Economic development of the British coal industry 1800–1914

This book provides a full account of the economic development of the British coal industry from 1800 to the First World War. It estimates and analyses the growth in the capital employed and the labour force in British collieries; their output, costs, productivity and markets; the changes in technology; and wage bargaining and determination in the industry. Figures are given for the United Kingdom as a whole, and also for all the major coalfields and some of the minor ones. It therefore provides the most complete study available of the economic determinants of the industry and by looking at the underlying economic relationships contributes to our understanding of the factors which motivated and determined the industry's expansion.

Economic development of the British coal industry 1800–1914

B.R. MITCHELL
Lecturer in Economics, University of Cambridge,
and Fellow of Trinity College, Cambridge

The right of the University of Cambridge to print and sell all manner of books was granted by Henry VIII in 1534. The University has printed and published continuously since 1584.

CAMBRIDGE UNIVERSITY PRESS
Cambridge
London New York New Rochelle
Melbourne Sydney

Published by the Press Syndicate of the University of Cambridge
The Pitt Building, Trumpington Street, Cambridge CB2 1RP
32 East 57th Street, New York, NY 10022, USA
296 Beaconsfield Parade, Middle Park, Melbourne 3206, Australia

First published 1984

Printed in Great Britain at the University Press, Cambridge

Library of Congress catalogue card number: 83–27269

British Library Cataloguing in Publication Data

Mitchell, B.R.
Economic development of the British coal
industry 1800–1914.
1. Coal trade – Great Britain – History
2. Coal mines and mining – Great Britain –
History
I. Title
338.2′724′0941 HD9551.5

ISBN 0 521 26501 0

WD

CONTENTS

TABLES AND FIGURES

Tables

Figures

PREFACE

This book had its origins in a PhD thesis at Cambridge University which was begun over thirty years ago. It has consequently had an unconscionably long gestation period, and as Sir Alec Cairncross wrote in the preface to his *Home and Foreign Investment, 1870–1913*, which had a somewhat similar history, it has led something of a subterranean life in footnotes. The eventual outcome is very different from the original thesis, which was not only confined to the period 1870–1914 and to the inland coalfields alone, but which also asked rather simpler questions of its material, and treated some of that material more naively than anyone would be likely to do today. Apart from the extension of both geographical and temporal scope the thesis has been rewritten twice since it was accepted by the examiners in 1955, and not too much remains of the original.

Surprisingly little has been written about an industry which came to have such importance in the British economy during the nineteenth century. Its earlier history was dealt with by Nef and by Ashton & Sykes half a century or more ago; but whilst there is a number of scholarly articles and chapters, several books on miners' trade unions, and a good study of the South Wales coalfield in the middle part of the period, no comprehensive study exists of the industry in the nineteenth century. This book is an attempt to fill that gap partially by examining the economic factors which determined its development in the century or so up to the First World War. It is still not a comprehensive history of the coal industry in the period, for aspects of its social history are touched on only as they impinge on its economics, and attention to the actual selling of coal is only peripheral. Moreover there is no attempt here to look in any depth at the people who were prominent in the industry. They are, for the most part, treated as entrepreneurs, as viewers, or as union leaders, rather than as individuals, such as George Elliott, John Buddle, or Thomas Burt. In other words, this is a study in the economics of the industry, fairly narrowly defined.

The general framework of the book is an examination of the output and markets of the coal industry, of the factors of production which were required to produce the output, and of the relationship between output and inputs. This format is not intended to imply that the author is a devotee of neoclassical production functions. It simply happens to be a convenient scheme for a systematic investigation of the economic influences at work. There *is* an attempt – a very limited attempt – at a total factor productivity study in the last chapter; but it will be clear to the reader who gets there that this has not been a major goal of the book. That goal has been to describe how the rather minor coal industry of 1800 grew to be one of the four or five major economic activities of the country, and how it coped with the economic problems posed by that growth.

Over the long period of its gestation I have incurred obligations to many people on account of this book, and it is with real pleasure that I can at length acknowledge them. Foremost amongst them are my two earliest guides in economic history – both, alas, no longer with us – Ethel Hampson, who first inspired my interest in the subject, and Munya Postan, who was my research supervisor in 1952–5. I am deeply grateful to both of them for their inspirational qualities as well as for their comments on earlier drafts. Others who have helped with comments or discussion of all or parts of the book, and to whom I wish to express my gratitude, are Professor R.A. Church, Dr J.L. Eatwell, the late Denys Munby, and two anonymous publishers' readers.

A wide variety of people has helped in one way or another to ease my path through archives and libraries, to provide access to unpublished records, and to afford other kinds of help. Many of those who helped an impecunious research student in the 1950s are now dead, but my thanks to them, and to those who have helped me more recently, must be recorded. The following, I am sure, is not a complete list, and I hope that those whose names I have inadvertently omitted will forgive me: Mr J.K. Bates, Mr D.C. Bradshaw, Rev. H. Stewart Carter, Professor J.D. Chambers, Professor C.H. Feinstein, Mr Philip H. Gibbons, Mr W.H. Harrison, Mr H. Horspool, Mr A.E. Horton, Mr K.W. Humphreys, Mr D.M. Hunter, Mr Roger Lewis, Mrs Pauline W. Maclean, Mr & Mrs Lyle McNaught, Mr Michael Moss, Mr Edward Robertson, Miss C. Warburton, and Mr Paul Warrington. I am also especially grateful to Dr Anthony Slaven for his help when I was examining the Govan Colliery Records at Glasgow University, and for his generous permission to use material from his dissertation. I must also express my gratitude to the University of Cambridge and to Trinity College for a year's leave in 1976/7, during which the final version was mainly written, and for help with travelling expenses. Finally I would like to record my profound sense of obligation to my first wife, Barbara, without whom I should not have been able to embark on this study at all, and to my wife, Ann, whose persistence has been mainly responsible for its eventual conclusion.

GLOSSARY OF MINING TERMS USED

adit	a tunnel driven into a hillside for access, drainage, haulage, or ventilation.
bank	the pithead area. (The term was also used in Yorkshire and the East Midlands for a coal face.)
banksman	the man controlling cages and operating signals at the shaft top.
blackdamp	a mixture of carbon dioxide and nitrogen leaving insufficient oxygen to support life.
brattice	cloth or wood screens dividing the shaft of a single-shaft pit.
chaldron	a measure of volume, conventionally reckoned to be 53 cwt in the North-East and 25½ cwt in London.
chokedamp	blackdamp (q.v.).
corf	basket used to transport coal from the face to the pithead. By extension, it was sometimes applied to the tub, which superseded it.
darg	Scottish daily stint for the hewer, often that which was fixed by the union.
deputy	colliery official, responsible for the safety of workplaces.
drawer	putter (q.v.).
filler	man employed to lead corves or tubs at the coal face.
firedamp	inflammable gas, mainly methane, emitted from some coal seams.
goaf (or *gob*)	the area underground left behind by the removal of coal, partly filled with waste.
hurrier	putter (q.v.).
keel	boat used in North-East England for carrying coal from the staithes to sea-going ships.
pikeman	West Midland term for hewer.
putter	haulage worker, mainly employed to move corves or tubs from the face to the main roadways.
round coal	large lumps.

screens	devices for separating large coal from small.
slack	very small coal.
smudge	small slack, almost dust.
staithe	coal wharf, or, more restrictedly, the spout through which coal was loaded into keels or ships.
timberman	man employed to set pitprops, etc., underground.
tram	tub (q.v.).
trammer	putter (q.v.).
trapper	child employed to open and shut underground ventilation doors.
truck	payment of wages in kind.
tub	wheeled wagon for conveying coal from the face to the pithead.
viewer	mine manager in North-East England, becoming obsolete in the second half of the nineteenth century.

ABBREVIATIONS

CG	*Colliery Guardian.*
CIC	Coal Industry Commission (otherwise known as the Sankey Commission), *SP* 1919 XI and XII.
DC 1907	Departmental Committee on an Eight Hours Day for Miners, *SP* 1907 XIV and XV.
HOLM	Home Office *List of Mines*, published in HM Inspectors of Mines' *Reports* for 1872–83, in *MS* for 1854–71, and separately from 1888 to 1950.
MS	*Mineral Statistics*, first published as part of the *Memoirs of the Geological Survey*, initially under the heading *Records of the School of Mines*. From 1882 they were published in *Sessional Papers* of Parliament.
RC	Royal Commission.
RC 1871	Royal Commission on Coal Supply, *SP* 1871 XVIII.
RC 1903–5	Royal Commission on Coal Resources, *SP* 1903 XVI, *SP* 1904 XXIII, and *SP* 1905 XVI.
SC	Select Committee.
SC 1873	Select Committee on the Dearness and Scarcity of Coal, *SP* 1873 X.
SP	*Sessional Papers* of Parliament.
TFIME	*Transactions of the Federated Institute of Mining Engineers.*

In the tables, — = nil, ... = not available, and ----- = a break in continuity.

1

INTRODUCTION

The British coal industry came during the nineteenth century to be regarded as one of the three or four staple industries of the country. In 1914 it had been one of the major employers of labour for well over fifty years, and the value of its net output was exceeded only by the conglomerate group of metal-working industries at the first Census of Production in 1907. At the beginning of the nineteenth century, coal was a major input for the fast-growing, but still quite small, iron industry, and for a number of other minor manufactures such as glass, bricks, and non-ferrous metals. It was also used to some extent in the textile industries, principally in cotton spinning, and in brewing. But its main use, taking the country as a whole, was as a consumption good, in the heating of houses; and in that it still had to compete in some places with wood and peat. In no sense could the coal industry be regarded as one of the fundamental basic industries of Britain in 1800. By the 1850s that had changed, a fact which was highlighted by the economist, W.S. Jevons, when he published a book in 1864 sounding a (premature) alarm about the likely exhaustion of what had come to be an essential input for all the main productive activities of the country except agriculture, and – what was perhaps the most strategically important for the economy as a whole – for the main means of transport both within the country and for its overseas trade.[1]

This change in the significance of coal came about scarcely, if at all, because of anything that happened within the industry itself, but mainly because of technical change in other industries, above all those changes involving the greater use of metal and those often-related changes involving the application of steam power. These, in turn, reflected mainly a major change in the scope and nature of capital equipment, with the spread of machines, of engines, of rails, and of iron ships. This change eventually affected all trading regions of the world in some degree, but in the first half of the nineteenth century it nowhere progressed nearly so far as in Britain.

It is probable that a major reason for Britain's primacy in the metal and steam revolution was the nature of its resource endowments – the close juxtaposition of known fields of iron ore and of good coking coal, coupled with the relative dearness of charcoal. This at-one-time generally accepted view has come under something of a shadow in the last thirty or forty years, as countries less well-endowed have industrialised successfully, and as technical progress has come to seem to be the major source of economic growth.[2] In the nineteenth century, however, and especially in its first half, when transport costs were still much higher than they later became, the location of suitable resources was much more crucial to the development of heavy industries than it has become subsequently. So that until the discovery and development of the deeper coking seams of the Ruhr coalfield and the exploitation of the Connellsville field in western Pennsylvania, both effectively in the 1850s and later, no other country had nearly such advantageous coal and iron resources as Britain. Indeed, it was only those of Belgium which approached the same class. For fifty and more years, therefore, the heavy industry strand of the Industrial Revolution was evolved in Britain almost alone; and it was this, above all else, which drew coalmining into a leading place in the British economy, a place far higher than it occupied anywhere else except Belgium until into the last quarter of the nineteenth century. In 1850, for example, British coal output was from eight to twelve times as big as that of France, the German states combined, or even Belgium. Only in the last case was its proportion to total GNP remotely similar to that in Britain.

Whilst Germany, the United States, and, in so far as its limited coal resources allowed, France began to follow in the same path as Britain (and Belgium) after 1850, the costs of production of most British heavy industries, held down partly by economies of skill and of scale, ensured that for more than two decades they maintained their superiority, and were able to continue to expand at roughly the same rate as before. Another aspect of these relatively low costs was the price of British coal, which remained lower than that of Continental European producers, largely owing to more favourable natural conditions. The continued availability of adequate iron ore close to coking coal was another factor. All these influences ceased to have any real importance in the last quarter of the century. British skills in the manufacture of wrought iron were rendered redundant by the shift to Bessemer and open-hearth steel, and economies of scale were then more readily available in America and Germany.[3] The native supply of iron ore became less and less adequate; and though British coal continued to be relatively cheap, the margin was decreasing. Certainly it was not enough to enable the British iron and steel industries to maintain their former rates of growth. Contemporaneously with this slackening in the expansion of the major customer of the coal industry there went the development of considerable economies in the use of fuel, especially, but by no means only, in primary iron manufacture. Together, these things brought down the rate of growth of the coal industry to roughly half of what it had been for fifty years. Where the

Table 1.1 *GB/UK coal output and its rates of growth, 1801–1913*

	GB output (million tons)	Annual percentage increase		UK output (million tons)	Annual percentage increase
1801–5	12.96		1856	66.6	
1806–10	14.79	2.8	1861	83.6	5.1
1811–15	16.59	2.4	1867	104.5	4.2
1816–20	18.90	2.8	1873	128.7	3.9
1821–5	20.90	2.1	1883	163.7	2.7
1826–30	24.80	3.7	1891	185.5	1.7
1831–5	29.56	3.8	1900	225.2	2.4
1836–40	35.27	3.9	1907	267.8	2.7
1841–5	41.71	3.7	1913	287.4	1.2
1846–50	50.97	4.4			
1856	66.4	6.1			

Sources: 1801–5 to 1846–50 – Pollard (1980); 1856–1913 – *MS*.

average annual rate of growth of coal output had been of the order of 4% or more from the middle 1820s to the 1860s, it was below 3% subsequently, and usually substantially below. There is a clear watershed in this respect in the early 1870s.

The point of inflexion is brought out in table 1.1, which establishes in broad outline the framework of the growth of the industry. Official statistics for the United Kingdom as a whole and for the various individual coalfields are available only from 1854 onwards. Prior to that date estimates have to be made on the basis of exports, of London imports, of ironworks consumption, and of other scattered, data, including the guesses of contemporaries. Sidney Pollard has made the most elaborate and best attempt at distilling such information into a usable series, and it is his figures which are reproduced here for the period 1801–50. They are annual averages of quinquennia, which, in view of the effects of cyclical fluctuations, are not ideal for estimating rates of growth that can be compared meaningfully at different times. Nevertheless, they do not distort the picture of the rate of change over longer periods, and they do not tell an essentially different story from that provided by the cyclical peak figures in the second part of the table.

Over the period as a whole the increase in coal output was over 2,500%, but the rate of increase varied considerably from time to time. The fastest rate was, apparently, in the 1850s, and it does not seem likely that the appearance is a false result of the selection of dates. It is not difficult, in fact, to see it as a consequence of the new-found effectiveness of the railways in the distribution of coal over substantial distances. The slowest rate of growth was in the 1880s and after 1907. The overall pattern was of steady, modest growth into the 1820s, a step up to a rate of around 4% from then until the late 40s, when a further shift of gear occurred to the peak rate of the 1850s. In the 60s there was a return to the rate of the second quarter of the century, followed by the watershed already referred to in the early 70s. The decade or so around 1900 – more precisely the years 1896 to 1907 – saw a return to the rate of growth of the first quarter of the century; but this was followed by a plunge in the last few years before the First World War, and, as everyone knows, the course of output thereafter was downwards.

The watershed of the early 70s was, it must be stressed, one associated almost entirely with changed demand. There *were* considerable changes in the supply conditions of the industry at about the same time, most notably a major reduction in weekly hours of work; but, perhaps surprisingly, these had little or no long-run influence on output or costs, and they do not affect the pre-eminence of the demand-side changes. These changes, unusually amongst the established major British industries, were mainly in internal rather than export markets. The latter, in fact, grew at a rather faster rate in the period 1873–1900 (and again after the export tax was removed in 1906) than in the previous quarter century. In view of the fact that labour productivity in British coalmines reached its peak (until fairly recent years, at any rate) in the 1880s this may seem rather surprising, especially as labour productivity in most other countries continued to edge upwards. However, the absolute level of productivity in Britain remained above that in Continental Europe, and the higher levels in the United States were still insufficient to enable them to compete with British coal in third markets after paying their higher wage rates and for the much longer hauls to the ports. British exports rose faster after 1873 because no other country's coalfields were able to compete with them at coastal sites in much of the world, and because demand for coal rose faster than before in places like Hamburg, Copenhagen, St Petersburg, Le Havre, Naples, Alexandria, Buenos Aires, Rio de Janeiro, and Havana. Behind this lay the increased use of coal-burning steamships and the extension to other countries of the metal and steam technologies begun in Britain a hundred years before.

However, the faster expansion of export markets was not enough to offset the slower growth at home. The major fuel-using industries decelerated; fuel economy was pursued more vigorously, especially in the immediate aftermath of the high prices of 1872–3 and in the period of more sustained price increase after 1898; and perhaps, too, there was increasing income-inelasticity in the demand for a 'necessity of life' as standards of living for the majority passed the level at which they could fairly readily afford 'sufficient' of that 'necessity'. Whether or not there was also price-inelasticity of demand is very difficult to assess. G.C. Allen believed that there was, for he remarked of the prewar period that 'the demand for British coal was so urgent, both at home and in the export markets, that it was possible for the industry to pass on the burden of increased costs to customers'.[4] Against this, however, one can argue that it was not possible to do so *and* to keep up the same rate of growth as before. Surely, therefore, *some* demand was being choked off by rising relative prices? At the margin, more Berliners (say) may have bought Silesian or Ruhr coal rather than coal from Newcastle; and more ship-owners may have opted for oil-burning engines. Though it was not, perhaps, apparent at the time, it seems likely that 1907 was the end of the long period of rapid expansion of the coal industry in which the entre-preneurs had had no sustained reason to worry about the demand for their product.

The rate of growth of the coal industry slackened after 1872–3, but it nevertheless continued to be substantial enough for large absolute increases in output to be registered, at any rate after a flat period in the later 1870s. For a time these increases were achieved mainly by increased productivity and the fuller use of the increased capacity created in and shortly after the great boom of the early 70s. But after 1888, rising output had to rely more and more on bigger inputs of labour and capital and on changes in technology, none of which was sufficient to prevent falling pro-ductivity, especially after 1907. It seems highly likely that this contributed to restraining the growth of the British economy as a whole, and perhaps especially to the poor record of the last few years before the war, both directly and through its effect in raising the costs of other industries.[5]

As has already been implied, the coal industry was one which was sub-ject to cyclical fluctuations – not surprisingly in view of the importance of heavy industries amongst its customers. This is readily apparent for the period after 1854, when annual production figures are available, but it may be inferred for earlier years too. The fluctuations, however, were nowhere near so violent as in the iron industry, and, in particular, the downswings were marked rather by output being more or less stationary than by pronounced declines. The typical production pattern of the trade cycle in the coal industry was a slight fall – 1% or 2% at most – in the year immediately following the peak of a boom, followed by three to five years in which output edged up a little bit. There was not normally, therefore, a gradual descent to the trough of the cycle, though this does describe the course of the *value* of output in the downswings which bottomed in 1886, 1896, and 1905, in all of which price falls were sustained. In a number of other depressions, too, aggregate value bumped along the bottom for a few years rather than edging up with output. Then, one or two years before the next peak, there was a fairly sharp rise in production, with the actual peak year showing only a modest further increase. In the major

Table 1.2 *Trade cycles in the coal industry, 1825–1913*

Peak	Trough	Peak	Trough
1825	?	1873	1879
1836/7	1840	1883	1886
1844/6	1851	1890/1	1896
1854/6	1858	1900	1905
1861[a]	1862[a]	1907	1908
1866/7	1869	1913	

[a]This may be seen as a minor, intermediate fluctuation, rather than as a proper trade cycle.
Source: author's assessments on the basis of table 9.3 below and of shipment and output statistics in Mitchell & Deane (1962), pp. 112–16.

boom of 1873, though, and again in 1900, 1907, and 1913, aggregate value continued to rise rapidly in the year of peak output, and it almost always rose more than did output, for prices were generally more volatile than production at those times. The weak boom of 1883, when capacity was scarcely if at all strained, was an exception to this. This suggests, incidentally – and the profit performance which we come to in chapter 9 bears it out – that the 'Great Depression' was not entirely a myth so far as coal was concerned.

With the beginning of the publication of annual output figures it is generally easy to identify the peak years, though reference to price data suggests a slightly different timing in one or two cases. On price grounds, for instance, 1854 or 1855 has as much claim to be seen as the peak of the 1850s as does the year of maximum output, 1856; and 1890 can be preferred to 1891. For the bottom turning points, and for the peaks prior to 1854, it is on the price data that one must rely primarily. In the earlier period this is often ambiguous, and it is only rather tentatively that one can suggest the timing of some of these cycles. Indeed, it is not really possible to identify cycles at all before the 1830s, for the only annual data we have relates to the Northumberland and Durham coalfield, which was probably less affected by industrial demand than any other district, and, moreover, it was subject to monopolistic manipulation of prices and production which varied in effectiveness from time to time. It does seem clear that there was a peak in 1825, but the subsequent trough is obscure, and the next obvious peak is not until 1836 or 1837. Thereafter the cycles can be dated as in table 1.2. It will be noted that the peaks, and the troughs from 1858 onwards, correspond very closely with the general cyclical turning points identified by Rostow.[6] The behaviour of the coal industry conformed very closely to that of the economy as a whole.

Having outlined the overall performance of the British coal industry in

Table 1.3 *Output of coal by main districts, 1801–1913 (in million tons)*

	Durham & Northumberland	Yorkshire	Lancashire & Cheshire	East Midlands	West Midlands	South Wales	Scotland
1801–5	3.2	1.5	1.4	0.8	2.1	1.2	2.2
1806–10	3.4	...	...	...	...	...	...
1811–15	3.7	1.8	1.6	1.2	3.1	1.8	2.5
1816–20	4.1	...	...	...	...	...	...
1821–5	4.5	2.4	2.4	1.5	4.2	2.5	2.9
1826–30	5.2	...	...	...	...	...	...
1831–5	6.1	3.6	4.5	2.2	4.8	4.0	4.0
1836–40	7.8	...	...	...	...	...	...
1841–5	9.7	4.5	6.0	2.7	5.7	5.0	5.3
1846–50	12.6	5.5	7.5	3.3	6.5	6.5	6.0
1856	15.5	9.1	9.7	4.3	8.1	8.9	7.5
1861	19.1	9.4	13.0	6.5	[8.1]	11.4	11.1
1867	24.9	[9.8]	13.8	8.2	13.6	13.7	14.1
1873	29.7	15.3	18.4	11.5	16.5	16.2	16.9
1883	37.4	19.6	23.9	16.6	16.1	24.1	21.3
1891	39.1	22.8	26.0	21.6	16.0	30.0	25.4
1900	46.3	28.2	28.7	28.9	15.8	39.3	33.1
1907	54.0	35.2	30.0	37.2	16.2	50.0	40.1
1913	56.4	43.7	28.1	38.8	15.8	56.8	42.5

Figures in brackets are believed to be defective.
Leicestershire and Warwickshire are included with the East Midlands. The estimates for South Staffordshire in 1861 and for Yorkshire in 1867 are known to be too low. Each should be perhaps 1.5 million tons higher.
Sources: as for table 1.1.

table 1.1, it remains to do the same for its major component parts. This is done in table 1.3.

From this table it appears that, over the period as a whole, the group of coalfields which came to make up the Midland Inspection District and South Wales experienced the highest rates of growth, followed some way behind by Yorkshire. These three were the only ones which grew faster than the UK average. Lancashire and Scotland had a rate only a little below average, and even the long-exploited North-East Coast grew not so very much less quickly. Only the smaller coalfields of the West Midlands and the West Coast grew markedly more slowly.

Looked at over shorter periods of time, there were many variations in rates of growth, both between coalfields and between periods. The North-East experienced its fastest growth in the mid-century period, from the middle 1830s to the middle 1860s. This period included the breakdown of the monopolistic restrictions in the London market, and the discovery and opening of the Cleveland iron ore field. There was no shortage of reserves away from the old-exploited coastal and riverine districts, and some of the earliest railways were built in order to open these up. During these years the North-East's growth rate was above the national average. This was never again the case, though it was approached in the 1890s and early 1900s when the concealed coalfield of south-east Durham and the Ashington district was being opened up and export demand was booming.

The Yorkshire coalfield grew particularly fast for two or three fairly short periods, with intervals of relatively slow growth between. The first spurt was in the 1820s and 30s, and was based on local demand. The next was in the early 50s, and was also in response to local demand, especially for manufacturing. The last followed 1900, and was associated with the exploitation of the concealed coalfield east of Doncaster, which had been recently discovered.

Lancashire's peak rate of growth was early on, in the 1820s and 30s, but it was fairly well sustained until 1861. The immediate check then was from the side of demand, with the Cotton Famine; but growth rates were always below average subsequently, largely, if not entirely, because of the inadequacy of unexploited reserves. In the late 1900s, when this problem was compounded by the difficulties following the Eight Hours Day Act, output began to decline.

The West Midlands group of coalfields also experienced stagnation or decline, but setting in much earlier. Their growth rate was always below average after 1815, and virtual stagnation set in after the peak output was reached in 1873. There were, actually, four separate coalfields in this group, and their individual experiences varied a good deal. Shropshire, small and long-exploited, began to suffer from lack of reserves very early, and its peak output was reached before 1850s. The Black Country suffered similarly after about 1870, though the problems of the wrought iron industry under the competition of cheap steel very soon became an additional influence on its decline. North Staffordshire, in contrast, enjoyed respectable growth rates almost throughout. Though small, it had adequate reserves, mostly by going deeper rather than by geographical extension; and its idiosyncratic local economy generally shared in the overall national expansion. The last component, the Cannock Chase coalfield, was only opened up from the 1860s, and thereafter grew quite quickly, though not enough to outweigh the decline in the district to its south.

The East Midlands and South Wales coalfields both grew fairly steadily and at above the average rate throughout the period, or practically so, and no time of unusual performance can really be discerned. Both had ample reserves, and both had more than normally buoyant demand, though of very different kinds. The railway released the East Midlands from purely local demand, and from about 1850 the main stimulus to its expansion was competition with the North-East Coast in the London market. South Wales, on the other hand, was mainly drawn on by iron-works demand until 1873 and by exports thereafter.

Finally Scotland grew fairly steadily at about the national average rate, apart from a slow beginning to 1830 and a surge in the 1850s and 60s, associated with industrial expansion, especially in the iron industry. Reserves were generally adequate, overall, though the older-exploited districts near Edinburgh and Glasgow began to decline in the 1880s, and most expansion came in Fifeshire.

2

MARKETS

When one talks about the British coal industry in the nineteenth century in the singular it is often straining reality, for there was, in fact, a large number of British coal industries in separate districts, more or less loosely connected. The connections became stronger as the years passed, but the diversity of physical conditions and of local customs has ensured that, even today, despite much uniformity introduced through mechanisation, and after well over thirty years of national ownership, the industry is by no means homogeneous. Not least of the many differences between the coalfields are those in the markets which they serve. In the nineteenth century these differences were so pronounced that it often conceals more than it illuminates to discuss them solely in terms of a national demand for coal. For many purposes it is more revealing to recognise that there were several local demands, some of which were shared by two or more coalfields. These latter were generally external to the coalfields – for instance, London and the export markets – but they could be within the coalfields themselves in cases where a particular class of coal was required, such as metallurgical coke in the Black Country, which came from Durham and Yorkshire by the later nineteenth century, as well as from local mines. Because of the strongly differentiated local characteristics of demand each coalfield – or, at any rate, the major ones – will be considered separately. However, there is some advantage in taking an overall view first, though no general analysis will be attempted at this point. Let us begin, though, with a short survey of the institutional organisation of coal sales.

Basically, there were two main methods of organising the transfer of coal from the producer to the consumer. Normally, large customers, including merchants, bought direct from the collieries paying pitmouth prices, if for internal delivery, and f.o.b. prices, if for export.[1] Small consumers – whether household, commerical, or industrial – usually bought from merchants, except for some in the immediate vicinity of the pits. The merchants (including export shippers) were normally not involved in coal production for most of our period, but operated separate businesses, buying competitively from different collieries. The larger and more old-established of these firms tended to deal regularly with the same producers and customers, working on relatively low margins. These were sometimes described as 'legitimate middlemen', to distinguish them from others operating more riskily and on the basis of opportunism, who were designated 'speculative middlemen'. These latter tended to buy and sell in smallish lots, except in cyclical downswings, when there was scope for 'bearing' the market, selling forward and short. The separation of producing and merchanting was virtually complete prior to the early 1870s. But the confusion of that period of extravagant prices and difficulty in obtaining supplies led to a tendency, mainly in the exporting coalfields and above all in South Wales, for selling and producing to draw closer together. By 1914 most of the larger South Wales colliery companies, and some in other coalfields, were selling direct to shippers, and in some cases to foreign consumers.[2] This particular change had largely taken place in the 1890s and 1900s. In the previous twenty years the convergence had been from the other direction, with some large merchants, like Cory's, buying up or buying into colliery firms, primarily in South Wales.[3]

The merchants and other large consumers who bought direct from the collieries usually did so on a regular basis, getting most of the coal they wanted on contracts, which lasted from a few weeks to as much as three years, though the latter was very unusual. Most contracts were for periods of six months to a year, with a tendency to be shorter when prices were relatively very high or very low, and longer at times when prices appeared to be relatively settled. Obviously neither party wanted to be caught out by a large movement of prices. The advantage of a contract to the merchant was security of supply and certainty of cost, both of which had more appeal on the upswing of a trade cycle. To the coalowner, on the other hand, contract sales had a rather more constant appeal. Their great advantage from his point of view was that they ensured regularity of working, which is technically desirable since it reduces the cost of underground repair work, and economically desirable both because of that, and because it helps to ensure fuller employment of resources and consequent reduction of unit overhead costs. The value of this was such that, at times, coalowners were even willing to accept contracts at prices which did not fully cover costs, so long as they made some contribution to overhead costs, or, perhaps, at the extreme, so long as they did not involve losses which exceeded the costs of closing down the pits.

The proportion of coal sold by contract varied considerably, both by district and over time. Early on in the nineteenth century the contract was a device found mainly in the North-East of England, and it always tended to be more prominent in those coalfields which shipped a large part of their output. But large consuming industries like iron also used contracts (where they did not own collieries, and so tie up the coal output even more securely). Railways, the Admiralty, the large shipping companies,

Table 2.1 *Estimated consumption of UK coal by uses, 1816–1913*

	1816		1830		1840		1855		1869		1887		1903		1913	
	m.tons	%	m.tons	%	m.tons	%	m.tons	%	m.tons	%	m.tons	%	m.tons	%	m.tons	%
Exports	0.2	1	0.5	2	1.6	5	4.8	7.5	10.2	9.5	23	14	45	20	73.5	25.5
Iron & steel	1.9	11	3.6	16	7.1	21	15.5	24.5	26.5	26.5	27	17	28	12.5	37	13
Non-ferrous metals	0.2	1	0.3	1	0.4	1	0.5	1	0.6	0.5	1	0.5	1	trace	1.5	trace
General manufacturing	4.9	28	7.0	29	9.7	29	18.0	28	30.0	26	46	28.5	62	27.5	68	23.5
Domestic	9.3	53	10.7	45	11.5	34	15.0	23.5	19.5	18	28	17	32	14	35	12
Railways	—	—	—	—	0.3	1	1.4	2	3.2	3	6	4	13	5.5	15	5
Steamships	—	—	0.2	1	0.6	2	1.6	2.5	3.3	3	9	5.5	13	5.5	17	6
Gas & electricity	—	—	0.5	2	1.0	3	2.7	4	6.3	6	10	6	18	8	23	8
Collieries	0.9	5	1.2	5	2.0	6	4.2	6.5	7.6	7	11	7	15	6.5	18	6

most gasworks, and some of the larger factories were others who did the same. Probably more than half the output of most coalfields was sold by contract by the later 1870s, the scarcities of 1872–3 having induced more consumers to adopt the practice. Earlier, some of the inland coalfields probably sold little in that way, at least until after the coming of the railways brought both railway company contracts and business from the large London merchants. Even in 1914, less was sold by contract in these districts – especially the smaller ones with a large proportion of house-coal – than in the major exporting coalfields or those with big ironworks. South Wales collieries were reckoned to sell over two-thirds of their output in this way, sometimes even as much as 85%.[4]

The rest of the output which was not sold by contract was usually disposed of in comparatively small parcels to small retail merchants, to local consumers, to make up a shortage in some other colliery's contract deliveries, and, of course, to those large customers who had unexpected extra demand or who might be holding off from making contracts because of their expectation of falling prices. Usually, these spot sales of 'free' coal were at somewhat higher than contract prices – perhaps even a good deal higher during the later stages of upswings.

Turning now to the question of how much coal went into various uses at different times, there is bound to be a good deal of guesswork in the attempt to assess this. We begin by examining the overall national picture first, and table 2.1 gathers together what seem to be the best of the avail-

Notes to Table 2.1

Sources: 1869– Royal Commission on the Coal Supply (*SP* 1871 XVIII), appendices to Report of Committee E (see note after 1913 source).

Years before 1869 – *ibid.*, adjusted by reference to data in Mitchell & Deane (1962), chs. 4–10; in Meade (1882); in Mushet (1840); in Scrivenor (1854); in McCulloch (1859); in Porter (1847); and in evidence to the Select Committees on the Coal Trade (*SP* 1830 VIII and *SP* 1836 XI).

1887 – Price-Williams (1889), amended for overestimation of coal used in steamships.

1903 – Reports of the Royal Commission on Coal Supplies (*SP* 1904 XXIII, p. 57, and *SP* 1905 XVI, p. 17).

1913 – Report of the Coal Conservation Committee (*SP* 1918 VII, p. 624).

The estimates of Committee E of the 1871 Royal Commission have been followed in all except two respects. The more important of these exceptions concerns coal used in the iron and steel industries. The normally accepted figures for 1869, quoted by Committee E, is 32.2 million tons – 16.3 million tons in pig iron manufacture and 15.9 million tons in subsequent stages. However, when iron works consumption is calculated by districts (see table 2.2) on the basis of the various different ratios of coal to iron given in the sources referred to above (including evidence to the 1871 Royal Commission), it becomes clear that the figure of 15.9 million tons of coal used in finished iron manufacture is markedly too high. The combined district figures have therefore been used in this table. (It is perhaps worth pointing out that the ratio of coal used in pig iron to coal used in finished iron manufacture in 1840, suggested by G.R. Porter, is much closer to the one implied here than it is to that of Committee E.)

able estimates. They must not be taken as exact, of course, though the margins of error vary greatly. At one extreme, the statistics for exports are quite precise. At the other extreme, the division between 'general manufacturing' and 'domestic' demand is very hazy for the years before 1869. Broadly speaking, the estimates for 1869, based on the detailed enquiries of Committee E of the Royal Commission on the Coal Supply of 1871, may be taken as having quite a low margin of error; and those for 1903 and 1913, which also derive from official enquiries, are reasonably reliable. Price-Williams's estimates for 1887 were based mainly on projections of the 1871 Royal Commission's figures, and also probably have only a moderate margin of error. The estimates for the years before 1869 are more variable. Those for the more specific markets are probably quite good, since there are fairly firm bases of estimation – e.g. the output statistics for iron and for other metals coupled with contemporary expert opinions about the amount of coal used per ton manufactured; tonnage of steamships registered, again coupled with consumption per unit estimates; train mileage and passengers carried. Much the greatest degree of unreliability is in trying to allocate the output remaining after these specific uses have been assessed. The attempt has been made here on the basis of a variety of information about various industries – the amount of beer brewed; the horsepower of steam engines employed in the textile industries; glass and paper production; and so on. Domestic consumption has been taken as the final residual. It must be admitted that these estimation procedures are pretty close to guesswork, and the two categories 'general manufacturing' and 'domestic' use may be as much as 50% out. It is believed, however, that 20% is a more likely margin.

No full analysis of these national estimates will be made here, since this will emerge from the consideration of the individual coalfields. But five major trends that can be discerned in table 2.1 should be noted. These are: (1) the steady increase in overseas markets; (2) the rise and fall of the relative place of the iron and steel industries; (3) the rise of demands for the conversion of coal into more convenient sources of light and power; (4) the rise of the demand for steam transport; and (5) the steady decline in the relative importance of house-coal.

Turning now to the major coalfields or groups of coalfields, it seems that the best method of comprehending the variations in their markets is to put the estimates of the uses to which their coal was put in tabular form. There is a danger in doing this that an impression of accuracy will be conveyed which is altogether false. But it seems better to risk this, whilst giving warning, rather than write in terms which imply specific magnitudes without revealing the estimates themselves. The margins of error are, of course, proportionately a good deal greater than was the case with table 2.1. There is far less contemporary material to go on for individual coalfields than for the national aggregate. And there are numerous markets which can be assessed quite accurately on an overall basis, but which are not sufficiently specific to particular districts for anything like the same degree of precision to be achieved. But, as with table 2.1, some estimates are more reliable than others, and, broadly speaking, it is the same ones in the local as in the national case, though the superiority of the 1869 figures to any others is worth stressing. The margin of error of some of the earlier estimates of 'general manufacturing' and 'domestic' use is extremely high – probably between 50% and 100%, but possibly even higher. Even the twentieth-century estimates for these items may not be very much better, for the existence of more information is offset by greater movement of coal between coalfields. At the other extreme, figures of exports and coastwise shipments have a high measure of accuracy.

The importance to Durham and Northumberland of non-local markets throughout the period is readily apparent from this table. At the beginning, this coalfield had an entrenched position in the major market of the day – that for house-coal in London. In fact it was so firmly established that its owners operated a cartel, the Limitation of the Vend, which was broken for comparatively minor periods only during the whole of its formal existence from 1771 to 1845.[5] In 1816 London imported 1.6 million tons of coal, and practically all of it came from the ports of North-East England. About as much again was shipped from those same ports to other destinations, and since only 159,000 tons of this went abroad, we may reckon roughly that a third of the entire output of Durham and Northumberland went for consumption in London and the areas served from it, and another one-third went to the ports of the east and south coasts from Aberdeen to Plymouth. The greater part of this coal, we may safely assume, was for household use, with brickmakers, brewers, and a host of small workshops taking most of the rest. The 1.5 million tons or so of North-Eastern output which was not shipped in 1816 was probably split fairly equally between local household use and industrial uses in the same area. The salt pans, glassworks, and the nascent chemical industry were the main industrial customers, whilst the mines themselves used proportionately more at that time than in other districts where the pits were less deep, less fiery, and generally less up-to-date.

In the 1820s there was little change in the market pattern, but in the following two decades trends began which were to be significant for the rest of the century. One was the rise of the export trade – by 1840 overseas shipments took over 11% of output, and by 1855 this had reached 18%. The other was the increasing importance of local manufacturing consumption, a development which depended overwhelmingly on the start and rapid expansion of the iron industry based on the Cleveland ores. In

Table 2.2 *Estimated consumption of the output of the main British coal-fields by uses, 1816–1913 (in million tons)*

	1816	1830	1840	1855	1869	1887	1903	1913
Durham & Northumberland								
Exports	0.2	0.2	1.2	2.3	5.0	7.4	14.0	23.0
Coastwise[a]	2.8	3.5	4.5	5.4	5.5	7.1	10.9	8.0
Ironworks	—	—	0.1	1.7	7.9	9.6	9.8	12.6
Local general manufacturing[b]	0.8	1.0	1.5	2.6	2.7	3.4	4.0	4.5
Local domestic	0.7	1.0	1.2	1.6	1.8	2.1	2.6	2.8
Railways	—	—	0.1	0.4	1.1	3.1	3.8	3.5
Steamships	—	—	0.5	0.7	1.2	2.6	3.2	3.6
Collieries	0.2	0.3	0.6	1.1	1.6	2.1	3.1	3.4
Yorkshire								
Exports	—	—	—	0.1	0.4	1.9	3.1	6.8
Coastwise[a]	—	—	0.2	0.1	0.1	0.4	1.1	2.4
Ironworks	0.2	0.3	0.5	0.9	1.4	3.4	4.4	6.1
Local general manufacturing[bc]	0.6	0.9	1.4	3.7	5.3	6.9	7.3	10.9
Local domestic	0.9	1.1	1.5	1.9	2.2	3.0	3.7	4.3
Manufacturing & domestic railed south[b]	—	—	—	0.2	0.5	2.0	4.0	5.4
Railways	—	—	—	0.2	0.4	0.9	1.4	2.2
Steamships	—	—	—	—	—	0.5	1.9	3.2
Collieries	0.1	0.2	0.3	0.5	0.6	1.2	1.6	2.4
Midland Inspection Division								
Exports	—	—	—	—	—	—	0.2	2.1
Coastwise[a]	—	—	—	—	—	—	—	0.2
Ironworks	0.1	0.1	0.2	0.6	1.1	1.0	1.2	3.1
Local general manufacturing[b]	0.3	0.5	0.7	0.8	1.7	3.4	8.3	9.9
Local domestic	0.7	1.0	1.1	1.0	1.6	2.6	3.5	3.7
Manufacturing & domestic railed south	—	—	—	1.2	3.2	8.3	12.9	15.0
Railways	—	—	—	0.1	0.4	1.0	1.2	2.0
Steamships	—	—	—	—	—	—	0.2	0.5
Collieries	0.1	0.1	0.2	0.2	0.6	1.0	1.9	2.2

Table 2.2 (*cont.*)

	1816	1830	1840	1855	1869	1887	1903	1913
Lancashire & Cheshire Cumberland & North Wales								
Exports	0.1	0.1	0.1	0.5	0.8	0.6	0.6	0.8
Coastwise[a]	0.3	0.4	0.6	0.9	1.1	1.5	2.8	2.7
Ironworks	0.1	0.2	0.2	0.3	0.8	0.9	0.9	1.0
Local general manufacturing[b]	0.7	1.3	2.4	4.5	7.9	12.7	13.9	13.9
Local domestic	0.7	1.1	1.1	2.6	3.6	5.1	5.5	5.5
Railways	—	—	—	0.1	0.2	0.3	0.6	0.7
Steamships	—	—	—	0.2	0.4	1.0	1.3	1.3
Collieries	0.1	0.2	0.3	0.6	1.4	1.9	2.0	2.0
West Midlands								
Ironworks	1.7	2.6	3.3	5.8	6.8	4.5	4.5	5.2
Local general manufacturing[b]	0.5	0.6	0.7	1.2	4.5	5.7	6.3	5.9
Local domestic	0.6	0.7	0.8	0.8	1.4	2.9	3.2	3.5
Collieries	0.2	0.2	0.3	0.5	1.3	1.4	1.3	1.2
South Wales								
Exports	—	—	0.1	1.1	3.2	11.2	19.9	29.8
Coastwise[a]	0.4	0.9	1.4	1.8	2.5	3.2	4.0	3.4
Ironworks	0.6	1.3	2.1	3.6	3.5	3.5	3.6	4.2
Local general manufacturing[b]	0.2	0.2	0.4	0.8	1.2	2.5	3.3	5.2
Local domestic	0.5	0.6	0.6	0.7	0.8	1.2	1.7	2.1
Railways	—	—	—	0.2	0.4	1.6	2.3	2.2
Steamships	—	—	0.1	0.7	1.4	3.6	4.8	6.8
Collieries	0.1	0.1	0.3	0.5	0.8	1.6	2.7	3.3
Scotland								
Exports	—	—	0.2	0.4	1.2	3.1	7.2	10.4
Ironworks	0.1	0.3	0.9	2.8	4.4	4.0	4.0	4.7
Local general manufacturing[b]	0.4	0.7	0.8	1.4	3.8	6.5	11.2	13.1
Local domestic	1.3	1.7	1.9	1.3	2.6	3.7	5.3	5.9
Railways	—	—	—	0.4	0.8	1.5	3.0	3.4
Steamships	—	—	—	0.1	0.4	1.2	1.9	2.1
Collieries	0.1	0.2	0.2	0.4	1.2	1.5	2.3	2.7

[a]Includes some coal also included under 'railways' and 'steamships'.
[b]Includes gasworks and electricity works.
[c]Includes Lancashire.
Sources: as for table 2.1.

1840 ironworks still took less than 1% of North-Eastern output, but by 1855 this had risen to 10%. Whilst these two trends developed in the mid-century period, the old markets for coastwise shipments and for local domestic and general manufacturing coal continued to grow quite substantially in absolute terms up to the 1850s, though thereafter they virtually stayed still for the next fifteen years or so. But as a result of the rapid growth of the other markets, their share of the output fell. Each constituted only a third in 1855 and a fifth in 1869.

One small market for Northern coal which grew up during these mid-century years was that for railway coal (or, more precisely at this time, coke). In 1855 it came to perhaps 100,000 tons locally, together with something like 250,000 tons of the coastwise shipments, which went to fuel the locomotives of the eastern and south-eastern lines. The demand from gasworks also grew rapidly, and Durham was one of the major suppliers on a national scale of the best appropriate type of highly bituminous coal, contributing around one million tons, most of which was shipped coastwise.

The main changes in the demand for North-Eastern coal in the second half of the century were two. First came the massive increase in the iron and steel industries as a market, and after this there was a vast expansion of exports. All the time, as a result of these two developments, there was a steady fall in the proportion going to other uses, even though there was a fairly steady growth in absolute terms.

The chief cause of the massive increase in ironworks demand continued to be the expansion of the industry based on the Cleveland ores, though by the end of the period these had to be heavily supplemented by imported ores. But the attention devoted by the railway companies to the development of their potential freight traffic brought the high-grade metallurgical coke of Durham into competition with local fuel in Furness, South Yorkshire, and the East Midlands, and even as far away as the Black Country. All told, by 1869 about a third of the North-East's output was consumed by the iron and steel industries, roughly seven million tons being used locally. This proportion was maintained during the following two decades, but then fell in the 1900s as the local iron industry ceased its rapid growth, and as Yorkshire coke came into more effective competition in other regions. In 1903, and again in 1913, it was under 25%.

The expansion of export markets for Durham and Northumberland coal, which began in the 1830s, continued steadily but unspectacularly for over half a century, reaching 20% or so of output by 1869 and staying around that level until the 90s. Then there began a very rapid increase in exports, checked but not halted by the export tax imposed from 1901 to 1906, and reaching a peak of over 40% of output in the record year, 1913. Some of these exports, especially in the earlier years, were the highly bituminous types – gas-coal and the kinds preferred for the British domestic open fire – but the great majority was steam coal for use in factory boilers, in the Continent's closed domestic stoves, and to some extent in steamships bunkering at overseas ports, though this latter was nothing like so important as in the case of South Wales's exports.

The expansion of the demand from steam transport for North-Eastern coal can be seen from the table. In the early days, this coalfield was the main national supplier, though by the end of the 1850s South Wales had surpassed it so far as shipping consumption was concerned. Even so, demand from these markets was never of great proportional importance for the district as a whole, varying from about 5% to under 15% of output. It was, of course, vital for some individual collieries.

For the rest, local general manufacturing and household markets grew steadily but slowly, as they had done earlier, and continued their gradual decline in relative importance: by 1913 they only took 13% of output. Coastwise shipments, the other major remaining market, also lost ground relatively, but behaved more erratically. In the 1850s and 60s, as rail-borne supplies from Yorkshire and the East Midlands bit deep into the South-Eastern markets, coastwise shipments from the North-Eastern ports stagnated and fell to around 20% of output by 1869. In the last quarter of the century, however, they staged something of a comeback, and this was continued into the 1900s as the export tax encouraged some of the seasale collieries to seek home markets for their expanding output. The effects of this are reflected in the high figure for 1903, which was 23% of output. There followed a rapid decline, both in absolute and in relative terms, when the export tax was removed. Seasale collieries diverted their shipments overseas, leaving a greater share of the South-East's markets to the East Midlands collieries and, above all, to the rapidly growing small coalfield of Warwickshire, which was the nearest geographically to London.

South Wales, the largest British coalfield in 1913, was another, like Durham and Northumberland, which expanded during our period mainly on the basis of demand from ironworks and from overseas, though the timing of its expansion was generally different. In 1816 its markets were primarily local, with only about a fifth of output being shipped out of the district – mostly across the Bristol Channel, above all for use in the tin and copper mines of Cornwall and Devon. The biggest single user in South Wales was the iron industry, a position it maintained until the end of the 1860s, though it was by then thirty years beyond the peak of its relative importance. This was reached around 1840, at a time when South Wales produced over a third of all British pig iron and had an extensive finished iron industry as well. By then 42% of output was used by local ironworks – a growth from 32% in depressed 1816, but only marginally above the proportion consumed in the more prosperous years from the 1820s through to the 50s. From then onwards, South Wales pig iron output, though fluctuating considerably with boom and slump, was more or less stagnant. And though the finished iron and steel industries continued to expand, partly on the basis of pig iron shipped from Furness and Cumberland, there were large economies introduced in blast furnace working which ensured a practically stationary overall demand for coal from this source. As a result, the proportion of output used by the iron and steel industries continued to slip – to 26% in 1869, 14% in 1887, 8.5% in 1903, and only a little over 7% in 1913.

The major developments in demand for South Wales coal after 1840 were an accelerating increase in exports and in the use of the famous smokeless steam coal in ships. J.H. Morris and L.J. Williams (1958) have analysed the early stages of this expansion, which took exports from 63,000 tons in 1840 to nearly four million tons in 1874, and which saw Welsh coal overtake that of the North-East as the main source of bunker fuel, especially for the Royal Navy. For the dates we are using here, exports rose from a negligible proportion of output in 1840 to 13% in 1855, 24% in 1869, and 44% in 1887. Thereafter the rate of increase was slower, though in absolute figures the growth was enormous, and by 1913 exports accounted for over half the district's output. At the same time there was a parallel growth in sales for ships' bunkers – a not unconnected development, since a substantial part of exports went to overseas bunkering stations. By 1869, as much as 10% of output went into ships' bunkers in British ports, and over the next four or five decades the proportion hovered around 12–15%. Perhaps as much again was so used in foreign ports.

Beside the huge export and shipping demands, and the still substantial consumption of the iron and steel industries, other markets for South Wales coal were relatively insignificant in 1913, though in the early part of our period they had exceeded half the total. Local industrial and household markets continued to grow steadily throughout, but by the 1870s they were overshadowed, taking only 15% of output, a proportion which fell even further, to 12% or 13%, by the twentieth century. The main local industrial consumers apart from iron and steel were the non-ferrous metal smelters of West Wales – industries with a rather low rate of growth overall. Markets for South Wales coal that were supplied by shipment coastwise were even less important than this by 1913, though for the anthracite producers of West Wales they did have considerable significance. At the start of our period, the tin and copper mines had been the main market supplied in this way, and this probably remained so until the middle of the century. But soon thereafter the mines began to close, under the influence of competition from overseas, and by the 1880s their demand for Welsh coal was negligible. Meanwhile, there was a gradual increase in ordinary household and general manufacturing uses supplied from the South Wales ports, and, in particular, anthracite came to be used much further afield in uses where its smokeless quality was of special value – hop-drying and malting above all. As a proportion of total South Wales output, however, these coastwise shipments fell steadily, especially in the 1870s and 80s, and again after the abolition of the coal export tax in 1906.

The Scottish coalfields' markets were always to a large extent isolated from those further south, though North-Eastern coal (especially house- and gas-coal) did compete effectively in the Scottish East Coast ports throughout our period; and Scottish coal itself became a competitor with English and Welsh coal in export markets. The bulk of sales, however, were local to Scotland and not *directly* affected by non-Scottish competition. (Indirectly, of course, there was a considerable effect, especially through competition between Scottish and English or Welsh iron.)

There are three main trends in the development of markets for Scottish coal in our period, and these are similar to those we have already seen in South Wales. There is the rise and fall in the relative place of the iron and steel industries; there is the increase in exports; and there is the growth of demand from steam transport. One major difference, however, is very significant. This is the relative steadiness of the place of general manufacturing and domestic uses.

The iron industry in Scotland was long-established but small prior to Nielson's discovery of the hot blast techniques in 1828. The local coal was not very suitable for ironworks, and the amount used per ton of iron made was exceptionally high – 8.5 tons around 1800 according to Mushet,[6] and still about 8 tons in the 1820s according to Scrivenor.[7] As a result, consumption in this small industry took only about 10% of output. Then came Nielson's invention, which proved especially valuable in Scotland, where it became possible to use the 'splint' coal of Lanarkshire in furnaces without previous coking, and yet needing only about a third as much coal per ton of pig iron as previously. The hitherto little-used blackband ores were exploited rapidly in the following two decades, and Scottish pig iron output climbed from 37,500 tons in 1830 to 241,000 tons in 1840 and 828,000 tons in 1855. Meanwhile, the proportion of coal output going into this industry, and to the much smaller finished iron industry represented by the old Carron works and newer works at or near Glasgow, grew to 23% in 1840 and 38% in 1855. By the early 60s, when pig iron output was nearly 1.2 million tons, a proportion of 40% was probably exceeded. This marked the peak of the relative importance of the ironworks market to the coal industry, though in absolute terms it continued to grow until 1870, and even, after a period of stagnation, resumed its expansion in the

1900s. Its share of output, though, fell to a little over 30% in 1869, under 21% in 1887, and only just over 10% in 1903 and 1913.

The two markets which took over some of the iron industry's share of Scottish output were steam transport (especially the railways) and exports. Scottish coalfields had a virtual monopoly in supplying the Scottish railway companies as a result of their relative isolation from coalfields to the south.[8] And, in addition, both Lanarkshire and Fifeshire produced steam coals which could compete for bunkering contracts in Scottish ports with Newcastle or Welsh coal. Welsh smokeless might be preferred for transatlantic liners sailing from Glasgow or for the Navy, but cheaper Scottish coal was good enough for the majority of freight steamers. Once again, distance helped to reduce competitors' power to push into Scottish markets. Demand from steam transport, therefore, built up more quickly than in other coalfields not producing special types of coal, and was up to 7% or 8% in the 1850s and 60s, and to 13% or so by the 1880s, at which level it stabilised.

The growth of exports as an outlet for Scottish coal began to be significant in the 1860s, and continued to grow right up to 1913, with the largest expansion in the 1890s. By 1887 they accounted for 14% of output, and when the coal export tax was imposed in 1900 they had grown to take almost a quarter, a proportion which fell during the currency of the tax but was recovered by 1913. The chief reason for this considerable increase in the last quarter of the nineteenth century was the extension of the Fifeshire coalfield, coupled with the creation by the North British Railway of a new port at Methil. The new collieries of this district were not very well placed to supply existing markets in the Central Lowlands; the potentialities for developing markets for coastwise shipments were limited by Durham and Northumberland competition to the south and by geography to the north; they were virtually forced, therefore, to seek outlets in Northern Europe – which they did with considerable success.

Finally, there is the large demand for Scottish coal for local domestic and general manufacturing uses. As everywhere, except North-East England and the iron areas of the West Midlands, these markets were easily the most important ones at the start of our period, and suffered a proportionate decline as others were developed. But unlike the cases we have considered so far, this decline had halted by the middle of the nineteenth century, and thereafter the share of output going to these uses remained stable at between 45% and 50%. Within this aggregate there was a large decline in the relative place of household consumption and a corresponding rise in industrial demand.

Once again, as with the railway and steamship demand, the high share of output going to local domestic and general manufacturing uses was largely a function of the isolation of the Scottish coalfields from southern competitors, given that household and general manufacturing demand in Scotland grew at approximately the same rate as demand in general. Except in export markets, Scottish coal did not compete to any significant extent with that from other coalfields; and, conversely, other coalfields only competed to a minor degree in Scottish markets. Consequently, the general expansion of industry in Scotland, the still-increasing use of steam engines, the improvement in living standards, and the extension of railways to new areas were all reflected in demands for Scottish coal. It is also fair to point out that the whole of Scotland is here treated as 'local' – a much larger area than that pertaining to any other coalfield, though the high degree of concentration of population and industry in the Central Lowlands practically offsets this.

We come now to the group of coalfields which, towards the end of the nineteenth century, came to be known as the Federated Area, from the fact that they were the districts in which the Miners' Federation of Great Britain was formed and first operated. They can also be called the inland coalfields, both because they were (with insignificant exceptions) some distance away from the coast, and, more importantly, because their markets were almost entirely inland, except for Yorkshire and the East Midlands in the last two decades of our period. Several of these markets were shared, to a greater or lesser extent, between two or more of these coalfields. One major consequence of the internal nature of their markets was that their prices were more stable than those of the major exporting coalfields, and this meant that wages also were steadier. As a result the miners in these districts were able to throw off sliding scales earlier than elsewhere, and this, together with the fact that longwall was the standard method of mining, thus limiting the degree of variation of local customs, formed a bond between them and resulted in 1888 in the formation for the first time of a stable union covering more than one district – the Miners' Federation of Great Britain. However, this is anticipating. For the moment let us examine the markets of these inland coalfields, bearing in mind that, as time went on, they became increasingly competitive one with another.

Until the 1840s, the Yorkshire coalfield was of only moderate size, meeting local demands from households and industries – wool textiles primarily, together with the metal trades of Sheffield (which, according to Porter,[9] took over 300,000 tons a year in the mid-1830s), and the blast furnaces around Bradford and Leeds which smelted the iron ores found in the coal measures. Rapid expansion began in the 40s, but the structure of markets was little altered until the last quarter of the nineteenth century. The iron and steel industries provided a small but growing market for Yorkshire coal from the very beginning of our period. Apart from a brief spurt in the 1830s, however, its growth did little more than keep pace with coal production up to 1869, in part because Durham coke was able to compete in the high-quality iron and steel works of South Yorkshire. By the 1860s, however, the practice of making coke from the small coal of the rich Barnsley Bed seam was established, and the great iron boom of the early 1870s stimulated this development further, both through the acquisition of collieries by Yorkshire iron and steel firms and through the opportunity offered to independent coalowners by the absorption of Durham coke-producing capacity in supplying its own local markets. Coupled with these was the growth of pig iron manufacture on the newly opened Lincolnshire and Northamptonshire ore-fields, both of which could be supplied economically with coke from Yorkshire. By 1873 it was said that Durham had lost the West Riding and Lincolnshire markets almost entirely.[10] The result, for a time, was a substantial, but not spectacular, increase in the proportion of output going to the iron and steel industries – a development contrary to that in all other coalfields at the time (except the North-East, where the proportion was stable), and caused almost entirely by the new works on the Jurassic limestone belt. By 1887 the figure was 17% of output. Thereafter there was a slight decline in the proportion as export and transport demands grew rapidly; but nothing like the fall in the relative position of iron and steel occurred in the Yorkshire coalfield's markets as elsewhere.

Reference has already been made to the increase of exports. These were of negligible importance until the surplus capacity after the great boom of the early 70s drove the coalowners to seek fresh outlets. That they should have looked to exports and coastwise shipments was largely the result of the fact that much of the recently created capacity was in the eastward extension of the coalfield, nearer to Hull, Goole, and Grimsby than the older-exploited mines. As time went on this tendency was accentuated, and the sinkings, in fact, began to be made with export markets primarily in mind. This was the case with most new developments after about 1890, the latest sinking before the outbreak of the First World War taking place within ten miles of Goole. New docking facilities were constructed at all three ports, together with a completely new coal-shipping terminal at Immingham, near Grimsby. Because of the nature of the export markets, the steam coal seams tended to be the ones on which exploitation was concentrated. This, in turn, encouraged the expansion in the use of Yorkshire coal by shipping and the railways, which together had reached 12% of output by the twentieth century.

Until the early 1870s, the transportation of Yorkshire coal outside the district to internal markets was as unimportant as were exports. In 1869 coastwise shipments were negligible, and only about 5% of output was railed southwards. But in 1871 there was a lengthy rate 'war' between the Great Northern Railway and the Midland Railway, as a result of which there was a considerable expansion of coal railed out of Yorkshire, so that by 1872 practically 10% of the output went southwards on these two railways. For a time, much of this went to London; but soon normal economic considerations reasserted themselves, and in 1878, for example, the Barnsley correspondent of the *Colliery Guardian* could say that 'the business doing with London has been anything but brisk, and now the Great Northern is working extensively in the Derbyshire coalfield it is not likely to be very much better, for the company . . . will prefer carrying the coal the short in preference to the long distance'.[11] However, the proportion railed out of the district was maintained and even increased. The quickly found and quickly lost London markets were replaced by others in East Anglia (greatly helped by the building of the joint GN/GE line south-east from Lincoln in the 1880s), and the south-west Midlands along the Midland Railway's line to Bristol.[12] These continued to grow faster than total output until into the 1900s, when, under the influence of the very rapid expansion of the Warwickshire pits, further growth was partially checked, and the proportion of Yorkshire's output railed southwards fell back to about 12% in 1913. But in the meantime the growth of coastwise shipments, mostly of house- and gas-coal for London,[13] had increased by enough to hold the total share railed and shipped to the south at 17–18%.

With the expansion of all these other demands – iron and steel works at first, exports, steam transport, and southern and eastern manufacturing, gasworks, and household uses – local household and general industrial consumption took a steadily declining proportion of Yorkshire coal from the 1870s onwards. Up until that time these two uses had been dominant, the growth of output to over 11 million tons in 1870 being borne primarily upon the expansion of the textile industries, upon their mechanisation, upon the increased numbers employed in those and dependent trades, and upon the rise in their standards of living. According to the 1871 Royal Commission's estimates, 4.2 million tons of coal were used in general manufacturing in the West Riding.[14] Steam power in the woollen and worsted industries itself probably accounted for over a fifth of this,[15] to which should be added up to 0.5 million tons for the linen industry of Leeds and Barnsley and the cotton factories of towns bordering Lancashire. Then there were dye-works and other non-engine requirements. In addition, textile engineering works were important coal users. Then there were the gasworks, the waterworks, the brewers, millers, bakers, and so on, catering for the textile workers and all the rest. Though clearly the most important influence, the West Riding textile industries were not the only ones bringing about the expansion of Yorkshire coal output. The iron and steel industries have been dealt with already, but there were others as well. Textile engineering was not the only sort of engineering in the district; there were brickworks to the east of those districts to which Pennine stone could be easily transported; there was a small glass industry; lead was mined and smelted; there were tanneries; and, of course, the workers in all these and other industries, and in agriculture too, required beer, bread, shoes for the horses that carried the things they made and bought, and so on. Furthermore, the West Riding was not the only 'local' area served by Yorkshire collieries: parts of the other two Ridings and of Lincolnshire were also within their orbit. And even more important, they sent at least 0.75 million tons, probably over 1 million, via the Lancashire and Yorkshire or the London and North-Western Railway into South Lancashire.[16]

All told, local (including Lancashire) domestic and general industrial consumption accounted for nearly 70% of output at the time of the 1871 Royal Commission, but only 50% in 1887, 40% in 1903, and 35% in 1913. This fall was, of course, more a result of greater changes in other demands than of influences in these markets themselves. It is worth noting, however, first that the Leeds and Barnsley linen industry contracted in the 1870s and 80s and by the end of that decade flax spinning in Leeds was finished;[17] and second that the woollen and worsted industries' growth, whilst maintained up to the early 1890s, was checked then, and scarcely resumed until a few years before the First World War. It can also be pointed out that there was little, if any, increase in the amount of Yorkshire coal railed to South Lancashire until the 1900s, when it may have risen to not far short of 2 million tons.[18]

Though the Derbyshire and Nottinghamshire coalfield is geologically the same as that of Yorkshire, there are good reasons for considering them as separate fields. Perhaps best of all is the fact that the statistics were kept separately, whilst the small Leicestershire and Warwickshire coalfields were sometimes lumped in with the East Midlands, with which for a good part of the period they formed the Midland Inspection Division. But there is also a major quality difference which goes some way to justifying separate treatment of the East Midlands and Yorkshire. This is the fact that the Barnsley Bed seam cokes much more easily than its continuation southward into the East Midlands, where it is known as the Top Hard seam. This is reflected in differences in markets between the two districts; as also, of course, is the relative closeness of the coalfields of the Midlands Inspection Division to London.

Like Yorkshire, the output of the East Midlands and the two small coalfields to its south was for a long time disposed of almost entirely locally. Until the coming of the railway only canals and rivers allowed the carriage of coal over any considerable distance without a prohibitive addition of costs of transport; and even they were unable to take any substantial quantity to the major southern market, London, at prices which could compete with the North of England seacoal. The most brought by canal to London in any one year was 72,000 tons in 1844, most of which probably originated in Warwickshire. Two years later, however, there came the first tentative beginnings of a railborne supply for London – 8,000 tons in 1846, rising to 55,000 tons in 1850. Then, with the completion of the Great Northern main line, growth became very rapid. By 1855, 1,138,000 tons were railed into London, and it is almost certain that this came entirely from the East Midlands and Warwickshire, constituting not far short of a third of the combined district's output. After that the rate of growth was slower; but the 2 million ton mark was passed in 1864, 3 million tons only three years later, 4 million tons in 1871, 5 million tons in 1873, and so on to a maximum of nearly 8 million tons in the late 1880s. Except for the brief interlude of fierce competition between the GNR and the Midland[19] in 1871, with its aftermath stretching on to the opening of the GN line into Derbyshire in the mid-1870s, most of this railborne coal came from the East Midlands, with a little from Warwickshire, and small amounts of special coals from other districts. In addition, the East Midlands and Leicestershire sent substantial amounts of coal into Central Southern England, the South-Eastern Midlands, and East Anglia.[20] It was these markets which kept up the increase in coal railed out of the coalfield districts after 1890, when the London market for railborne coal almost ceased to expand as rail freight rates rose relative to those of coastal shipping. As a proportion of output, however, distant markets reached by rail had peaked around 1887, when they were 48% for the group as a whole and well over half so far as Derbyshire and Nottinghamshire were concerned. Thereafter there was a steady relative decline, to 39% in 1913, though this was made up of a larger fall in the case of the East Midlands (to around 35%) and an increase in Leicestershire and above all in Warwickshire. The opening of new, deeper pits, tapping richer seams, enabled these two coalfields to break into markets that had been virtually closed to them until the 1890s, and by the mid-1900s they provided London with the bulk of its kitchen cobbles, bakers' nuts, etc., and had an appreciable stake in the house-coal market there.[21]

Of local uses for any coalfield's output, that which can be most easily identified is the ironworks demand. So far as the Midland Inspection Division group of coalfields is concerned this was never of outstanding importance, though for the East Midlands component it had a period of considerable significance in the middle of the nineteenth century, during which it accounted for up to a fifth of production – the figure for 1855, 16% for the whole group, was probably around the peak year. This was in the heyday of smelting with raw coal from seams which bore some similarity to the Lanarkshire 'splint' coal. By the 1870s, however, this method was out-of-date, and the small proportion of output which continued to be taken by the iron and steel industries was coked in orthodox fashion.

As coking methods were devised, in the 1900s, which could use the smudge of the Top Hard seam, this proportion began to creep up a little, and by 1913 it was 8%.

Until the 1890s this group of coalfields remained entirely concerned with internal markets. However, the building of a direct railway link from Chesterfield to Grimsby in 1891 opened up the possibility of exports, though this was not taken up immediately to any significant extent. But, as in Yorkshire, the eastward extension of mines turned coalowners' attention to finding new markets at the same time as it made transport to the staithes cheaper. Until 1906 the export tax inhibited much development along these lines; but in the following year about a million tons of East Midlands coal was exported through each of Hull and Grimsby. By 1913, though the Hull figure had dropped, shipments through Grimsby and the new docks at Immingham had more than doubled.[22] At 6% of East Midlands output this was a modest but valuable addition to the district's markets. Coastwise shipments, it is worth noting, remained completely negligible, for it was cheaper for all but one or two collieries to send coal to markets in the South-East by rail.

Once the burning of raw coal in railway engines replaced coke, which was more usual in the middle of the nineteenth century, railway company contracts became a matter of importance to some East Midlands collieries, especially those furthest away from Derby, Nottingham, and the area of industry south of Sheffield. The Midland Railway normally took most of its supplies from this district, and the Great Eastern and Great Northern took a large proportion of their requirements thence from the 1870s and 1880s onwards respectively.[23] So far as the Midland Inspection Division as a whole was concerned, the share of output going to railways was around 5 or 6% at all times from 1869 to 1913.

There remain the local domestic and manufacturing uses which took the bulk of the district's output in the first part of our period, and which continued to grow in absolute terms right to the end. It will be seen from table 2.2 that the amounts involved were not large as late as 1869. The 1871 Royal Commission estimated that 1.2 million tons of coal were used in general manufacturing in Derbyshire and Nottinghamshire, 150,000 tons in Lincolnshire, Northamptonshire, Leicestershire, and Rutland, and 120,000 tons in Warwickshire (excluding Birmingham, where 600,000 tons were used).[24] In addition, a small amount was sent to the cotton towns on the Cheshire, Lancashire, Derbyshire border,[25] and there were the local gasworks. The total came to about 20% of output, which was not very much less than it had been earlier in the century. Domestic consumption, on the other hand, suffered a very drastic relative decline over this period, as other markets were found, whilst the scope for local house-coal sales was restricted to the rising population and living standards of a fairly small area by the existence of competitors to the north and the west. In 1830 it was still over half of output; by 1855 it was down to little over a quarter. In 1869 the figure was less than a fifth; 15% by 1887; 12% by 1903; and only a tenth by 1913. Local general manufacturing demand had meanwhile done rather more than hold its own in proportional terms, and by 1913 took over a quarter of output. The most significant factors in this increase were continued growth and mechanisation of the traditional Midlands hosiery and leather trades; larger sales of coal in Lancashire, where local production was almost incapable of further expansion; the success on a national scale of Burton beers; and the growth of engineering industries, especially in Derby and Coventry.

The most uncompromisingly inland of all coalfields were the group in the West Midlands – Shropshire, North Staffordshire, the Black Country, and Cannock Chase. Their markets were entirely in England, and to a very large extent within a few miles of the collieries, right through our period. Indeed, for the first half of that period much the greater part of their output was consumed in local furnaces, foundries, and forges, very often operated by the owners of the mines. From 1816 to the 1860s this proportion varied between 56 and 70%, partly according to the ups and downs of the trade cycle. Thereafter, factors operated to reduce it to around a third by the 1880s, at which level it was subsequently maintained. The first of these factors was the discovery and opening out of the Cannock Chase coalfield, which produced primarily house and gas varieties of coal, with none suitable for making metallurgical coke. A second factor was the gradual exhaustion of most of the Black Country pits, which also reduced local supplies of iron ore. Thirdly, there was the growth of coal consumption by other industries in the area, notably by those using the products of metallurgy (above all engineering), but also including others catering for a rising population with an improving standard of living. The main reasons for the maintenance of the iron and steel industries as important customers for the region's collieries were the continued expansion of the later stages of manufacture, where there was scope for special skills,[26] in all the areas concerned, using pig iron brought in from elsewhere (especially Cumberland and Furness), and the stability of pig iron output in North Staffordshire, which – along with Cannock Chase, but in contrast to Shropshire and the Black Country – was a (slowly) expanding coalfield. One can add that, of course, the whole *raison d'être* of some collieries in the region was the supply of the iron industry, and when the latter contracted (for whatever reason), the mines were closed.

Apart from iron and steel, other local industries provided the main market for the West Midlands group taken as a whole from the 1830s onwards, reaching a third of output in 1869 and around 40% at the later dates. For the Cannock Chase component this was certainly never the case; and very probably not for Shropshire either, where most of what was not used in ironworks must have gone into household consumption. North Staffordshire was the district where other industries than iron were most important. The potteries were large coal users at all times, taking probably over 600,000 tons in 1869,[27] or between 15 and 20% of output, and a similar proportion throughout. But the Black Country, too, had other important industries – the best examples are glass (using 200,000 tons in 1869), engineering and other metal-using trades, and leather working.

Household consumption never formed a really major market for the West Midlands coalfields as a whole during our period, though it was always important for Cannock Chase. At the beginning it took a fifth of output and had climbed back to a little over that figure by the end. In the meantime it had been as little as 10% in the middle of the nineteenth century. For most of the period, the collieries of the region supplied only very local areas with house-coal, and it was only with the development of Cannock Chase after 1870 that there was ever much to spare for dispatch further afield. Even then, very little went farther than the immediately surrounding counties.

The last group of coalfields with which we are concerned is that in the North-West of England – Lancashire and Cheshire, and the two small outlying districts of Cumberland and North Wales. The last two do not share many of the characteristics of the main component, and it is unfortunate that the compilers of nineteenth-century statistics often lumped them all together. For the group as a whole, table 2.2 makes clear the overwhelming importance throughout the period of local domestic and general manufacturing demand. There is also an unusually high degree of stability in the rate of expansion of all markets. The unspecified local markets never took less than 70% of output; and if the small shares going to steam transport, and to the iron industry are combined with them, the range is from 78% (in 1840) to 86% (in 1869). The only non-local markets of any significance for the group were those reached coastwise – and, besides Ireland, which was the main one, these included some not very distant places on the Lancashire, Cheshire, and North Wales coast.

The figures for the group as a whole are, of course, dominated by Lancashire, where the growth of markets depended largely on the expansion of, and technical progress in, cotton textiles and connected industries (especially engineering); though there was a number of other important coal-using industries in South Lancashire, and after the 1870s they gradually closed up on textiles. It was estimated[28] that in 1869 steam engines in cotton factories used nearly 2.5 million tons of coal, out of a total of 8.5 million tons used in the two counties of Lancashire and Cheshire, exclusive of Liverpool, where milling and other industries used a further 570,000 tons. In 1861, at the height of the boom before the cotton famine, factory engines had taken over 3.5 million tons, which makes the estimate for 1869 seem a trifle too low. A further 580,000 tons in 1869 was used in the alkali industry; probably half of the 770,000 tons used in salt production in Britain as a whole; and probably as much again in brickworks. Gasworks consumption must have been exceptionally high in this densely populated area, and there was also a concentration of consumer industries such as brewing. There is also included under this heading – quite wrongly, but for the sake of convenience – a small amount of coal which was railed to London. This came to about 200,000 tons in 1869, and consisted almost entirely of cannel, a special gas-coal, suitable for naked flame lamps (that is, fish-tail burners). With the invention of the incandescent mantle in 1885, this demand ceased to grow, and very soon began to decline. In addition to the manufacturing uses, in a district where the delivered price of coal was comparatively low and family incomes were relatively high, there was a large domestic consumption of coal.

The great majority of the small proportion of the North-West's output which was exported (as of that used in ships' bunkers) came from the Lancashire coalfield. With the more important coastwise shipments things were rather different. Up to the 1850s, approximately three-quarters of the total amount involved came from the Cumberland coalfield, which at that time had no other major markets. Even in 1869 this proportion was little lower, despite expansion in the meantime of the Cumberland iron industry. By the 1880s, however, Cumbrian shipments had fallen off a little, whilst those from Lancashire had risen to about 1 million tons. Thereafter shipments through the Cumbrian ports stagnated, whilst by 1913 those through Chester and the Lancashire ports had increased to 2.5 million tons, some part of which, however, consisted of Yorkshire coal.[29]

Having completed these surveys of individual markets, it is perhaps appropriate to say something more about demand in general. Reference was made earlier to five major trends, and we can now say a little more about how these developed. The most important of them, in the end, was the expansion of exports. These began in earnest in Durham and Northumberland in the 1830s, in South Wales a decade later, in the 1860s in Scotland, and the late 1870s in Yorkshire. Whilst their rate of growth was highest in the earlier stages of expansion, there is no doubt that their influence on the supply side of the industry was greater when the absolute amounts involved were large – after 1870 and, above all, in the 1890s and after the repeal of the coal export tax in 1906.

The fluctuation in the importance of demand from the iron and steel industries matches the rise and fall in the world position of Britain in these industries, and is not unconnected with it. Another factor of considerable importance in the 1870s, especially, was the development of fuel economy in these industries. In the period up to 1840, too, there was a useful reduction in the amount of coal used to manufacture a ton of pig iron, and this was coupled with a very substantial economy in the amount of fuel used in the finished iron and steel trades. Such was the expansion of iron sales at this period, however, that the effect of fuel economy on the demand for coal was completely swamped. But it was a different story after the early 70s when the very high coal prices provoked a search for fuel-saving innovations, which not only had a good deal of success, but had it at a time when Britain's comparative advantage in wrought iron was being rendered obsolete by the advent of cheap steel. Between 1869 and 1887 coal used per ton of pig iron made in the United Kingdom fell from 4.9 to 3.6 tons, and that used per ton of finished iron and steel fell from 3.2 to 2.2 tons. Fortunately for the coal industry there was little further progress before 1914.[30]

Whatever the long-run influence of other markets, one can say with some certainty that cyclical fluctuations arose primarily from these first two markets. This is especially evident in the two greatest booms of the century. The first, in the early 1870s, was set off, as the Select Committee of 1873 had no difficulty in concluding, by demand from the iron industry. The second, in 1899–1901, had as its dominant feature 'the demand for steam coal for navigational purposes (assisted no doubt by the requirements of transports for the war in South Africa) coming on top of depleted stores as a consequence of the prolonged strike in South Wales'.[31]

The impact of the rise in the consumption of gasworks, and at the end of the period electricity stations, cannot be identified very particularly, though it is clear that Durham, Yorkshire, and Lancashire were the main beneficiaries of expanding demand for gas-coal, and it is likely that the two former plus the East Midlands had the largest share of the power station market, since they supplied the biggest consuming areas in the South. However, neither gas nor electricity demand had a crucial influence on any coalfield at any time before the First World War, though gas-coal was of great importance for some individual collieries. The same is true of the steadily increasing demand for coal for steamships and railway engines, except so far as South Wales and its famous smokeless steam coal for bunkering was concerned.

Nevertheless, the steady increase was clearly of value to most coalfields, especially after 1873. By that time expansion in the rail network and in the intensity of its use had reached a point where a significant demand for coal existed, and its subsequent expansion was not hindered by the development of fuel economy. Railway engines may have been made marginally more efficient users of coal, but this was more than offset by the striving for higher speeds and by the proliferation of less intensively used services. Coal consumption per thousand train-miles appears to have reached a minimum of about 2 tons in the 1850s, to have changed very little from then until the end of the 80s, and then to have risen rather sharply to a new equilibrium of 3.5 tons by the early 1900s. The course of demand for coal for steamships, whilst following a similar course to that for railways, did so by very different means. The earliest marine engines were very greedy of coal, and though annual consumption per net register ton of steamships was brought down from 6.7 tons in 1830 to 3.5 tons by 1869, it was not until the developments of the late 70s and early 80s that steamships broke out of the confines of short-distance freight traffic plus the Atlantic passenger trade. The expansion of these trades had led to a significant demand for coal by the 1870s, but the very size of the demand for each marine engine inhibited the overall development of the market. By 1887, annual coal per NRT was down to 2.2 tons, and by 1903 to 1.5 tons, and the release of cargo space which this implied led to a virtual tripling of total coal requirements between the late 1860s and the late 80s, and a further doubling by 1914.

The fifth trend identified was the steady decline in the importance of house-coal, as its markets grew but slowly. This was an almost universal experience amongst British coalfields, with a small exception in South Staffordshire, resulting from the discovery of the Cannock Chase coalfield and the high degree of exhaustion of the Black Country seams. Perhaps the Midland Inspection Division, with its large rail consignments to London, was the nearest to constituting a major exception. The slower growth of house-coal demand than of demand for other types was almost entirely a reflection of what was happening in other markets. It certainly was not in any way a result of the preference of coalowners, for one major virtue which the market for house-coal had from the point of view of the entrepreneur was its relative freedom from short-term (though not from seasonal) fluctuations. Of course, demand was not completely insensitive to price, but it was far from elastic. To the upper and middle classes, good fires in more than one room in the house were a necessity. To the better-off working man, at least his living-room fire must be well-stocked. And if poorer people were compelled to use coal more sparingly the higher its price, at least it came second only to soup kitchens in the list of means of charitable distribution.

Finally, another general development may be noted, which was at times the most important market influence of all. This was the solid chunk of demand from factories, workshops, breweries, and so on, which grew steadily in amount throughout the period, and only suffered any significant proportionate decline towards the very end. This was of major importance in all the inland coalfields and in Scotland, especially in the last thirty years or so of the period. It had been very important by the 1850s in Yorkshire and from the very beginning in Lancashire. Even in South Wales and the North-East it had some significance in the end.

The aim of this chapter so far has been to delineate the main features of change in the markets of the major coalfields and in the overall consumption pattern of the country. Now seems to be the point at which to examine the significance of demand as a whole to the British coal industry.

In the last two decades there has been a number of studies which have attempted to 'explain' the growth of industries, in the sense of attributing this much to demand-side factors and that much to supply-side factors.[32] These attributions often have a somewhat artificial air, since they cannot satisfactorily take account of the feedback effects of demand on supply and *vice versa*.[33] Another problem is their dependence on accurate estimation of the elasticity of demand. For the nineteenth-century British coal industry it is this which makes it extremely hazardous to use the model concerned. Annual data are available for both output and prices for a large part of the period, but it is clear that these cannot be used to estimate elasticity since the changes in the two series are often positively correlated, indicating that shifts of the demand curve must have swamped the influence of any movements which occurred along existing demand curves. Even if one attempts to remove the effects of cyclical influences by taking the average changes in output and in price between successive trade cycles, and furthermore, one deflates the coal prices by an index of general prices, the result is a nonsense so far as the estimation of elasticity is concerned. The largest rate of growth of output is actually associated with the largest rate of increase of prices, whilst the next largest rate of growth of output is associated with the largest rate of decrease of prices. Perhaps it would be sensible to conclude that what this tells us is that, by the second half of the nineteenth century (when the data permit this sort of calculation), demand was very inelastic over the range of prices which actually obtained; that the supply-side conditions were sufficiently stable; that the prices outside this range, where a different sort of elasticity of demand might have been experienced, were not reached; and that, given these conditions, the expansion of output was largely determined by the expansion of a demand that was, for practical purposes, independent of price levels.

This is probably going rather too far, but there is a good deal to be said in support of this line of reasoning, particularly so far as the second half of the nineteenth century is concerned. Reasons have already been offered why elasticity of demand for house-coal should have been low, though these must be qualified by stressing that they apply in their full force only to (a) places close to the pits or within easy reach of cheap water transport and unaffected by monopolistic restrictions on the part of coalowners; or (b) elsewhere, to the period after prices had been brought down, mainly by the existence of the railways, to levels at which employed members of the working class could readily afford to buy reasonable amounts of coal. The effects of these qualifications are to suggest that in the middle of the century, from the 1840s to the 1870s, there probably *was* more price elasticity than later in the biggest house-coal market of all, London; and that this was also the case in most of the myriad of small, inland, rural markets. Perhaps it was also true of London before the 40s. The Limitation of the Vend seldom remained in tight control of the London market for very long owing to the expansion of output by interlopers, who had to be accommodated within the cartel sooner or later if it was to survive.[34] But until the coming of competition from rail-borne coal, it did have periods of considerable success in keeping up prices. This implies that there was a fair degree of inelasticity of demand over the range of prices concerned, or else, surely, the temptation to chisel would have been too great. But experience after the coming of rail-borne coal suggests that there may have been more elasticity at prices below those which actually obtained earlier.

The hypothesised changes in the pattern of elasticity of demand for house-coal are consistent with the data on house-coal sales given in table 2.1. Combining these with national population estimates[35] we can deduce rough figures of consumption per head. These remained pretty well stationary from 1816 to 1840, rose by about a quarter to 1855, by 17% from then to 1869, and by something over 20% from then to 1887, where-after they remained virtually unchanged. Since this pattern in no way reflects any estimate of national income per head, and since it is only in the 1840s that it can be said to reflect a fall in prices in London, it seems reasonable to suppose that the extension of the railway system was the main influence.[36]

So far as industry is concerned, except in iron and steel and one or two others of much less importance – and, of course, gasworks – coal was not a really large proportion of the cost of production for any manufacture. This implies that elasticity of demand was likely to be low except where a technology using another source of power could be substituted without much difficulty or a significant change in costs. Such technologies *were* available in many of the major manufacturing industries of the early nineteenth century, especially in textiles until the 1840s or 50s, according to von Tunzelman's study of the diffusion of the steam engine.[37] However, in the light of the historical record of the relative growth of steam and water horsepower,[38] it is hard to believe that a sufficient supply of suitable sites for cheap waterpower existed, even by the 1830s, to make this technology truly competitive with steam at that time so far as the *expansion* of production was concerned – though doubtless it had been twenty years earlier. One qualification must be added, and that is that there clearly was some measure of long-run elasticity in the demand for manufacturing coal, at least in the first half of the nineteenth century. This is attested by the search for more economical engines. Watt's chief advantage over the Newcomen and earlier engines lay in fuel saving, and the same was true of the later inventions of McNaught and others in compounding, and of the Cornish engine improvers whose work von Tunzelman analyses so carefully.[39]

Coal for locomotives and steamships was, as with general manufacturing, a proportionately fairly small item in the cost of transport by rail or by sea, at any rate for the period after steam shipping demand became significant. Admittedly, the last qualification implies that, earlier, coal costs might have been instrumental in restricting the expansion of steam shipping; though actually, it seems, this was only a smallish part of the explanation of the slow transfer from sail to steam. A more major influence – again similar to the case of the stationary steam engines in factories – was a technical one, namely the effect of the very high rate of consumption of coal by the early marine engines in cutting down the available cargo space.[40] Together, these ensured that the tonnage of steamships remained below that of sailing ships until into the 1870s.[41] However, competition between sail and steam was normally confined to a limited number of trades and routes at any time, the majority being firmly in the hands of one or the other. Consequently, it may be reckoned that the demand for coal for steamships operating in those trades and routes in which they had established a clear advantage at any time was generally pretty price-inelastic. As for railways, their considerable superiority for passenger carriage was established quite rapidly, and as early as 1844 they were reckoned to be without effective competition in this aspect of their business.[42] A little over twenty years later the same was said of most freight transport,[43] but for many trades and routes it could have been said long before that. Even such an efficient and well-placed canal system as that of the Bridgwater Trustees found that 'some trades were easily lost', and though the quantity of goods they carried rose from *circa* 700,000 tons in the early 1830s to 1,500,000 tons by about 1850 and 2 million tons by 1860, the net profit of the canal, though fluctuating greatly, was never again as high on average as it had been in the 1820s.[44] On the whole, then, it seems reasonable to conclude that transportation's demand for coal was mostly price-inelastic, especially and increasingly from about 1850 onwards.

What, then, about the industries for which coal was a substantial element in costs? Iron and steel were by far the most important of these, and it is on them that we shall concentrate our attention. The first point to make is that, throughout our period, there was no effective substitute technology to that using coal. In other words, over the range of coal (and coke) prices which obtained at any time after 1800, charcoal was always too expensive to capture a significant fraction of the market. Discussion of the elasticity of demand for coal for use in ironworks, therefore, largely boils down to discussion of the elasticity of demand for iron itself (including steel). This is readily apparent from the very close correlation between coal and iron prices,[45] implying that ironmasters could recoup their outlays on coal if they went up. (Actually, it is more likely that the sequence was the other way round, and that it was the rise of iron industry demand for coal in a boom which resulted in a rise in the price of coal.) This is not to say that there was never any long-run price elasticity in the iron industry's demand for coal. Clearly the same considerations applied as applied to economy in steam engines, though it would seem that they did not apply very strongly until the 1870s, except perhaps in the very early part of our period. The amount of coal used per ton of pig and finished iron appears to have gone down by around 15% between 1816 and 1830, but thereafter to have fallen only very slowly. By 1869 it was only 8% below the 1830 level. But then the very high coal prices of the early 70s proved a tremendous stimulus to economy in the use of coal, especially in blast furnaces.[46] The result was little change in the ironworks demand for coal between 1872 and the following cyclical peak, despite continued growth in iron and steel production, with a consequent persistence of excess capacity in coking coal pits and of low prices. The low prices do not appear to have had any significant influence on the quantity demanded, though it is possible that this is misleading if the demand for iron itself was extremely inelastic. The high degree of volatility of iron prices does not indicate this, however, reflecting, as they probably do, a high degree of volatility in domestic capital formation.[47] It is also pertinent to add that coal economy in ironworks scarcely improved at all between the 1880s and the First World War.

Gasworks, like iron and steel, used large amounts of coal, which was one of their major costs of production. However, the secular rate of growth of gas consumption seems to have been quite strongly positive until around the end of the nineteenth century, suggesting that output was relatively impervious to price changes, being determined principally by a mixture of the gradual extension of the availability of gas supply, and of technical progress which made possible new ways of using gas – in gas fires, in ovens, and so on. Certainly the quality of illumination provided by gas (particularly when the Welsbach mantle came into use), and the degree of convenience associated with its use for cooking and, to a less extent, for space heating, meant that it was often selling in markets where its superiority ensured that there were no close substitutes. Moreover, there was a close correlation between coal prices and the price of the coke which gasworks produced as their main by-product, so that the gas undertakings were partially insulated from the effects of rises in raw material costs. There does not appear, even, to have been any effective attempt to economise the use of coal in gasworks until the 1900s, when, though coal prices were rising, the main influence was almost certainly the onset of serious competition from the nascent electric supply industry.[48]

When we come to examine the elasticity of demand in export markets it is, of course, not quite the same thing at which we are looking. In considering internal markets we have been treating the British coal industry as a unit, and ignoring the existence of many competing firms, for each of which demand was far more elastic than it was for them when taken together. In some export markets, however, there was competition from foreign producers, and hence the demand curve facing the British industry as a whole must have been a bit more like those facing individual coalowners in home markets. But this did not apply to every overseas market. Most of those situated on coasts, which included the bunkering stations that took a large proportion of exports, could be more cheaply supplied from Britain than from anywhere else, since no foreign coalfield of any size was situated close to a sea coast. And, in fact, it seems probable that the majority of British coal exports did go to such markets. Certainly the high degree of volatility of export prices – even higher than of home prices often – does not suggest particularly elastic demand curves. Rather, it probably reflects the bodily shifting demand curves of a highly cyclical situation, coupled with a lagged supply response and comparatively low elasticity of demand; for if demand had been elastic, price fluctuations would surely have been damped.

In conclusion, the upshot of this discussion is that elasticity of demand for the British coal industry taken as a whole was low, especially in the second half of the nineteenth century. This was only just beginning to change, with the advent of oil as a competitor, in the last few years before 1914; and, at that time, it was only the steamship market that was directly affected to any serious degree. Such price elasticity as there was in markets for coal was mainly to be found where prices were relatively high owing to the absence of cheap means of transportation for a good with such a high weight-to-value ratio as coal. In addition, there was some long-run elasticity in some industrial markets and in shipping, but this was mostly made evident by the search for means of fuel economy in steam engines. Such economies, which superficially may seem to have a tendency to reduce demand for coal, in fact did the opposite by encouraging the diffusion of coal-using technologies. That did not apply, however, in those cases where there was no viable alternative technology to those using coal. These became increasingly prevalent from the 1840s or 50s on into the twentieth century, but included the iron industry throughout our period.

3

CAPITAL

Responses to the rising demand for coal resulting from the industrialisation first of Britain, then of other parts of the world, and from the spread of steam transport, entailed the exploration and exploitation of new areas of the coalfields, the sinking of new pits, the introduction of new techniques and machinery, and the bringing together of a bigger labour force. As Morris and Williams say of the South Wales coalfield in the middle of the nineteenth century, 'the lucrative opportunities the industry offered, or seemed to offer, not only stimulated men of enterprise within the coalfield itself to embark upon or to extend production but also attracted men from outside into the ranks of the coalowners. The hope of profit stimulated capital investment.'[1] In this chapter we shall be concerned with these entrepreneurs and their response to the hope of profit. That is, we shall examine the amount of capital involved in the coal industry and the course of its growth, and the sources from which it was obtained. These will be discussed in turn.

Capital formation

The estimation of the capital stock and of capital formation in the coal industry present peculiar difficulties, as well as the more general problems of lack of sufficient records. In particular, there is the fact that the shafts and roadways underground, which constituted around two-fifths of the cost of a colliery in 1914 and a good deal more up to the 1870s, are always unique creations. There is thus no possibility of imputing a standard price to them, even if it was always possible to categorise them according to depth and size of colliery, which is far from being the case. The only way, therefore, in which it would be theoretically possible to arrive at accurate figures of the historical cost of gross capital formation in coal mines would be by examining the accounts of the various colliery concerns. And it would have to be every single one of them, since there is no way of relating the figures pertaining to a sample group to the whole satisfactorily.[2] This is, of course, impossible in practice, since most colliery records have not survived, and a number were never properly kept in the first place, probably the great majority in the first half of the nineteenth century. Often, indeed, the owner-manager of a one-man business – and these were by far the most numerous, if not the most important, up to the 1880s, and probably had contributed a really substantial proportion of output up to the middle of the century – had no accurate knowledge of the amount of capital he employed. He had the traditional figure, probably, of the cost of sinking and opening out; but beyond that he often only had the figure on which he had paid probate duty, for there was seldom any proper separation in the books of capital account and revenue account, and investments as well as maintenance were included in the 'costs of getting'. If and when he turned the business into a limited company, a more or less accurate valuation might be made, its accuracy perhaps depending on whether or not it was intended to bring in any outside capital. But, clearly, this is no basis on which to make worthwhile estimates of capital stocks and flows.

Since all idea of approaching the problem from the side of the individual concerns – other than as a check in certain cases – must be abandoned, there remains only the approach from the side of the industry as a whole. Even here there is no easy answer. The only aggregate statistics which exist are those of annual output, published in *Sessional Papers* for the years from 1854 onwards, with more or less reliable estimates for earlier years, and those of numbers of men employed, which go back to 1851 on an annual basis, with earlier estimates based on the censuses. Using one or other or these as a means of estimating capital implies making assumptions either about the capital–output ratio or about the amount of capital per man. Whilst it is conceptually possible to use either, in practice there is no satisfactory basis for making the latter assumptions. Capital–output ratios, on the other hand, were not without interest to a number of contemporaries, and it seems possible to make use of their ideas. Before doing so, however, it is as well to get clear exactly what the capital stock of the industry, which we are trying to estimate, consisted of.

As has just been indicated, the shafts and underground roadways were a major part of the industry's capital at all times, more especially so when mechanical equipment was limited. By the beginning of the nineteenth century there was usually some sort of winding and haulage machinery, though this might involve nothing more than ropes and winches, and this sort of primitive equipment was clearly still quite common at the time of the Children's Employment Commissioners' enquiries in the early 1840s, as is testified by some of the splendid line-drawings which illustrate their reports. However, by that time, steam engines were normal in collieries of any substantial size, for pumping as well as for winding; mechanical haulage was beginning to spread; and mechanical ventilation was only a decade or two ahead.[3] These were the main capital requirements on the equipment side by the end of the century, together with the buildings and headgear to house them; though by then coal preparation plant had also become a sizeable item, and in some cases coke ovens were also included in a colliery's equipment. In addition, there were various smaller items, of which the most important was probably corves or tubs.

Apart from the pit, underground equipment, and pithead gear – and often the land on which the latter was erected – colliery firms usually had to finance other capital outlays which were less directly connected with the productive process. Foremost amongst these were investments in means of transportation. In the case of large collieries in the early nineteenth century, this may have consisted of wagonways or railways, staithes or spouts, wagons and horses, keels or other vessels. Later, railway sidings, wagons, shunting engines, etc., became the main items of this kind. In some cases – for example, Cramlington Colliery in Northumberland[4] – very substantial proportions of a firm's capital might have been laid out on this sort of thing – on a private railway in the case of Cramlington, which was also used by other collieries at rates which one at least thought were excessive.[5] In the estimates which follow, based on capital–output ratios, it may be reckoned that, in principle, the more spectacular examples of this sort of investment are excluded, only the 'normal' amount of sidings, etc., necessary to connect with public means of transport being included, together with the standard complement of 'private owner' wagons. Similarly, colliery firms' investments in workers' houses, and in such things as welfare institutes, chapels and churches, schools, and so on, are excluded; even though in the case of colliers' houses, they might be a necessary investment in order to attract a workforce to some particular individual pits.[6]

In making capital stock estimates based on capital–output ratios, one thing it is certainly not safe to do is simply to take the annual output statistics and apply a ratio. For one thing, fluctuations on the demand side of the industry invalidate that procedure, and for another, the contemporary estimates of capital–output ratios were related to output capacity rather than to the realised output of any particular year. The best that can be done is to make estimates of the stock of capital for the years of high demand, when output was getting pretty close to capacity, but when overmanning of the pits, in an attempt to raise output when productivity was falling, had not yet set in. Such estimates, together with the stages by which they are arrived at, are shown in table 3.1. To them have been added estimates of the implied capital per man (at constant prices), using the labour force statistics presented in chapter 4 below.

The estimates of the capital–output ratio are the crucial ones in this table, and require justification. The most accurate are probably those for the later years, so that it will be easier to work backwards. In the last few

Table 3.1 *UK capital stock in coalmining at intervals, 1816–1913*

	Output (million tons)	Capital–output ratio (shillings per ton)	Capital stock at current prices (£m.)	Index of colliery opening costs (1886 (= 100)	Capital stock at constant prices (£m. at 1886 prices)	Capital per man at constant prices (£ at 1886 prices)
1816	17.4	5.0	4.4	112	3.9	43
1825	21.9	5.5	6.0	125	4.8	45
1836	31.0	6.0	9.3	117	7.9	59
1846	46.0	6.25	14.4	112	12.9	67
1856	66.6	6.5	21.6	119	18.2	74
1866	101.6	6.75	34.3	121	28.3	88
1872	123.5	{7.5 9.0}	{46.3 55.6}	144	{32.2 38.6}	{77 93}
1882	156.5	7.5	58.7	112	52.4	104
1889	176.9	8.0	70.8	107	66.2	118
1899	220.1	9.0	99.0	121	81.9	112
1906	251.1	9.5	119.2	119	100.2	114
1913	287.4	11.0	158.1	129	122.6	108

Sources: output, based on table 1.1 above, and the sources used there. Index of Colliery of Colliery Opening Costs; table 3.A1 below; for remainder, see text.

years before 1914, the generally accepted figure for the average capital in Britain's coal mines was 10s per ton of annual capacity.[7] This figure, however, was certainly not meant to apply to every coalfield separately. In the districts where coke ovens were often part of the equipment of collieries it should be a good deal higher, and likewise in those coalfields where there had been a high proportion of expensive recent sinking, notably South Yorkshire, the East Midlands and Warwickshire. In some other coalfields, the true figure was probably below 10s a ton, especially where there had been little recent new sinking, such as Lancashire and the West Midlands. However, the importance of new sinking should not be exaggerated, for there was undoubtedly some increase in the amount of capital involved in many of the older mines owing to the adoption of technical innovations. Between 1907 and 1913, for example, a large, old-established Nottinghamshire firm, with five relatively old collieries producing 1 million tons a year, spent £34,000 on the installation of mechanical cutters and £20,000 on the electrical plant they required, a total of over 1s per ton of capacity. At the same time normal replacements of such things as tubs were maintained.[8] On the whole, then, there is good reason for supposing that the accepted figure of 10s per ton was on the low side for the period from say 1910 to 1914; whilst when what we require is a figure for the year 1913 itself, when both equipment prices and sinkers' wages had risen significantly since 1910–11, it is certain that it should be higher than 10s. On the assumption that 10s 6d per ton would be appropriate for the quinqennium 1910–14, 11s would seem correct for 1913, and this figure has been adopted in table 3.1. It is possible, however, that it is still a little too low.

The 1899 ratio is the next, working backwards, for which there is any general evidence. In this case, there is a series of estimates of the amounts of capital needed in collieries of various sizes by a well-known North-Eastern mining engineer, J.H. Merivale.[9] These were as follows:

collieries up to 80,000 tons a year capacity – 12s per ton
collieries up to 150,000 tons a year capacity – 10s per ton
collieries up to 400,000 tons a year capacity – 8s per ton
collieries over 400,000 tons a year capacity – 7s per ton.

He gave an average overall figure of 9s per ton. So far as the larger collieries are concerned, the figure receives confirmation in the evidence of Emerson Bainbridge to the Royal Commission on Coal Supplies. Bainbridge, another eminent mining engineer from the North-East, who had become a large shareholder and a director in colliery companies in Yorkshire and the East Midlands, reckoned that £300,000 was needed to open a colliery of 3,000 tons a day capacity, unless exceptional conditions occurred in sinking.[10] There seems to be no reason why the estimates for the smaller pits cannot be accepted too. Indeed, the accounts of a number of medium-sized collieries in the inland coalfields confirm them, though it is doubtful whether a figure of 12s a ton can be taken as a fair average of all pits producing less than 300 tons a day. In fact, it probably applied only to the best-equipped. Nevertheless, the average of 9s per ton seems reasonable enough, despite the fact that it is so much lower than the estimate for 1913. In the interval, output per man-year fell substantially,[11] and as that falls so, *ceteris paribus*, does output per unit of capital. In addition, the mild inflationary trend must have had some effect in increasing the capital–output ratio at current prices.

The 1906 estimate of 9s 6d per ton is simply an interpolation between the two firmer estimates for 1899 and 1913. It is placed much closer to that of the former year than the latter both because of the faster fall of productivity after 1906 and of the faster rate of rise of prices.

The basis of the 1889, 1882, and 1872 estimates is various colliery accounts, papers on individual collieries in the *Transactions of the Federated Institution of Mining Engineers*, and Bainbridge's evidence to the Royal Commission on Coal Supplies already referred to, coupled with evidence about the cost of sinking, which will be discussed shortly. That the 1889 figure should be as much as 1s a ton lower than ten years later, whilst being itself only 6d a ton higher than 1882, is not surprising when one considers the history of the industry at this time. The years after 1888 were a period of technical progress in many pits, when attempts were made to compensate for the slowness of the previous ten or fifteen years of depression. The boom of 1889–91 was a very satisfactory one from the coalowners' point of view, unlike that of the early 80s, and provided substantial profits. The opportunity was taken by most firms to plough back a fair proportion of them.

The figure of 8s per ton of annual capacity in 1889 derives in part from a generalisation of G. Lewis, an experienced mining engineer, who reckoned that at that time it took £100,000 to open an average colliery, 'though some cost twice as much'.[12] Since this figure of £100,000 was certainly too low for the larger collieries being opened at that time in South Durham, South Wales, Yorkshire, and the East Midlands, it must be assumed to apply to medium-sized pits – indeed, the phrase 'average colliery' suggests just that. In other words, it applied to pits employing from about 600 to 1,000 men and producing from 200,000 tons to 300,000 tons a year – that is, around 8s per ton of annual capacity. The majority of the colliery accounts which we have examined show ratios of 7s 6d to 7s 9d per ton, but these were of rather above average size, and the one smallish colliery among them had a ratio as high as 9s 3d per ton.[13]

That the 1882 ratio is estimated as lower than that of 1889 is a result of two things. First, there was a definite, though slow, spread of capital-using innovations in the 1880s;[14] and second, and more important, there was the closing down of the great majority of really small mines on the outcrop of coal seams, which had begun in the later 1870s, and went very much further in the following decade. These mines, unlike the smaller ones to which Merivale referred in his later book, did not have the higher-than-average capital–output ratios. They were shallow, and usually very lightly equipped. This same reason would tend to make the 1872 ratio lower than that of ten years later. However, it is clear that there were offsetting factors, though not exactly how important they were. One of these was the relatively low output per man-year in 1872, when wages had already gone up a very great deal, though they were to reach even greater heights in 1873, and productivity was to plunge much lower. It is probable, therefore, that actual output in 1872 was some way short of theoretical capacity. It is also possible that, compared with later statistics, output was underestimated in (and up to) 1872. In addition to this, sinking and equipping costs were at an unusually high level in 1872, and the assumed capital–output ratio of that year must make allowance for this. Because of the imponderable nature of both these influences, two estimates are made here for 1872: a low one assumes the same capital–output ratio as in 1882, whilst a high one multiplies this ratio by the difference between sinking costs at the two dates. The truth probably lies somewhere between the two estimates, but is more likely to be nearer the higher than the lower figure.

Evidence for periods before the 1870s is very slight and scattered, and the ratios assumed for 1866 and earlier years are not much more than informed guesses. The earliest landmark we have is an estimate by the famous North Country viewer, John Buddle, in 1829, reckoning that the total capital in Durham and Northumberland was £2.1 to 2.2 million.[15] This amounts to 7s 3d per ton of actual output, but the ratio per ton of capacity may well have been more like 5–6s, since the Limitation of the Vend resulted in a good deal of underemployment of capital.[16] However, during the operation of the Limitation, theoretical capacity is not the relevant measure for the large North-Eastern coalfield; though it does suggest that a lower figure than 7s 3d per ton would be appropriate for the other coalfields during the 1820s and 30s. Lesser depths of working in all these districts except Lancashire reinforce this suggestion, though the smaller average size of collieries (not counting the tiny ones on the outcrop) was possibly some offset. After weighing all these considerations, an overall national figure of 6s per ton has been adopted here for the mid-30s. The two earlier figures are simply guesses based on the gradual adoption of more expensive equipment, with the falling cost of labour (i.e. labour for sinking and opening out), especially in the decade after the end of the Napoleonic Wars, as an offsetting factor to be taken into account. Similarly, the increases between 1836 and 1866 are assumed to come with the elaboration of equipment, partially offset by rising overall productivity.[17]

Before leaving the subject of capital–output ratios, it is worth pointing out how comparatively low they were relative to most manufacturing industries – and, of course, relative to the late twentieth century. A capital–output ratio of 11s per ton in 1913 implied a ratio in value terms of virtually 1 to 1. This compares with an economy-wide ratio of 3.8 to 1 on Feinstein's estimates.[18] In the course of a century or so the capital–output ratio in coalmining had risen by about 75% in real terms, and the rise from 1856 to 1914 was 54%. The latter can be compared with an economy-wide *fall* of 24% estimated by Feinstein.

The estimates of the stock of coalmining capital, so far as possible at times when output may be supposed to have been close to unstrained capacity, derived from the capital–output ratios just discussed are shown in column 3 of table 3.1 at the current prices of the years concerned. In order to make meaningful comparisons between them it is necessary to express them in terms of constant prices, and this has been done in the last two columns by using an index of the costs of sinking and equipping collieries specially devised for this purpose. The last column, showing capital per man, is included here because it seems the most appropriate place to show how it is derived, and also because it serves as some sort of check on the plausibility of the capital stock estimates. Comment on it, however, is reserved for chapter 10, when we discuss productivity.

The index of colliery opening costs was constructed from six price series – hewers' wages in the major coalfields (weighted by the numbers employed at each census or by interpolations between censuses) to represent the labour cost of sinking and opening out; wood and brick prices, taken to represent the cost of materials other than machinery; and engineers' wages and bar iron and steel prices, taken to represent the cost of machinery.[19] The weights given to each series were changed in the following way, to allow for the increasing importance of machinery both in sinking and in equipping pits:

hewers' wages	– 60 up to 1850, falling by 0.5 per year to 1910
wood prices	– 10 up to 1880, falling by 0.25 per year to 1900
brick prices	– 10 throughout
engineers' wages	– 10 up to 1850, rising by 0.25 per year to 1890
bar iron prices	– 10 up to 1850, rising by 0.25 per year to 1880, by 0.5 per year to 1885, and then falling by 0.5 per year to 1910 and by 1.0 per year subsequently
steel prices	– first coming into the index in 1886 at 1, rising by 1.0 per year to 1890, by 1.25 per year to 1900, and by 1.0 per year subsequently.

The full index is given in the Appendix to this chapter.

The difference between the stock of capital at constant prices at any one of these cyclical near-peak years and the next such year represents the value of new capital formation at replacement cost in the interval, on the assumption that the value of mines and equipment scrapped was exactly balanced by expenditure on renewals. This would clearly be an invalid assumption for any individual year, but it is probably not far from the truth over the course of a whole trade cycle. There is no need to make any allowance for depreciation, since the capital stock figures (being, in effect, estimates at replacement cost) already have such allowance built into them. These figures, shown in table 3.2, are the closest we can come to reasonably reliable estimates of capital formation in the industry. There just is not sufficient information available to put them onto an annual basis except in a very speculative fashion. The only thing one can say with any certainty is that the annual average over each trade cycle is seldom likely to have applied to any individual year, for capital expenditure tended to be relatively high in the later stages of a boom and on into the downswing (e.g. in 1872 to 1875), whilst being relatively low in the slump and the early stages of a boom (e.g. in 1868 to 1871). For this reason the final column of table 3.2 has not really got a great deal of significance, and it serves mainly to show that, until the twentieth century, there was rather little difference in colliery development costs from cycle

Table 3.2 *UK new capital formation at replacement cost over trade cycles, 1817–1913 (in £ m.)*

	Total at 1886 prices	Annual average at 1886 prices	Average value of colliery opening costs (1886 = 100)	Total at average current prices of trade cycle concerned	Annual average at current prices of trade cycle concerned
1817–25	0.9	0.10	122	1.2	0.12
1826–36	3.1	0.28	103	3.2	0.29
1837–47	5.0	0.45	104	5.2	0.47
1848–56	5.3	0.59	101	5.4	0.60
1857–66	10.1	1.01	109	11.0	1.10
1867–72 (a)	3.9	0.65	116	4.5	0.75
(b)	10.3	1.72		11.9	2.00
1873–82 (a)	20.2	2.02	122	24.6	2.46
(b)	13.8	1.38		17.2	1.72
1883–9	13.8	1.97	103	14.2	2.03
1890–9	15.7	1.57	115	18.1	1.81
1900–6	18.4	2.62	122	22.4	3.20
1907–13	22.4	3.20	122	27.3	3.90

Sources: this table is derived from tables 3.1 and 3.A1. The alternative figures for 1867–72 and 1873–82 are based on the alternative estimates for 1872 in table 3.1.

to cycle. From 1900, however, capital formation at current prices is noticeably larger than it is on a constant 1886 prices basis.

It is difficult to draw conclusions from this table since they have, in effect, been built into the figures by our assumptions about capital–output ratios. Perhaps the best that we can do is to point to one or two implications of those assumptions which are revealed or emphasised by the table.

So far as absolute amounts of capital expenditure in the country as a whole are concerned, it is readily apparent that there was a fairly smooth increase in the annual average right through to the twentieth century, with a pause only in the 1890s. If these outlays are compared with estimates of total domestic fixed capital formation,[20] a fluctuating relationship is observed. At average current prices of the cycle, colliery capital expenditure formed 1.5% of the national total in the first period for which comparison is possible, 1837–47. In the next period the proportion fell a little, to 1.4%, then rose to 1.6% in 1857–66. It then either fell substantially and recovered in the next two cycles, or, more likely, fell a little in each of them. In 1883–9 it rose to its highest point, at 2.2%, fell substantially to 1.2 and 1.1% in the next two cycles, and finally recovered to 1.7% in 1907–13. It is clear, however, that the greater part of the explanation for these fluctuations must be sought outside the coal industry, in the behaviour of other items of capital formation, notably those which were much more voracious for capital, such as housing, railways, and other public utilities.

Turning from the absolute amounts of colliery capital expenditure to the rates of change, it seems that there were three trade cycles when the annual average outlay showed rapid increases. The first was in 1837–47, when the average was nearly 100% above the previous cycle; the second was in 1857–66, when the average was also almost 100% above the previous cycle; and the third was in the period 1900–6, when the average was over 60% above the 1890s. There was only one cycle when capital expenditure fell off, namely 1890–9, when it was 20% below the 80s. The other trade cycles all showed modest increases in capital outlays. Of all these figures, the ones which seem most to call for explanation are the increases in the 1900s and the relatively small increases of 1867–72 and 1873–82. The reason for the increase in 1900–6 compared with the 1890s lies in two factors: on the one hand was the greater elaboration of colliery equipment and the greater difficulty of sinking; on the other hand, there is the incentive effect of the high prices of the boom of 1899–1900. The latter reason, however, prompts the question as to why the even more remunerative prices of 1872–4 did not produce a similar large rise in the rate of colliery investment. The answer to this is that it did, but that it was spread over two trade cycles, 1867–72 and 1873–82. In the former, high investment in 1872 was offset by particularly low investment in the middle of the cycle; in the latter the opposite occurred: the effect of the high prices was seen in continued new sinking into the middle 1870s, but this was followed by a period when capacity was clearly excessive and profits were low (or non-existent).

The differing progress of capital formation in the various major coalfields is brought out by tables 3.3 and 3.4, though it must be emphasised that the margins of error involved in these tables are quite high, and the figures can only be regarded as rough approximations. The methods used are the same as in the previous two tables.

These regional tables call for little further comment. The main local peculiarities distinguishing individual coalfields from the aggregate behaviour are five in number. Taking them in chronological order they are as follows:

1. The relatively high level of investment in Yorkshire in 1847–56 and 1867–72, with a very low level in the intervening period. The explanation of this probably lies partly in the behaviour of demand, since the 1850s were a period of rapid mechanisation in the woollen and worsted industries, whilst 1867–72 saw an especially large expansion of coking coal output in the district. But there were also supply-side factors in the beginnings of exhaustion of mines on the western outcrop, which was becoming apparent in the 1860s and was not compensated for by vigorous expansion in the east until the book of the early 70s.
2. The relatively low level of capital outlays in Scotland, South Wales and the

Table 3.3 *Capital stock in coalmining by coalfields, 1816–1913*

	Output (million tons)	Capital output ratio (shillings per ton)	Capital stock (£m.)	Index of colliery opening costs (1886 = 100)	Capital stock at constant prices (£m. at 1886 prices)
Durham & Northumberland					
1816	4.8	6.5	1.56	110	1.42
1825	5.3	7.0	1.86	121	1.54
1836	8.0	7.25	2.90	109	2.66
1846	11.0	7.5	4.13	109	3.79
1856	15.5	7.5	5.81	115	5.05
1866	25.2	7.5	9.45	116	8.15
1871	29.1	8.75	12.78	115	11.11
1882	36.3	7.75	14.07	111	12.68
1889	39.1	8.25	16.13	104	15.51
1899	46.1	9.25	21.32	119	17.92
1906	52.1	9.75	25.40	118	21.53
1913	56.4	11.5	32.43	128	25.34
Yorkshire					
1816	1.8	4.75	0.43	103	0.42
1825	2.2	5.25	0.58	114	0.51
1836	3.2	5.75	0.92	112	0.82
1846	4.2	6.0	1.26	105	1.20
1856	9.1	6.25	2.84	116	2.45
1866	9.7	6.5	3.15	122	2.58
1872	14.6	8.0	5.84	142	4.11
1882	18.5	7.5	6.94	111	6.25
1889	22.0	8.0	8.80	108	8.15
1899	26.9	9.0	12.11	122	9.93
1906	32.5	9.5	15.44	120	12.87
1913	43.7	11.25	24.58	128	19.20
Lancashire & Cheshire					
1816	2.0	5.0	0.50	118	0.42
1825	2.7	5.5	0.74	126	0.59
1836	4.2	6.0	1.26	122	1.03
1846	6.8	6.25	2.13	106	2.01
1856	9.7	6.5	3.15	124	2.54
1866	13.2	6.75	4.46	124	3.60
1872	18.4	8.0	7.36	145	5.08
1882	23.0	7.5	8.63	114	7.57
1889	24.6	8.0	9.84	108	9.11
1899	27.6	8.75	12.08	122	9.90
1906	28.4	9.25	13.14	119	11.04
1913	28.1	9.75	13.70	126	10.87
Midland Inspection District					
1816	1.2	4.5	0.27	101	0.27
1825	1.6	5.0	0.40	113	0.35
1836	2.0	5.5	0.55	107	0.51
1846	2.4	6.0	0.72	103	0.70
1856	4.3	6.25	1.34	115	1.17

Table 3.3 (*cont.*)

	Output (million tons)	Capital output ratio (shillings per ton)	Capital stock (£m.)	Index of colliery opening costs (1886 = 100)	Capital stock at constant prices (£m. at 1886 prices)
1866	8.0	6.5	2.60	121	2.15
1872	10.7	7.75	4.15	142	2.92
1882	15.6	7.0	5.46	112	4.88
1889	19.7	7.75	7.63	108	7.06
1899	27.6	9.0	12.42	124	10.02
1906	33.2	9.5	15.77	123	12.82
1913	38.8	10.5	20.37	130	15.67
West Midlands					
1816	3.0	4.75	0.71	118	0.60
1825	3.7	5.25	0.97	131	0.74
1836	4.6	5.75	1.32	121	1.09
1846	6.3	6.0	1.89	124	1.52
1856	8.0	6.25	2.50	126	1.98
1866	13.4	6.5	4.36	119	3.66
1872	15.7	7.5	5.89	142	4.15
1882	15.9	7.0	5.57	111	5.02
1889	15.6	7.25	5.66	108	5.24
1899	15.5	7.75	6.01	123	4.89
1906	14.9	8.5	6.33	120	5.28
1913	15.8	9.0	7.11	127	5.60
South Wales					
1816	1.9	5.0	0.48	109	0.44
1825	2.6	5.5	0.72	126	0.57
1836	4.8	6.0	1.44	117	1.23
1846	6.6	6.25	2.06	119	1.73
1856	8.9	6.5	2.89	120	2.41
1866	13.8	6.75	4.66	120	3.88
1872	15.2	8.0	6.08	137	4.44
1882	22.2	7.5	8.33	113	7.37
1889	28.1	8.0	11.24	105	10.70
1899	39.9	9.0	17.96	117	15.35
1906	47.1	9.5	22.37	118	18.96
1913	56.8	10.5	29.82	129	23.12
Scotland					
1816	2.0	5.0	0.50	118	0.42
1825	2.4	5.5	0.66	146	0.45
1836	3.5	6.0	1.05	126	0.83
1846	5.0	6.25	1.56	110	1.42
1856	7.5	6.5	2.44	116	2.10
1866	12.6	6.75	4.25	124	3.43
1872	15.4	8.0	6.16	156	3.95
1882	20.5	7.5	7.69	114	6.75
1889	23.2	8.0	9.28	109	8.51
1899	31.1	9.0	14.00	128	10.94
1906	39.2	9.5	18.62	119	15.65
1913	42.5	10.5	22.31	132	16.90

Note that for Durham and Northumberland the year 1871 is used instead of 1872.
Sources: as for table 3.1.

Table 3.4 *New capital formation over trade cycles by coalfields, 1816–1913 (in £ m. at constant 1886 and current prices)*

	Durham & North-umberland		Yorkshire		Lancashire & Cheshire		Midland Inspection District		West Midlands		South Wales		Scotland	
	con.	cur.	con.	cur.	con.	cur.	con.	cur.	con.	cur.	con.	cur.	con.	cur.
1816–25	0.12	0.30	0.09	0.15	0.17	0.24	0.08	0.13	0.14	0.26	0.13	0.24	0.03	0.16
1826–36	1.12	1.04	0.31	0.34	0.44	0.52	0.16	0.15	0.35	0.35	0.66	0.72	0.38	0.39
1837–46	1.13	1.23	0.38	0.34	0.98	0.87	0.19	0.17	0.43	0.57	0.50	0.62	0.59	0.51
1847–56	1.26	1.64	1.25	1.58	0.53	1.02	0.47	0.62	0.46	0.61	0.68	0.83	0.68	0.88
1857–66	3.10	3.64	0.13	0.31	1.06	1.31	0.98	1.26	1.68	1.86	1.47	1.77	1.33	1.81
1867–72	2.96	3.33	1.53	2.69	1.48	2.90	0.77	1.55	0.49	1.53	0.56	1.42	0.52	1.91
1873–82	1.57	1.29	2.14	1.10	2.49	1.27	1.96	1.31	0.87	−0.32	2.93	2.25	2.80	1.53
1883–9	2.83	2.06	1.90	1.86	1.54	1.21	2.18	2.17	0.22	0.09	3.33	2.91	1.76	1.59
1890–9	2.41	5.19	1.78	3.31	0.79	2.24	2.96	4.79	−0.35	0.35	4.65	6.72	2.43	4.72
1900–6	3.61	4.08	2.94	3.33	1.14	1.06	2.80	3.35	0.37	0.32	3.61	4.41	4.71	4.62
1907–13	3.81	7.03	6.33	9.14	−0.17	0.56	2.85	4.60	0.32	0.78	4.16	7.45	1.25	3.69

Note that for Durham & Northumberland the years 1871 and 1872 are used instead of 1872 and 1873.
Sources: derived from table 3.3.

West Midlands in the period 1867–72. In large part, this reflected the fact that the iron industries of Scotland, South Wales, and the Black Country were no longer in a position to take much advantage of the huge expansion of demand in 1870–2, whilst the great days of British coal exports were still just around the corner.
3. The comparatively high figure for capital formation in the North-East in 1867–71, and its obverse in the following cycle. To a large extent this is probably the result of our method of estimation overstating the true figure in the earlier and understating it in the later period. However, there certainly was rather less new sinking in this coalfield relative to its size in the boom of 1872–3 and its aftermath than in the others. Comparative shortage of proven reserves readily accessible with current technology seems to have been the main reason for this.
4. The slackening of capital formation in the West Midlands after 1882. This resulted mainly from disinvestment in the old-exploited Black Country mines and in Shropshire, where many of the best seams were approaching exhaustion or were being flooded. Except in the 1890s, the development of Cannock Chase, the slow growth of the Potteries coalfield, and the elaboration of equipment, especially in the larger collieries, were more than enough to offset this.
5. The slackening of investment in Lancashire in the 1890s, becoming disinvestment after 1906. This was also mainly the result of closure of pits as they reached exhaustion, especially in the eastern part of the coalfield.
6. The especially high capital expenditure in the last cycle in Yorkshire. The opening of the large new mines in the Doncaster area, resulting from geological discoveries coupled with improved sinking techniques, was the main factor responsible for this, aided by the sustained growth of export markets.

Finally, what general conclusions about investment in the British coal industry may be drawn from this attempt to delineate its statistical framework? None with any certainty, of course, since so much depends upon assumptions. But it would seem reasonable to suggest that, whilst increasing capital-intensity, resulting both from more elaborate equipment and from the need to overcome more difficult natural obstacles, was steadily raising the investment requirements of the industry, the main reason for rising capital formation was the demand for coal. The major bursts of investment in the mid-century period, in 1837–47 and in 1857–66, were both associated with rising demand from the iron industry, whilst the first of these was also linked with the expansion of markets consequent upon the establishment of the railway network. The investment of the boom of the early 1870s was similarly connected with the iron industry. Only the rising capital expenditure of the years after 1900 seemed to have owed as much to supply-side influences (i.e. capital-using technical changes) as to demand; and, even then, increasing exports and renewed expansion of steel output (after a pause in the 90s) were essential to the successful opening of many new collieries in South Wales, Durham, South Yorkshire, and Fifeshire. It does seem reasonably safe to conclude that rising demand was the mainspring of investment in British coalmining during our period, rather than technical innovations or geological discoveries, though both these played a part, and the latter had an important influence on the distribution of investment between the different coalfields.

Sources of capital

In some respects it is even more difficult to say anything about the sources of investment in the coal industry than it is to estimate how much capital was involved. For if the statistics of the latter are few, there is even less to indicate whence it came, especially in the first half of our period. Until Robert Hunt, Keeper of Mining Records of the Geological Survey, published *Mineral Statistics*, including a list of mines and their owners, in 1854, there is no comprehensive survey of the ownership of collieries; and even that list and its successors[21] only give the trading names of the businesses concerned. However, these lists do enable us, in conjunction with the *Return of Joint Stock Companies*,[22] to say something about the relative importance of limited companies, and about the place of iron and steel concerns in the ownership of the coal industry. But before the 1850s it is impossible to be at all precise about where capital came from.

Nevertheless, it is easy to make too much of this problem, for it is clear that, in the days before limited liability companies with London Stock Exchange quotations – and for a long time afterwards, for that matter – by far the greater part of capital came from sources local to each coalfield. And it is also clear that the principal source of funds must have been the savings of those local people who had incomes large enough to provide a sufficient margin for saving. This group comprised landowners, businessmen in whatever industries there were in the neighbourhood, merchants, professional men, and existing coalowners, who were certainly the most important component of the group.

The first of these, the landowners, had been the main original developers of coal mines; but by the beginning of the nineteenth century the heyday of their active involvement in colliery enterprises was past. Many were still active, of course, and some of them continued to be large coalowners[23] right through to 1914 – the Duke of Bridgwater's Trustees in Lancashire, the Marquis of Lothian in Scotland, Earl Granville in North Staffordshire, the Earl of Dudley in the Black Country, the Marquis of Londonderry in Durham, to name only a few. But though these 'feudal' enterprises might catch the eye, they were increasingly outnumbered by other types of entrepreneur; and by the end of the nineteenth century some of the more important amongst them had converted themselves into limited companies, and made use of the stock market to bring in extra capital. As J.T. Ward has said, 'there were many reasons for landowners' withdrawal', and there is little to add to his discussion of them.[24] The growing complexity of mining technology, difficulties with agents, accidents of inheritance, and increase in capital and managerial talent available from outside the landowning class, the growing need for personal supervision at a time when competition was growing keener – all these were influential, perhaps the last above all, for it emphasised what Ward calls 'the comparative tranquillity (sometimes more apparent than real) of a *rentier* class.'[25]

By 1800, the majority of coalfield landowners, and the owners of the great majority of coal-bearing land outside Durham and Northumberland, were content to leave the profit and risk of developing the pits to others, and to lease their minerals in return for payment of royalties; though later in the century, when joint stock companies became important, many landowners became shareholders in them, often substantial ones; and at all times royalty owners were apt to be drawn into a more active role if their tenants went bankrupt, or if there was a natural disaster which required recovery expenditure beyond their capacity.[26] By 1860 only the Londonderry and Durham (Vane-Tempest) estate enterprises remained as major landlord ventures even in the North-East. The rest were run by lessees.

The coalowners who took the leases on royalty were not in any significant way a homogeneous group. For the most part they were local men, though there were exceptions to this in most districts at most times, especially in periods and places where expansion was unusually rapid, such as South Wales in mid-century, and the eastern part of Yorkshire and the East Midlands in the 1890s and 1900s. They differed very greatly in their scale of operations, and, usually in the same proportion, in their social class. They sprang from various occupational backgrounds, but (leaving aside the colliery capital which was invested by iron-making firms) the most important single source of recruitment was the coal industry itself. By the beginning of the nineteenth century, the old-exploited Durham and Northumberland coalfield already had a substantial group of people who could best be described as professional coalowners – families whose main income was derived from the working of coal mines, and had been so derived for at least a generation, such as the Joiceys, the Lambs, and the Taylors (who, of course, bought themselves landed estates when they had been successful). There were similar but smaller groups in the other, less intensively exploited coalfields, such as the Nadins in Leicestershire, the Knowles in Lancashire, or the Barbers in Nottinghamshire. As the century went on, this class was greatly enlarged, and it is almost certainly true to say that at all times after 1850 (and earlier outside the North-East) it provided the majority of coalowners and the majority of colliery capital.

It is difficult to pick out any sources of recruitment to this class as clearly more significant than others, since they were so varied and numerous. An impression of this variety can be gained by looking at a few individual cases. A particularly unusual one is the Clay Cross Company, in Derbyshire, which was started by George Stephenson and the Liverpool railway financiers when coal seams were found in cutting a tunnel for the North Midland Railway. One of the 'Liverpool party' eventually became the sole owner, in 1871. He was Sir William Jackson, and he may then be said to have joined the ranks of the professional coalowners.[27] Another example from the same area is H.D. Pochin, an industrial chemist and inventor, who began in the 1860s to use his savings to buy up coal and iron firms and turn them into public companies, in which he and his friends took a controlling interest. Both the Staveley Coal and Iron Company and the Sheepbridge Coal and Iron Company were owned by this group, which included Charles Markham, who became Staveley's managing director and one of the leading professional coalowners in the East Midlands and Yorkshire. Originally Markham had been an engineer, educated at Oundle and at Edinburgh University, but his family origins were in the legal profession.[28] Yet another example, and perhaps a more common one, is the Bairds of Gartsherrie, whose origins were in farming, as leaseholders in rather a big way. Their original move into coalmining was when the farmer, Alexander Baird, leased Rochsolloch Colliery in 1816; but all except one of his sons were rapidly involved in collieries as their main business, branching into iron in 1828 as one of the pioneers in the use of Neilson's hot blast technique.[29]

Morris and Williams found that in South Wales in the third quarter of the nineteenth century 'a source of recruitment for owners that stood out . . . was the growing class of mining engineers and colliery managers', and this is confirmed by R.H. Walters. They also refer to the 'many minor local capitalists' who were attracted into colliery enterprise.[30] Both statements can be applied to the other British coalfields up to the 1870s.[31] Indeed, an examination of shareholders' registers of the limited companies which, after 1870, increasingly became the main means of organising the development of new collieries (and often old) suggests that both statements held good right the way through to the end of private ownership of the industry in 1947.[32] Up to the 1870s, a Merthyr grocer like Samuel Thomas, or a draper like David Davis, or viewers like William Hedley and George Elliott of Northumberland, could embark on coalowning comparatively easily owing to the small size of the unit of enterprise. By the 1890s, or earlier in some districts, some form of joint stock organisation was almost essential. But it was still people of much the same background as before who started most of the new firms, namely established coalowning families, up-and-coming mining engineers, and local small capitalists of various kinds – some manufacturers, but even more merchants and traders, and professional men. Three examples will suffice. One is the Cannock and Rugeley Company, established in 1868 to operate in a newly discovered part of the South Staffordshire coalfield. Of its 800 shares at that time, 288 were held by the Harrison family, who owned collieries in the Black Country; 237 were held by local gentry, whose exact status is partially hidden by the description 'gentleman'; 120 were held by local tradesmen; and 75 by local solicitors, clergymen, and accountants. Another is Manston Colliery Company, opening a new mine in South Yorkshire in 1872. Of its 2,400 shares in 1874, Robert Tennant, a Leeds flax spinner by origin (and still so describing himself), held 720; a Leeds maltster held 520; two local coalowners (Akeroyd and Johnstone) held 650 between them; and a 'gentleman' from Penrith held most of the rest. The third example, which shows that there was some measure of regional mobility of capital, is the Bolsover Colliery Company, opening up a new area in the East Midlands in 1890. The prime mover, Emerson Bainbridge, has already been referred to.[33] He was a well-known mining engineer from the North-East, who was a witness at most of the parliamentary and other official enquiries into aspects of coalmining in the following two decades. He held 100 out of the company's initial 1,500 shares, and others already connected with him in the coal and iron industries of the North-East held another 540 shares. The Knowles family, large coalowners in Lancashire, held 235 shares; shipowners from Hull held 100 shares; local mining engineers held 89 shares; and other local men, mostly in the professions, had another 65. The remaining 371 shares were owned by a variety of people from outside the district, mostly described as 'gentlemen', and including a number known to have been connected with Bainbridge in other businesses.

These three examples are reasonably typical of new firms started from the 1860s onwards, though there were many with a different emphasis – with more or less local bias, with more or less existing coalowners, with more or less participation by landlords, with more or less involvement of enterprising mining engineers, and perhaps, as time went on, with more rather than less London-based businessmen with some immediate connection with the coal industry.

But this period also saw joint stock companies of very different composition joining the ranks of coalowners, which, whilst not so typical, were certainly not insignificant. These were companies whose owners were, with minor exceptions, new to the coal industry, and often came from outside the coalfields. They can be divided into two kinds. The more common of the two had comparatively few shareholders, and its shares were placed privately by the promoters. The other kind was the genuine public company, floated by way of appeals to the public to subscribe to the shares. Both kinds were, of course, only possible following the companies legislation of 1844, 1855–7, and 1862. But coal company promotion was not something for which there was an enormous demand only awaiting the release of the legislation. The new companies were very largely the product of boom conditions in the coal industry when those conditions came along after the legislation had been passed. In fact, the second of the two kinds referred to was almost entirely confined to the great boom of the early 1870s, though *existing* businesses were floated as genuine public companies at other times. Both kinds were means of bringing outside capital into the coal industry, and this meant that they were mostly to be found in those districts where the rate of expansion of demand was too great to be met from capital resources internal to the local coal industry – above all in South Wales in the early 1860s and the early 1870s, and in Yorkshire in the early 70s. Morris and Williams refer to several examples of the first kind of company in South Wales, and conclude that most of the capital brought in via this route was external not only to the coal industry, but also to the district.[34] An examination of the shareholders' registers of some of the same kind of company in other districts confirms this view. For example, the Wrexham Colliery Company (1863) was mainly owned by the contractor Thomas Brassey's family and by 'gentlemen' with addresses in the North of Scotland; whilst of the 8,000 shares in Thorp's Gawber Hall Company (1872), operating near Barnsley, all but 178 were held by a small number of 'gentlemen' resident in the London area. The public flotation of the second kind of company, on the other hand, was a means of mobilising relatively small savings, which were very largely local in practice, though this was probably not the intention. About half of the Lofthouse Company's £5 shares were taken up by residents of the West Riding in blocks of less than 100 shares. The holders of more than 100 shares each were, overwhelmingly, local businessmen, and there were no really large holdings amongst these.

So far we have been concerned with people and firms who invested in the coal industry for the sake of the profits to be derived directly from it. In most coalfields, and at most times, this covered the great majority of coalowners. But there were also investments in coal mines, and often quite substantial ones, which were made by way of businesses which were primarily concerned in other industries. By a long way the most important of these was the iron and steel industry, and the associated shipbuilding industry of the late nineteenth century; but there were also one or two others involved in this way. In the West Midlands and in South Wales there was a number of engineering firms which, at one time or another, bought or developed collieries to ensure a regular and stably priced supply of fuel – Nettlefolds, the Patent Nut and Bolt Company, the Patent Shaft and Axletree Company, and Bagnall & Sons are examples of this. Never proportionately very significant, they tended to become less and less important towards the end of the nineteenth century, largely as a result of the approaching exhaustion of much of the Black Country coalfield. Another industry for which coal was an important raw material, brick and firebrick manufacture, was also drawn to invest in coal mines. But, once again, it was never proportionately very significant in this respect, although it owned a considerable number of small pits in the Black Country, Cumberland, and Yorkshire, and a fair number in most other coalfields. The largest colliery owned by a brickmaking firm in 1894 employed only a hundred men, however, and the great majority had less than thirty men. More often than not, it seems, the coal output of these mines was really a by-product of mining for fireclay or ganister.

Other coal-using industries also had a small stake in coalmining at various times, particularly textiles. Naturally, it was in Lancashire and West Yorkshire that these were most important, if that is not too misleading a word to use about such an unimportant element in the coal industry. Since the demand for coal from an individual mill was not all that high – generally to be measured in hundreds of tons a year, rather than thousands – its ownership of a coal mine was dependent on the possibility of opening (or buying) a mine small enough not to involve it in an elaborate adventure into the coal trade. This possibility was still quite strong until around 1880, when the Pennine outcrops neared exhaustion. But because of its relatively small demand, an individual textile firm did not have much inducement to become a coalowner except in times of especially high prices. This meant, in effect, the early 1870s, and a year or two around 1900.

We may conclude, then, that industrial consumers of coal, other than iron and steel works, were not of more than trifling significance as investors in collieries. This, of course, applies to industrial firms as such: as private individuals, many people concerned in textiles and other industries were also quite heavily involved in coal companies.

The stake of the iron and steel industry in coal mines naturally varied a great deal from district to district. At one extreme were the small coalfields of Leicestershire and Warwickshire, where the coal was unsuitable for coking, and no mines at all were ever owned by iron firms.[35] At the other extreme was another small coalfield, Shropshire, where over 40% of pits was owned by such concerns from 1854 onwards, and almost certainly back into the eighteenth century. The other coalfields came in between, and the relative importance of the iron industry naturally depended on the state of development of that industry in the neighbourhood. Though we have no comprehensive statistics of any sort before 1854, it is clear that before that time, iron firms were heavy investors in the coal industry in the West Midlands and in South Wales before the beginning of the nineteenth century, and in Scotland since the 1830s. To some extent, too, there had been iron industry involvement in coalmining in the East Midlands and Yorkshire, back in the eighteenth century in the case of the Bradford and Leeds districts and parts of Derbyshire. It was just beginning to happen in the 1850s in Durham, following the recent discovery of the Cleveland iron ores.

By the 1870s, the movement of iron firms into the coal industry had virtually come to an end. Those which were already involved stayed in, unless they themselves closed down or were forced out by exhaustion of seams in their area, like the Leeds and Bradford firms. Most iron firms did, in fact, increase their investment in collieries; but they came to be of decreasing relative importance, as more purely coalowning businesses expanded far faster. The approximate proportions of coal mines owned by iron firms in those coalfields where they were significant are given in table 3.6 below.

One characteristic of the iron firms which distinguished them from purely coalowning concerns virtually throughout our period, with only a few exceptions, mainly towards the end, was their size. The efficient unit of production in iron was always bigger than in coal, and this was reflected in the capital involved in the average firm. This means that there was less scope for newcomers, and it also meant that the public limited liability company early became much more significant, and that it drew capital from much more widely scattered sources than the purely coal company. On the whole, therefore, the capital brought into the coal industry by way of iron firms tended to be owned either by wealthy individual ironmasters, like the Crawshays of Cyfarthfa, the Guests of Dowlais, the Darbys of Coalbrookdale, or the Wrights of Butterley, or else by something like the modern type of portfolio investor, largely ignorant of the industry, and concerned only with his dividends and share prices.

Whatever the origins of the men who owned collieries, there is no doubt that, except in the case of some public limited companies (of which more later), the majority of the capital which they invested was derived from the profits of the coal industry itself, either through the direct ploughing back of retained earnings, or from savings out of distributed profits. Evidence about the extent to which ploughing back occurred is, naturally, elusive, owing to the scarcity of surviving accounts, especially those of private concerns. Morris and Williams cite the Ocean Coal Company, which between 1865 and 1875 distributed only £204,000 out of total profits of £493,000, and had transferred £27,000 to capital account, £52,000 to reserves, whilst £182,000 had been used in paying calls on new shares created for opening additional pits.[36] Griffin states that Bentinck Colliery in Nottinghamshire was largely sunk in the early 1890s out of the profits of New Hucknall, and that the Sherwood Colliery Company was started with the profits made at Hucknall Colliery according to the

Table 3.5 *Profits and distributed profits of Henry Briggs, Son & Co. Ltd, 1867–1913 (in £)*

	Total profits	Distributed profits
1867–73[a]	187,873	98,350
1874–82	168,770	136,980
1883–91	193,071	134,446
1892–1900	377,394	328,353
1901–7	445,696	419,607
1908–13	386,598	281,427

[a]The figures for 1872 are missing.
Note: the Briggs financial year ended on 30 June.

company's officials.[37] Barber, Walker and Company sank Bentley Colliery near Doncaster in the 1900s out of the profits of its older, Nottinghamshire pits.[38] Yet more evidence is provided by the accounts of Henry Briggs, Son & Company,[39] a West Yorkshire publicly quoted company, control of which was retained by the original owning families. Table 3.5 shows their distributions out of total profits for the period 1867–1913, subdivided into trade cycle periods which fit the fluctuations in profits. Over this whole period retentions came to over £360,000. Meanwhile, the value of the company's assets and investments had risen from £88,000 to £573,000, and £7,000 of debentures had been redeemed. Only a little over £200,000 of this had been raised by calls on shares. Clearly, if this sort of amount of ploughing back occurred in companies which had, or could easily have had, access to the national capital market because of their size and obvious profitability, there seems little reason to doubt that lesser firms must have relied on profits to secure capital for growth to at least as great an extent. And the prominence of existing coalowners amongst the subscribers to newly formed companies simply reinforces the view that the coal industry's expansion was financed largely from its own past activities.

Hitherto we have talked about the sources of capital in general, but what has been said applies in the main to fixed as opposed to working capital. The latter would seem to have been financed to a very considerable extent by local banks, even early in the nineteenth century. Most of the accounts we have examined show fairly substantial debts to banks, and others have noted the same thing.[40] No doubt some of the banks' short-term loans were converted, deliberately or otherwise, into long-term ones, and occasionally one finds banks becoming owners of a colliery when the borrowers defaulted. Banks were never important financiers of fixed capital, however.

The general remarks so far made about coalowners and the sources of

Table 3.6 *Percentages of collieries (1854, 1873, and 1890) and of workers (1894 and 1913) according to categories of owners*

	Publicly quoted iron companies	Other iron companies	Publicly quoted coal companies	Other coal companies	Partnerships and individuals
Durham & Northumberland					
1854	3		97		
1873	8	5	4	16	67
1890	12	3	6	40	41
1894	12	3	7	60	18
1913	15	4	21	56	4
Yorkshire					
1854	5		95		
1873	6	5	3	32	53
1890	6	8	3	39	45
1894	13	4	10	51	22
1913	9	1	23	.62	5
Lancashire & Cheshire					
1854	4		96		
1873	6	2	5	33	56
1890	8	1	11	42	38
1894	12	1	13	56	19
1913	10	3	15	61	11
East Midlands					
1854	6		94		
1873	8	37	3	13	39
1890	8	42	5	24	21
1894	10	60	7	16	7
1913	17	62	8	11	2
North Staffordshire					
1854	17		83		
1873	13	12	0	28	48
1890	10	12	0	36	42
1894	29	8	3	47	14
1913	38	7	5	39	11
South Staffordshire					
1854	23		77		
1873	4	26	1	18	53
1890	2	15	2	27	53
1894	3	15	10	40	22
1913	0	4	15	75	8
South Wales					
1854	20		80		
1873	9	16	4	22	47
1890	7	7	11	41	35
1894	9	10	41	34	6
1913	14	5	51	27	3

Table 3.6 (*cont.*)

	Publicly quoted iron companies	Other iron companies	Publicly quoted coal companies	Other coal companies	Partnerships and individuals
Scotland					
1854	26		74		
1873	3	17	3	11	66
1890	6	20	4	35	35
1894	9	21	14	37	19
1913	17	11	34	31	6
All coalfields					
1854	13		87		
1873	6	11	3	23	57
1890	7	8	6	40	39
1894	10	9	16	50	16
1913	13	5	27	50	5

The Earl of Dudley's enterprises are counted with 'other iron companies'.
Sources: see text.

their capital are probably all that can be said for the period up to 1854, when any sort of comprehensive evidence on the ownership of collieries is lacking. For later years a slightly more exact picture can be formed, though even then much still remains hidden. Using the *List of Mines*, the *Return of Joint Stock Companies*, and the *Stock Exchange Year Book*, table 3.6 has been constructed to show the changes in ownership by major categories of owner. This is a hybrid table, and it scarcely needs stressing that there is no direct comparison between the first three figures for each district and the last two. In fact, just about the only thing that a comparison between the figures for 1890 and 1894 does tell us is that the average size of mine owned by individuals and partnerships was much smaller than the average size in other categories. But despite the fact that there is a discontinuity and that there is a number of qualifications to be made about the dividing lines between categories, table 3.6 does make some things clear. First, there was a considerable increase in the importance of publicly quoted companies in all coalfields from the 1870s onwards, with the bulk of the increase in most districts taking place after 1894. This applied to the four biggest coalfields even more than to the smaller ones, and it also applied most particularly to companies not directly involved in the iron industry. Indeed, with the exceptions of Scotland, the East Midlands, and North Staffordshire, such increases in the relative importance of publicly quoted iron companies as did occur were very modest. And in all these exceptional cases, some of the main expanding iron companies were opening coal mines which were not primarily intended to supply their ironworks, but which were self-contained colliery enterprises.

The exact significance for a study of the sources of capital in the coal industry of the increased relative importance of publicly quoted companies is far from clear. Reference has already been made to the importance of ploughed-back profits, even for public companies. And another cause of the bigger place they had in the industry by 1913 was, quite simply, the conversion of existing private businesses. The remainder, which was drawn from subscriptions by 'the general public', varied greatly in size between different districts, but was nowhere anything like the whole source of public company expansion. Some measure of the relative importance of these three things is provided by a survey of the accounts of the public coal companies in two coalfields with very different characteristics – Lancashire and Yorkshire – during the period 1894 to 1913.[41] In Yorkshire, the capital of these companies increased over the period by £5.2 million, of which only £0.6 million was contributed by the expansion out of profits of companies already publicly quoted in 1894. About £1.7 million came from the establishment of completely new companies, and about £2.8 million through the conversion of existing private concerns.[42] Of this last, approximately £1.7 million was a mere book-keeping transfer, whilst £1.1 million was newly subscribed. Thus something like 60% of the total increase in public coal companies' capital in Yorkshire between 1894 and 1913 came from subscriptions by 'the general public'.[43] In complete contrast was the Lancashire coalfield, where there was no reliance at all on public subscriptions. There was very little new sinking in this district after 1890, and there was certainly no individual sinking comparable in size to those in the Doncaster area of Yorkshire. Consequently there was not the same need to attract capital from the public. Coalowners felt that they could get enough from their own resources, so that there were no flotations of new public companies, and there was no introduction of outside capital by the old-established ones. The increase of about £750,000 in capital owned by public coal companies came entirely from the ploughing back of profits.

A cursory inspection of the *List of Mines* and the Stock Exchange publications suggests that most of the smaller coalfields resembled Lancashire in the unimportance of public subscriptions, whereas South Wales and Scotland had a similar experience to Yorkshire. Northumberland and Durham came somewhere in between.

So far the phrase 'the general public' has been used without any probing into what exactly it meant. In fact, 'the general public' which subscribed for the shares of colliery companies was not, typically, all that 'general'. As has already been suggested with reference to the great mass of private coal companies, it was drawn mainly from people who were already involved in the coal industry, though not necessarily in the same district. In the case of new shares issued by established companies, this would happen in the normal course of events, through 'rights' issues to existing shareholders. But even newly formed companies were, for the most part, owned by existing coalowners. A few examples will illustrate this. Horden Collieries Ltd, established in Durham in 1900, was largely owned by families already heavily involved in the coal and iron industries – Bells, Dormans, Peases, Stobarts, and so on. Over half the shares in Bolsover Colliery Company, begun in 1890 in Derbyshire, were owned by Durham and Northumberland coalowners, and the Knowles family of Lancashire had another substantial slice. Shirebrook Colliery Company, in the same area at about the same time, had a more mixed bag of shareholders, but about a fifth of the shares was owned by the Joicey family and other North-Eastern coalowners, and the Nettlefold family, which had engineering and colliery interests in the West Midlands and South Wales, was also involved. Desford Coal Company, started in Leicestershire in 1900, was mostly owned by Lancashire coalowners. The Warwickshire Coal Company was financed by Scottish coalowners, some of whom were connected with the Coltness Iron Company. Finally, East Cannock Colliery Company was mostly owned by the men who ran the big iron, steel, and shipbuilding firm, John Brown & Co. and the Sheepbridge Coal & Iron Co. of Derbyshire.

A second conclusion that can be drawn from table 3.6 is that, despite the conversion of existing companies from private to publicly quoted status, there was a compensating accession to the ranks of companies from those of partnerships and individuals. In fact, up to 1890, there was more than just an offset, there was an increase of considerable dimensions in the private companies. In part this seems to have resulted from mere changes of name. The dividing line between the two categories is very blurred, and it appears likely that, as time went on, more businesses carried on their activities under a company name, even though there was no change in their ownership. For all that, there was clearly a considerable increase in the importance of genuine private joint stock companies, especially in the 1870s and 80s. Equally clearly, there was a decline in the position of individuals and partnerships. This is explained in part by the exhaustion of most outcropping seams, which could be exploited easily by small mines requiring little capital, and in part by the gradual realisation of the attractions of limited liability.

A third conclusion from table 3.6 is that, over the country as a whole, iron and steel firms were no longer increasing their relative stakes in the coal industry after 1873, but that they more or less retained their place, which implies quite a large increase in their absolute investment. Different coalfields, however, had different experiences in this respect. In

South Staffordshire there was a very large decline in the involvement of iron firms, caused entirely by the exhaustion or flooding of most Black Country mines. In Yorkshire there was a similar cause for the reduction of iron industry investment in coalmining in the western parts of the coalfield, and this was by no means offset by the continued investment by the Rotherham and Sheffield iron and steel companies. The same sort of thing happened in parts of Scotland, Lancashire, and the East Midlands, and the offsetting expansion of other collieries owned by iron firms in these districts was not always connected closely with their ironworks business, as has already been mentioned.

Finally, in brief summary, one can say that, although the joint stock form of organisation was increasing in the coal industry from the 1860s onwards, and although some genuine public subscription capital came in, especially from about 1890 onwards, the great majority of capital in 1914, as at the beginning of the nineteenth century, was owned by people long connected with the industry as owners and managers, and that the source of their investments was overwhelmingly the profits of the industry itself. In 1800 this group of people had still included a good many of the old landed classes, but by 1860 most of these had withdrawn, and the group was composed of what we have described as professional coalowners – entrepreneurs with experience of the industry, quite often professionally involved as mining engineers. The origins of this group go back into the eighteenth century, especially in Durham and Northumberland, and a major source of its recruitment was always its own sons. But newcomers came in in every generation – from other trades and industries, and a few from amongst the working miners, such as John Warburton, who had charge of laying out Lofthouse Colliery in the 1870s.[44]

APPENDIX

Table 3.A1 *Index of colliery opening costs, 1816–1914 (1886 = 100)*

	Hewers' wages	Bar iron	Engineers' wages	Wood	Bricks	Steel	Weighted index
1816	68.7	246.4	79.8	270.1	111		112
1817	66.4	188.4	79.8	270.7	111		105
1818	65.5	279.7	78.9	292.7	111		116
1819	63.6	268.1	80.7	250.0	111		109
1820	63.0	236.2	78.9	231.7	111		104
1821	64.0	192.8	78.9	267.1	111		103
1822	69.0	172.5	78.9	224.4	111		100
1823	74.0	182.6	78.9	219.5	111		104

Table 3.A1 (*cont.*)

	Hewers' wages	Bar iron	Engineers' wages	Wood	Bricks	Steel	Weighted index
1824	79.2	188.4	78.9	228.0	111		108
1825	86.4	300.0	78.9	256.1	111		126
1826	77.0	236.2	78.9	193.9	111		108
1827	73.5	214.5	78.9	200.0	111		105
1828	72.0	192.8	78.9	184.1	111		100
1829	71.7	166.7	78.9	198.8	98		97
1830	74.1	142.0	78.9	181.7	98		95
1831	80.0	133.3	78.9	211.0	115		102
1832	77.7	129.0	78.9	204.9	105		98
1833	77.1	152.2	80.7	212.2	110		102
1834	79.5	160.9	80.7	213.4	110		104
1835	81.5	149.3	80.7	212.2	110		104
1836	85.8	230.4	80.7	229.3	113		117
1837	87.4	198.5	80.7	222.0	110		114
1838	84.9	204.3	80.7	215.9	105		112
1839	85.5	213.0	80.7	215.9	105		113
1840	82.4	181.2	83.5	215.9	105		108
1841	76.8	156.5	83.5	213.4	100		101
1842	72.2	130.4	83.5	203.7	90		94
1843	68.6	107.2	83.5	178.0	86		87
1844	72.7	124.6	83.5	169.5	110		92
1845	80.1	191.3	83.5	173.2	115		104
1846	90.4	204.3	83.5	173.2	120		112
1847	89.0	208.7	83.5	173.2	120		112
1848	75.0	150.7	82.6	145.1	110		94
1849	75.2	134.8	80.7	137.8	110		91
1850	75.7	126.1	80.7	125.6	90		88
1851	76.0	117.4	82.6	120.7	86		86
1852	78.3	134.8	83.5	124.4	81		90
1853	84.4	198.6	85.3	161.0	110		107
1854	96.1	214.5	86.2	172.0	110		117
1855	103.1	182.6	88.1	163.4	110		117
1856	106.6	189.9	89.0	152.4	110		119
1857	97.0	176.8	89.0	141.5	110		110
1858	90.2	150.7	87.2	123.2	100		101
1859	90.2	144.9	88.1	140.2	93		102
1860	94.3	139.1	88.1	156.1	95		105
1861	97.4	129.0	88.1	140.2	100		105
1862	98.9	134.8	89.0	140.2	100		107
1863	99.3	150.7	89.9	137.8	100		107
1864	105.7	176.8	89.9	136.6	100		116
1865	111.1	166.7	91.7	139.0	100		118
1866	118.9	155.1	92.7	130.5	111		121
1867	107.9	144.9	92.7	123.2	100		112
1868	97.2	136.2	91.7	120.7	100		105
1869	96.8	144.9	91.7	122.0	100		106

Table 3.A1 (*cont.*)

	Hewers' wages	Bar iron	Engineers' wages	Wood	Bricks	Steel	Weighted index
1870	100.8	158.0	92.7	123.2	100		110
1871	110.1	163.8	93.6	106.1	122		117
1872	137.6	236.2	99.1	113.4	133		144
1873	166.1	268.1	100.9	143.9	156		169
1874	146.1	220.3	101.8	152.4	133		150
1875	125.2	179.7	102.8	130.5	122		131
1876	115.6	155.1	102.8	130.5	122		122
1877	109.1	144.9	102.8	134.1	122		118
1878	101.3	120.3	100.9	117.0	111		107
1879	96.8	123.2	98.2	102.4	93		102
1880	96.7	144.9	98.2	120.7	100		108
1881	100.5	123.2	100.0	118.3	93		105
1882	105.3	134.8	100.0	120.7	111		112
1883	106.1	123.2	100.0	113.4	100		108
1884	103.5	110.1	100.0	106.1	100		104
1885	102.1	104.3	100.0	106.1	100		102
1886	100	100	100	100	100	100	100
1887	98.9	100.0	100.9	95.1	100	97.8	99
1888	100.1	104.3	101.8	102.4	100	88.1	101
1889	112.1	105.8	101.8	113.4	100	83.0	107
1890	131.9	136.2	103.7	107.3	133	89.1	124
1891	141.4	120.3	103.7	97.6	122	87.0	122
1892	136.9	117.4	103.7	98.8	122	88.5	119
1893	130.3	107.2	102.8	97.6	111	76.8	112
1894	132.4	104.3	102.8	98.8	111	73.7	112
1895	123.8	104.3	103.7	95.1	111	74.1	108
1896	121.3	107.2	108.3	100.0	111	67.7	108
1897	122.6	113.0	109.2	107.3	111	67.8	109
1898	127.2	117.4	110.1	107.3	122	67.9	112
1899	137.6	155.1	110.1	109.8	133	69.7	121
1900	156.7	192.8	111.0	126.8	133	78.6	135
1901	167.2	139.1	111.9	117.1	127	73.0	130
1902	154.4	131.9	111.9	115.9	124	62.7	120
1903	150.7	134.8	111.0	117.1	128	66.3	120
1904	145.8	131.9	111.0	113.4	131	70.7	118
1905	142.1	139.1	111.9	120.7	119	74.3	117
1906	142.9	155.1	111.9	126.8	119	77.1	119
1907	158.0	160.9	113.8	120.7	117	80.0	124
1908	168.9	144.9	112.8	106.1	114	80.9	125
1909	158.5	139.1	112.8	97.6	111	76.6	118
1910	158.4	139.1	113.8	104.9	121	72.4	118
1911	159.4	136.2	114.7	113.4	131	70.9	119
1912	162.6	158.0	116.5	132.9	147	71.7	123
1913	178.3	166.7	117.4	120.7	151	75.5	129
1914	180.2	150.7	117.4	120.7	151	75.5	128

Notes to Table 3.A1

The weights are given in the text, p. 47.

Sources: hewers' wages: based on table 7.1; engineers' wages: Mitchell (1962), p. 349 to 1850, and Spicer (1928) thereafter; bar iron and steel: *London Mercantile Price Current* to 1850, and the Sauerbeck *Statist* index thereafter (Journal of the Statistical Society (1886 and subsequently)); bricks: Laxton's *Builders' Price Book* to 1862, *Report on Wholesale Prices* (*SP* 1903 LXVII) for 1863–1900, and *The Builder* (four quotations per year) thereafter; wood: Tooke (1838–58) to 1837 (Memel fir), *The Economist* (Memel fir, four quotations per year) for 1838–53, *Annual Statement of Trade* (import values of sawn fir) for 1854–1913, and Laxton's *Builders' Price Book* (sawn fir) for 1914.

4

TECHNOLOGY

As the demand for coal increased in the nineteenth century, entrepreneurs responded by investing more capital in opening up new pits and new areas of coalfield. But the limits to expansion of output by these means are necessarily closely restricted, and in practice, most of the expansion of output had to come from more intensive exploitation of existing coalfields by improving mining technology. As A.J. Taylor has put it: 'In coalmining the role of the innovator is fundamental. Whereas in manufacturing industry invention has increased productivity and reduced costs, in coalmining it has been and is the indispensable condition of the industry's continued existence.'[1] Certainly, expansion of the amount achieved would have been impossible without technical change, and it is with the nature, motives and effects of this change that this chapter will be concerned.

The eighteenth century had inherited methods of getting the coal at the face, and had modified and applied these in different ways in order to expand output. The nineteenth century continued this process. But the eighteenth century had also met and solved, at any rate in principle, numerous problems of its own of increasing output from a limited area.[2] Most of these sprang from the need to sink pits deeper, with its corollary that each individual shaft had to tap a wider area to be profitable. Amongst the problems for which the solution was known by the beginning of the nineteenth century were those of drainage; of raising coal from pits too deep for manhandling or horsepower to be feasible; and of underground haulage over considerable distances. These solutions, though known, were not always widely applied in 1800, mainly for perfectly sound economic reasons. Other problems still awaited solution. One of the most important of these was that presented by noxious gases underground; another concerned safety and speed in winding; and there had scarcely been any perception of the possibility of aiding the efforts of the hewer with his pick.

The nineteenth century saw the application and improvement of earlier inventions; it saw new technical discoveries of its own put into operation; and, like its predecessor, it experienced frustrations in overcoming its difficulties. In all these things there were economic, as well as strictly technological, influences at work, and it is these that we are mainly concerned to elucidate. Unfortunately, owing to the great variety of conditions in different collieries, and also to the lack of extant records of more than a handful of individual companies, it is impossible to be precise about all the economics of technical change. Often, indeed, it is impossible even to guess at orders of magnitude. But so far as one can go – and perhaps a little further – the attempt to quantify has been made here.

There were many hundreds of technical changes during the course of the period we are concerned with, and it clearly would be impossible to consider them all. In order to reduce the discussion to manageable proportions, all the minor changes will be ignored, as will those which had no direct economic impact even though they were important from some other point of view – safety, for example.[3]

Let us begin at the beginning, as it were – once the pit has been sunk – with systems of mining at the face. Basically, there are two methods of working coal mines – pillar and stall,[4] and longwall – though there are modifications of the former which make it close to the latter, such as the bank system of Yorkshire and the East Midlands. Essentially, pillar and stall consists of driving narrow headings into the coal and then delving into the sides of them at frequent intervals, thus forming the 'stalls', 'bords', or 'rooms'. Longwall involves working all parts of a long face of coal simultaneously. Usually the operations are performed outward from the shaft bottom, but occasionally the coal is worked backwards from the pit boundary.[5]

When the first mines, as distinct from quarries or bell pits, were sunk in Britain, in the Durham and Northumberland coalfield, the pillar and stall method was devised for exploiting them,[6] and it has persisted there, at least until very recently, partly because of its peculiar suitability to the geological conditions and partly because, as A.R. Griffin puts it, 'a whole socio-economic system had been built around bord-and-pillar'.[7] In other coalfields,[8] the first mines seem generally to have followed the established practice of the North-East – either because they were influenced directly or indirectly by the Northern engineers, or because pillar and stall is, in some sense, the most 'natural' method of working, the one which requires least organisation by management, and the one whose consequences on the surface need least forestalling. According to Ashton and Sykes 'longwall working originated in Shropshire, perhaps in the early or middle seventeenth century.'[9] By the beginning of the nineteenth century it was the normal method used in the West Midlands (with the very important exception of the Black Country's Ten Yard seam) and in the small Bristol and Somerset coalfield, but elsewhere it had only been used experimentally.

During the course of the nineteenth century, longwall gradually replaced pillar and stall working entirely, or almost entirely, in every district except Durham and Northumberland. The main reasons for this were outlined in a late nineteenth-century textbook by Caleb Parnely:

> The advantages of working the longwall where it is applicable are a better yield of large coal, less injury to upper seams as the intermediate strata settle gradually, simplicity of working, ease of ventilating, and greater economy, for the superincumbent weight reduces the labour of 'holing'. These advantages are so manifest as to indicate the desirability of working all seams of usual thickness situated 100 fathoms or more below the surface on the longwall system.[10]

In addition, a greater proportion of the coal in a seam was extracted. In the light of these advantages, it is pertinent to ask why the changeover was not faster and more complete.

There are, of course, always difficulties in changing from one system to another, which constitute costs that have to be met immediately, as well as causing uncertainty which requires a higher prospective profit than would otherwise be needed to induce the change. High amongst these difficulties are objections which may very likely be encountered from the labour force, though in periods of new colliery opening and recruitment of fresh labour, such as were common at frequent intervals in the nineteenth century, these objections probably were not a major obstacle to introducing longwall working. Sheer inertia – the absence of alert and forceful management – may well have been more important; but in the competitive conditions under which each coalfield worked at most times, even this does not seem to be very plausible as the major reason for slowness. However, both labour and management reactions[11] no doubt played a part, given what does seem to be the most important reason. This was the lack of any very great profit to be got out of the change in many mines until quite late in the century. There were two major disadvantages to longwall working, which have to be set against the advantages mentioned earlier. It resulted in more surface subsidence, especially when working the upper seams; and it required a greater expenditure on the maintenance of underground roadways, more particularly where the strata above the coal were soft. The first of these disadvantages diminished in importance as pits became deeper. The second continued to contribute to the maintenance of pillar and stall in Durham and Northumberland well into the present century; but elsewhere geological conditions were not generally so unfavourable to longwall as not to make it sooner or later profitable to incur the extra expenditure on timber in order to achieve a larger output and a bigger proportion of large coal. But as late as 1870, one of HM Inspectors of Mines for Scotland, W. Alexander, told the Royal Commission on Coal Supply that pillar and stall was 'perhaps the cheapest mode of working',[12] referring, of course, to prime costs only.

And even Truram, the South Wales advocate of longwall, admitted that it was no cheaper there than pillar and stall, except in so far as a 30% greater rate of extraction lowered unit overhead costs.[13] With increasing capital per man, the cost advantage of increasing the extraction ratio must have risen, so increasing the incentive to use the longwall system.

During the first half of the nineteenth century, the main areas where change took place were the East Midlands (including Leicestershire) and North Wales. The Reports of the Children's Employment Commissioners in the early 1840s[14] contain various references to mining methods, and these suggest that longwall was still only a major feature in its old original areas and in the East Midlands and Denbighshire. Even in the East Midlands, the bank system was still common, and the change to longwall was mainly confined to new collieries.[15] In the 1850s, however, the situation again became more fluid. The last remnants of the older methods of working were virtually wiped out in the Midlands,[16] except for the Black Country's Ten Yard seam, where a variant of pillar and stall continued to be more common than the alternative, longwall used in two layers, right up to its exhaustion in the 1900s.[17] Experiments with longwall were undertaken in South Wales, such as those already referred to at Dowlais, and in Scotland and Lancashire. By 1870 these districts were all in the course of transformation.

In Lancashire, longwall had been used for a long time in the Bridgwater and Balcarres families' pits and probably in others, even though the Commissioner in 1842 had had no doubt about saying that pillar and stall was 'general' there.[18] By 1870 a modified form of longwall was the most commonly used method, though the modification did not permit the long face typical of the system in the Midlands, a series of short faces in echelon being used instead.[19] Twenty or so years later pillar and stall survived in Lancashire only in certain seams whose roof conditions some of the more conservative mining engineers continued to deem unsuitable for longwall.[20] A bare handful felt like this as late as 1914.[21]

In Scotland, as in Lancashire, longwall had a long history, even though it was used at only a small minority of collieries before 1850,[22] and once the period of rapid changeover began, its pace was only a little slower. By 1870, HM Inspector for West Scotland said that pillar and stall was 'going out of use' there and that longwall was 'gaining ground fast'. He reckoned by then that almost 55% of all Scottish coal was mined by the newer system.[23] Twenty years later pillar and stall survived as little more than a curiosity.[24]

In South Wales, longwall had made considerable strides in general acceptance by 1870, having been little used before the Dowlais experiments of the mid-50s. It was still only recently adopted along the northern outcrop of the coalfield, but was more common in the south. However, pillar and stall remained the system at more than half the collieries in the district.[25] Thirty years later longwall had become general.

Of the remaining coalfields, Durham and Northumberland retained pillar and stall, with only a minor flirtation with longwall, mostly in the isolated, thin-seam Scremerston region.[26] Yorkshire, on the other hand, eventually went over to the Midland system, though it is not clear whether this had made much headway by 1870, or whether the main switchover took place in the following decades.[27] A minor variant of pillar and stall, the bank system, was certainly much used in 1870,[28] and continued to have its adherents as late as 1903.[29] However, some of the 1870 witnesses thought that longwall was more important by then.[30] Whether or not this was true, the bank system was certainly holding its own better than pillar and stall in Lancashire or than the East Midlands variant.[31] Since the main advantage of the bank system was held to be its effectiveness in draining off firedamp, it may well have required the improvement of ventilation (to which we turn in a moment) to overcome this. By the 1890s this improvement had taken place, and longwall was definitely predominant.[32]

In the absence of comparative cost data at frequent intervals from pits working the same seams by different methods, it is impossible to be exact about the effects of operating one method of mining as opposed to the other. All one can do is to repeat the judgments of contemporaries about the advantages, and attempt to see how they applied to different times and places. The general conclusion to be drawn from the lack of much change before the 1840s is that the longwall system had little advantage on balance outside the Midlands, though this begs the question of entrepreneurial efficiency, of course. Either roofs were too soft (Durham and Northumberland, and parts of South Wales), or firedamp was too readily accumulated (parts of Yorkshire, Lancashire, and Durham and Northumberland), or the depths of the seams worked were too shallow to give longwall an advantage. These reasons were reinforced by the very low ratio of overhead costs (above all capital charges) to prime costs, and to some extent by the probability of labour troubles on a change of method. After 1840, increasing average depths of pits and increasing capitalisation combined to overcome the resistance to change, and in the space of three or four decades longwall became predominant everywhere except in the North-East, where the geological difficulties were especially great and where the miners' resistance may have been expected to be especially strong.[33]

Just how much the change of system contributed to increasing the productivity of labour and the profitability of capital, or to holding back any decline, is beyond any attempt at measurement. The impression left both by contemporary discussion and by the existence of the two methods side by side is that the economic effects of the change were relatively small. This, however, is probably misleading; for whilst the effects of a switch in an existing pit may have been small, it seems quite likely that it was only by using longwall that some new deep pits could have been economically worthwhile at all.

Apart from the method on which his work was organised, the hewer experienced little in the way of technical change in his job, narrowly defined, until after the beginning of the twentieth century. Right up to 1914 there was little mechanisation of his work, and he depended largely on his own muscles and the unchanging pick. The only significant exceptions to this – apart from coal-cutters, to which we turn shortly – were the safety lamp and the use of gunpowder for shotfiring.

The story of the invention of the safety lamp has been told often enough not to need repetition here.[34] Suffice it to say that probably its main economic effect was to make available for exploitation a number of seams which had previously been regarded as too dangerous to tackle. As Ashton and Sykes say: 'The inventors had provided the miner with a weapon of defence: armed with it he was led forward to meet fresh perils. They had sought to bring security of life: they achieved an increase in the output of coal.'[35] It is doubtful, however, if they did this by raising the output per man. In so far as the very fiery seams were thicker or more easily got, labour productivity would tend to increase; but, to offset this, the weaker illumination provided by the safety lamp compared with the naked flame tended to reduce productivity. What the lamps did was to make possible the employment of more men in collieries where some seams had previously been ignored. That is, it increased the return on capital invested in shafts and surface equipment if these had not previously been used to capacity.

In contrast to the safety lamp, the use of gunpowder to help the hewer get the coal must have had quite a marked effect on his productivity, for an explosive charge could do in a matter of seconds what might have taken hours with picks and wedges. Even allowing for the time taken in preparing for shotfiring, there clearly should have been a considerable increase in the hewer's output per shift. But, of course, its use brought an increased risk of accident, and initially this must have seemed more frighteningly large than it in fact was. The first recorded use of gunpowder to break down coal at the face was at Hetton in 1813,[36] though it had been used in sinking and opening out throughout the eighteenth century, according to Ashton and Sykes.[37] Presumably a mixture of fear and inadequate ventilation was responsible for the delay in everyday use, which continued even after 1813 for another twenty years or so. However, once its use did become fairly common, it was adopted rapidly, indicating that its impact on productivity was substantial. But just how big this impact was is largely a matter of speculation, for there is little evidence to go on. It has been impossible to find any contemporary calculations on the economic effects of shotfiring before the very end of the nineteenth century, and this is all that there is to go on. A leading textbook of mining at the beginning of the twentieth century reckoned that the use of explosives reduced the cost of driving headings by a factor of four;[38] which, assuming that this figure can be applied to coal-getting and that hewers formed two-fifths of the labour force before the innovation, implies a rise in overall productivity per head of just over 40%.[39] This is, in fact, the figure which W.Y. Craig of Podmore Hall Colliery, North Wales, found in an experiment lasting two months, in which the use of powder was discontinued for the second month. In this month, 233.5 tons were produced compared with 328 tons using shotfiring by the same group of miners, working on day wages throughout.[40] Conditions in this case were probably very favourable for explosives, and it is likely that this figure of 40% can be regarded as an upper bound for the increase in productivity resulting from shotfiring. The average was more likely the 20 or 30% suggeted by T.J. Taylor to the Select Committee on Accidents in Coal Mines in 1854.[41]

One implication of the introduction of shotfiring was that, other things being equal, the proportion of men engaged in conveying coal away from the face would be increased. But in practice other things were not equal, and in the second quarter of the nineteenth century there seems actually to have been a decrease in the proportion of haulage workers.[42] In view of this it seems safe to say that innovations in transporting the coal were even more important than the use of gunpowder for shotfiring. In the fifty years after 1850, of course, there was no hewing innovation which could enter into comparison.

The response of eighteenth-century mine managers to the need for hauling coal over longer distances underground in the Durham and Northumberland coalfield had been the invention of the rail, on which wagons could be moved by horses or men (or boys – or even girls). Which was used depended on the height of the roadways and the weight of the load, and initially at least, it seems, that in many places 'the invention of cast iron tramways . . . brought back the system of tramming, and the substitution of manual labour for that of horses in the interior workings'.[43] By 1830 probably,[44] and by 1840 certainly, the eighteenth-century invention had come into general use in all British coalfields, but there were still survivals of older systems of haulage in many small pits, notably in areas where thin seams were worked. The Children's Employment Commissioners noted that there were still some 'bearers' in the Lothians,[45] carrying coal on their backs; and that sledges were still used in some mines in West Yorkshire, Shropshire, and Lancashire.[46] But these were exceptional, and the normal system of haulage by 1840 was by wheeled trams, usually (though not always) on rails. S. Scriven, one of the Commissioners reporting on West Yorkshire, described the system graphically as it applied in a thin-seam district:

> Hurriers are children who draw loaded corves or wagons weighing from 2 to 5 cwt., mounted upon four cast-iron wheels, of five inches diameter, without rails, from the headings to the main gates . . . As soon as the hurriers enter the main gates they detach their harness from the corf, change their position by getting behind it, and become 'thrusters'. The vehicle is then placed upon the rail, a candle is stuck fast by a piece of wet clay, and away they run with prodigious celerity to the shaft, pushing the load with their heads and hands.[47]

In districts where the seams were thicker, whilst the subsidiary haulage – from the face to the main road – was similar to that described by Scriven, the main haulage was done by horses[48] pulling a train of wagons, either on rails or on rolleys. (Though these latter were beginning to be abandoned, they were still to be found quite frequently in the early 1850s, when, however, they were said to be 'fast giving place to the edge railway'.)[49]

Whilst it had taken the first three or four decades of the nineteenth century to get the general adoption of the best underground haulage practice of the eighteenth century, there had in the meantime been further advances in technology. And, as in so many industries, the basis of these nineteenth-century improvements was the steam engine. The first recorded use of a steam engine for underground haulage was in 1804 or 1805 at Birtley, County Durham, by John Curr, one of the most ingenious and inventive mining engineers of all time.[50] He used an engine placed underground near the upcast shaft to haul wagons up inclines from dip-workings. This continued to be the only use of a steam engine in underground haulage for many years, and this limited its widespread application, though it was generally employed throughout the country for this purpose by the 1840s. When Matthias Dunn wrote his *Winning and Working of Collieries* in 1852, he could say that steam engines underground for inclines had 'long prevailed'; but at the same time he went on to say that 'attention is now being turned to the adoption of engine power to haul the carriages along the bed of the mine'.[51] The main problem to be overcome was that of transmitting power to the further reaches of the pits, over what might easily be two miles or more; and it was only with the coming of wire ropes in the 1840s that any general solution became possible. Even then, there needed to be improvements in the design of steam engines before mechanical haulage became an effective competitor with horses By the mid-1850s, however, this had come about; though it was almost another two decades before main road haulage was more commonly done by engines rather than horses. The main period of change seems to have been in and just after the boom of the early 1870s, though there had been a gradual spread of mechanisation at the more go-ahead collieries in the 1860s[52] – and, of course, at new ones of any size. By 1882, Galloway was saying: 'In the haulage of the coal carriages along the main underground roads, horses have to a great extent been superseded by machinery, usually consisting of wire ropes, of iron or steel, or chains, worked by steam engines placed either at the surface or in the mine.'[53] Two decades later, horses had 'practically disappeared' from the main haulage roads of even the technically rather backward Lancashire coalfield.[54] Where steam could not be used for any reason, compressed air engines were used instead. The first had been used at Govan Colliery, Glasgow, in 1849 for pumping purposes, but it was soon used for haulage, and the first large-scale application of compressed air to that use came only a few years later, at Dowlais in 1857.

As with other technical developments, it is impossible to say with any precision what were the economic consequences of the mechanisation of main haulage. The careful enquiry by the North of England Institute of Mining Engineers published in 1869 suggested that there was very little difference in costs between horse-haulage and mechanical haulage by tail-rope, but that the endless chain produced a saving in cost of about 25%: a quarter of a century later the more recently developed endless rope was reckoned to be considerably more effective still. Even in the mid-60s it was claimed in South Wales that mechanisation might halve the former costs of horse-haulage.[55] And a similar conclusion is suggested by calculations made by T.V. Hughes of North Staffordshire thirty years later.[56] But these are no more than rough guides, and it is completely impossible to lay down any general law about the cost-saving effects of mechanisation – so much depended on particular conditions in each and every case. The most one can say is that the larger the output to be moved, the longer the distance to be covered, and the steeper the inclines to be tackled, the more likely was mechanisation to be profitable.[57]

Mechanisation of the main road haulage, mostly in the third quarter of the nineteenth century, was followed, though far from universally, by mechanisation of the auxiliary haulage, between the face and the main roads. This was one of the main topics of discussion amongst mining engineers in the 1890s and 1900s; but for all that, there were relatively few pits even in 1914 where haulage was completely mechanised from face to shaft.[58] The main technical problem was that of motive power; but there were also two economic considerations. First, in some pits, especially where gradients were low and the pillar and stall system prevailed, as in the North-East, the flexibility of pony-putting kept its costs down. And second, in other collieries, the thinness of the seams made it too costly to make roadways that were high enough for anything but hand haulage. The technical problem was solved – or at any rate partially solved – in the

1880s by improvements in the efficiency of compressed air engines.[59] The complete solution came in the next decade with safe methods of electric power transmission and with the invention of the conveyor, though the use of the latter was almost entirely linked with the adoption of mechanical coal-cutters and will be discussed later. But the economic barriers to widespread mechanisation of subsidiary haulage remained after the technical problem was solved – or, if they did not, they certainly appeared to do so in the minds of the majority of managers. Probably they really did exist in a great many cases, and the less thrusting managers were not prepared to find out whether the obstacles applied to them or not. Certainly in thin-seam pits, particularly those without any long future ahead of them, it seems likely that the need for unproductive ripping of bigger roadways would have swallowed any potential saving in costs.[60] Probably, too, there were pits in the North-East, where the roads were so often level, where the cost of pony-putting was little, if any, higher than the cost of mechanical auxiliary haulage.[61] But in more cases than not there was a saving in cost to be had from the innovation in thick and medium seams, and in view of the low level of electric horsepower used in underground haulage of all types, it is probable that this was not being achieved everywhere as late as 1913.[62] The failure, however, was not on its own of very much importance, and since it appears to have been tied up with the speed of adoption of electricity in general, it is covered by our discussion of that.[63]

The second earliest application of the steam engine to mining – after pumping – had been for winding coal (and miners) in the shaft. This was well established in the bigger collieries when the nineteenth century opened, though horse-gins and even more primitive methods were still in use, and remained in use, until the second half of the century.[64] Eventually, however, steam winding became universal. The increasing depths to which most pits – and all of any size – were sunk made mechanical winding essential, and until the very end of the century steam had no competitor. Even then electric winding was rare in South Wales and in Durham in 1914, and almost unknown in other districts.[65] There were enormous improvements on eighteenth-century practice, of course; but the principle remained the same. Cages in which tubs could be raised replaced the old corf hooked onto the winding rope from the 1830s;[66] guide rails were quickly added to prevent crashes in the shaft; wire rope replaced hemp, very rapidly after the Wingate Colliery wire rope trial of 1844, at which its safety was established;[67] steam engines were improved time and again. As a result winding speeds multiplied many times, and loads were vastly increased. The motive behind many of these advances was safety, but still more important generally was the desire to move the ever-increasing output which the ever-deeper pits had to turn out if their larger capitals were to prove profitable. Winding capacities at many pits doubled and more than doubled in the 1830s and 40s; and if the rate of advance was slower after that, it was steadily maintained.[68]

The increasing depth at which coal had to be worked, together with considerations of safety, were also the reasons for another major group of innovations in the nineteenth century – those to do with ventilation. At the beginning of the century, ventilation by means of a furnace, either at the bottom of the upcast shaft or underneath a tall chimney at the top of the shaft, was the universal means where any form of artificial ventilation was necessitated by the size of the workings. Where these were really extensive by the standards of the day, the simple method of 'coursing the air' through the pit, invented about 1760 by James Spedding of Whitehaven, was increasingly proving dangerous, when gas picked up on the long journey through the workings passed over the furnace. Some time around 1810, John Buddle, perhaps the most famous viewer of all time, got around this difficulty by what he called 'compound ventilation', later known as 'splitting the air'. This involved subdividing the workings into a series of independent systems by the use of stoppings and doors – a simple idea in conception, though more complex in execution.[69] It came into use at all the bigger pits in most districts in the course of the following two decades, and it seems to have been general by the time of the Select Committee on Accidents in Mines in 1835, when one witness, Anthony Winship, a colliery wasteman, described it as 'the greatest and best improvement that ever was known'.[70] However, not everyone was satisfied with the furnace as the basic means of providing a current of air, and various experiments were made with other methods from the 1830s onwards. The first to achieve some measure of success was steam jets, which had a certain vogue in the late 1840s, often as subsidiaries to furnaces. As in most other things at that time, South Wales seems to have been backward, and as late as 1850 natural ventilation was still not uncommon there, and not all furnaces were associated with 'splitting the air'.[71] However, the frequency of explosions and the rapid development of the especially fiery seams of the Aberdare Valley, soon brought the coalfield into the van of experimentation with mechanical ventilation. The first of the experiments was at Eaglesbush Colliery, near Neath, in February 1849, following a disastrous explosion there ten months earlier. This was W.P. Struvé's air pump, which was quite frequently adopted in South Wales in the next decade or so. A more important development still also took place in 1849 in South Wales, when William Brunton's centrifugal fan was tried out at Gelligaer Colliery. For a time, fans proved insufficiently reliable for widespread adoption, but experiments continued, and in the 1860s there came three types, the Guibal, the Schiele, and the Waddle, which superseded furnaces eventually – by about 1900.

Once again, the period of most rapid advance was during and just after the great boom of the early 70s, partly because by then new pits, of which there were a great many at that time, tended to be established with fan ventilation. But there were many conversions at older pits too, notably in South Wales, Scotland, Yorkshire, the East Midlands, and North Staffordshire. By 1880, roughly half of all collieries in these districts were mechanically ventilated, and not far short of that proportion in the North-East. In Lancashire and South Staffordshire only about a quarter of even the larger collieries had gone over to fans, and this was probably true of most of the smaller coalfields.[72] In the 80s the changeover was much slower, both because the rate of new sinking slumped to a very low level and because neither the incentive nor the finance for new investment was strongly present. But it picked up again in the 1890s, and by the end of the century furnaces were used at large collieries only in a very few exceptional cases, mostly in Durham and Northumberland.[73]

Early protagonists of mechanical ventilation were mainly concerned with urging its technical superiority, above all its greater safety; and it was on the same grounds that most opposition was based, until the fans of the 1860s had proved their reliability. By the 1870s, their safety was accepted, and the ground of argument was shifted to the relative economy of the two methods; and then one of the main considerations was the value of the very small coal which was used in furnaces. In places where the coal was of a non-coking variety, and so unusable for making coke, the coal used in pit furnaces had little value (except in the extraordinary boom circumstances of 1872–3). Often, indeed, it was never wound to the surface at all, but was packed in the goaves. In the iron-making coalfields, however, this 'smudge' was readily saleable, and the running cost of a fan there was about a third lower than that of a furnace. This, more than anything else, accounts for the greater willingness to convert old pits from furnaces to fans in those districts – and, perhaps most significantly of all, for the greater willingness to convert in Durham than in Northumberland. The return on the investment of £2,450 in a fan in the 1860s might be around 10% in a pit producing a coking smudge.[74] In other places it was only under the influence of high boom that it was enough to induce conversions, though after 1870 it was almost always enough for it to be installed in all new collieries except the very smallest.

There is a general pattern to most of the innovations which we have discussed so far – invention and adoption at a few pits; a period of a few years or even a decade or so of experiments; then widespread acceptance during the course of another two decades. A major exception to this pattern is provided by coal-cutting machinery. The first experiment (leaving aside such oddities as Michael Menzies's patent of 1761, long before portable power plant was available) was in 1861 in West Yorkshire. Even then, the saving in costs claimed was as much as 10% in thin seams.[75] Yet fifty years later the proportion of British coal output cut by machine was still not much more than a twentieth, and less than half the owners of pits working seams of three feet thick or less were using machines.[76] Ten years after the first machine was introduced, *The Economist*, in its 'Yorkshire Coalfield Notes' reviewing the year 1871, reported that two mechanical cutters had been installed at Wharncliffe Silkstone and one each at Woolley and Thorncliffe Collieries. The pace of change may be judged from the accompanying remark that machinery 'had made rather more progress than . . . hitherto'.[77] The long period of depressed profits and wages which followed the boom which peaked two years after that removed most of the stimulus and some of the means for experimentation with labour-saving machinery, especially as, with frequent underemployment,[78] the miners were generally hostile. Some of the machines which had been introduced were withdrawn during those years.[79]

In the 1890s progress was resumed, but very slowly, despite much canvassing of the question in the trade journals; and by 1900 there were only 283 cutters at work in Britain. The proportion of output cut by machine was only 4% in Yorkshire, 2% each in Scotland,Lancashire and North Wales, and the Midland Inspection District, and less than one half of 1% in the West Midlands, South Wales, the North-East, and the small coalfields.[80] The proportion of firms using cutters in each district was almost certainly higher, for it was common at that time to try out the innovation by using one or two machines on their own. Unfortunately, this was hardly the way to get good results. After 1900 the rate of adoption of mechanical cutting accelerated, and the proportion of national output so got rose from about 1% to 5% by the end of the decade, and by 1913 it had topped 8%. This was still low, however – and, as table 4.1 indicates, it was low everywhere except in Scotland.[81] Even there the proportion was relatively modest.

Mechanical cutting seems to have made even less progress in most other countries until during and after the First World War. The only exception was the United States, where it went much faster than in Britain. Already in 1891, when the first statistic becomes available, over 5% of output was machine-mined. By 1900 the proportion was 25%, and by 1913 50%.[82] Similar figures are not available for Continental Europe until the 1920s, probably because the proportions were insignificant before the War. Great changes came with postwar reconstruction, and comparisons for the interwar period are not in Britain's favour. However, one should not be misled by this into thinking that this was true before 1914.

The reasons for the slow spread of mechanical cutting were partly tech-

Table 4.1 *Proportion of various districts' coal output cut by machine, 1907 and 1913*

	1907	1913
Scotland	10.9	21.7
Yorkshire & East Midlands[a]	6.5	9.2
North-East and Cumberland	3.6	6.0
Lancashire and North Wales	3.8	7.7
South Wales	...	1.1
North Staffordshire	4.5	n/a
South Staffordshire	1.1	n/a
Midland and Southern[b]	...	4.3

[a]Includes Leicestershire and Warwickshire in 1907.
[b]Comprises West Midlands, Leicestershire, Warwickshire, Somerset and Gloucestershire.
Source: *MS*.

nical, partly economic, and partly psychological. Hurd, a machine maker, said there were 'good and reliable' machines as early as the 1870s,[83] and the experience of the West Yorkshire collieries goes some way to bear him out. But they were 'good' only in a limited way. They could only undercut the coal for a distance of three feet, and they were undoubtedly cumbersome;[84] and their 'reliability' depended perhaps excessively on the skill of their operators and on their suitability for a particular seam – and, of course, skilled mechanics who would work at a coal face must have been exceedingly rare. However, by the 1890s these problems had been largely overcome, for those who cared to profit from the experience of others and recognise that 'the machine is not to be viewed as a mere incident in the general scheme of operations, where it is used it is the prime factor upon which the whole working of the section should be based'.[85] There really were good and reliable machines by 1900, and in sufficient variety to suit most conditions (outside South Wales) provided they were correctly used. Most owners and managers, however, seem to have looked only superficially at other people's experience, and were discouraged by what they saw as technical failures, but which were really management failures.[86]

There was no lack of contemporaries to say that British managers and owners were unenterprising about cutters. One mining engineer wrote in the journal of his profession in the late 1890s: 'Many colliery officials are intensely conservative and refuse to believe that any modification of their methods is either possible or necessary';[87] and J B. Atkinson, HM Inspector of Mines for East Scotland, told the Royal Commission on Coal Supplies in 1903 that he did 'not think that there is any objection simply to the introduction of machinery, but there is a sort of *vis inertiae* that you have to overcome at a great many collieries before you can get any new system introduced'.[88] Another inspector, J. Gerrard of Manchester, said much the same thing: 'people have been accustomed to handwork, and they are difficult to move from that. They think it is best for their conditions, and they do not care to go to the expense of giving £200, £300 or £400 for a machine, and spend £2,000 to £6,000 in putting down plant to make the experiment.'[89] As a final example, we may note the admission of a Yorkshire owner, G.B. Walker, that 'we are slow in making new departures'.[90] Yet for all that, the pre-1914 British coal industry, with its numerous independent entrepreneurial units, must have been one of the closest approximations to the model of pure competition that it is possible to get. And this should surely make us pause, and wonder how, if the many who failed to adopt mechanical cutting were as unenterprising as all that in failing to pursue cost-reduction (and profit maximisation) *à l'outrance*, they managed to get away with it. Of course, it is possible that an answer can be found in the great variety of natural conditions in coal mines and the many quasi-rents that abounded. Maybe these provided a cushion for the unenterprising, and only those in poor natural conditions felt the full spur of competition. But this is not really very plausible, for it still relies on irrational or lazy conservatism on a mass scale, which is an unlikely condition in an industry which newcomers could still enter fairly easily as late as the 1890s. For the same reason it is also implausible to suggest, as is sometimes done, that British coalowners were seeking short-term profits at the expense of long-term ones. For this allegation implies that most coalowners were adopting an inappropriately high rate of discount for future returns – mass irrationality once again. In fact, the answer to the question 'why was the rate at which the pioneers were copied so sluggish?' seems to lie mainly in limitations on the profitability of the innovation – which is just what another inspector of mines, A.H. Stokes, said in his *Report* for 1906. A close review of the writings of leading mining engineers of the day and of the evidence they gave to committees and commissions tends to confirm this, although it can never do so conclusively. As Arthur Taylor has pointed out, 'the economics . . . of machine mining are . . . elusive: it is doubtful whether they have been, or can be, assessed with full accuracy even in the modern, statistically minded era'.[91] The harder evidence that we have, however, all points in the same direction.

The cost-reduction claimed by advocates of cutting machinery in the 1900s was generally of the order of 6d to 1s per ton, with some rather higher claims for very thin or hard seams.[92] These figures included an allowance for the increased proportion of the more valuable large coal which was got by machinery.[93] The somewhat unsatisfactory replies to a circular sent out by the Royal Commission on Coal Supplies to the 122 known users of cutters tend to confirm these claims to some extent, though they also indicate that the machines did not do so well in all cases. It is not always clear whether the people replying to the circular took the greater proportion of large coal into account, nor is it always clear whether they were thinking in terms of gross saving (i.e. the prime cost, or even merely the labour cost at the face) or of net saving (i.e. the saving on prime cost *less* the increase in overheads). For what it is worth, however, of the forty-seven usable replies, seventeen indicated no cost saving, six said they saved 3d per ton, twelve saved 5d or 6d, four saved 9d, and seven saved 1s per ton or more; one reckoned the machines cost 4d per ton *more* than handwork.[94] A rather less optimistic view than would be justified by the Royal Commission's circular came from the report of a committee of the North of England Institute of Mining Engineers in 1903, which found that the average saving per ton in the twenty-six cases they investigated was only 6.99d *before* allowing for increased overheads.[95] All told, the general impression one gets from an extensive study of contemporary sources is that a figure of 6d per ton net saving (including the greater value of coal got by machine) was representative of seams from about three feet thick upwards where coal-cutters were actually used, but that this gradually disappeared above four feet thick, and had always disappeared by the time seams reached five feet in thickness.[96] For seams thinner than three feet, much depended on their hardness as well as their thickness, but a total net saving of 6d per ton was probably a minimum, and 1s a ton would be more representative.[97]

If one asks why the saving in costs resulting from machinery was not higher at any given time, there are two sides to the answer. First, the actual level of the increase in productivity achieved by the cutters; and second the actual level of wage costs. If the machines had raised productivity by more than they did, or if wages had been higher than they were, then more cutters would have been used. So far as the first aspect is concerned, one can probably say that the machines were about as efficient as the technology of the day allowed,[98] though management did not always so organise them as to get the best out of them. Properly managed, they were reckoned by the leading textbooks of mining to raise physical productivity by about a third in seams of three feet thickness or above;[99] but in very hard and thin seams they could do much more than that. In the twenty-five cases investigated by the committee of the North of England Institute, for example, nearly all operating in seams below two and a half feet thick, the average increase in faceworkers' productivity was from 2.45 tons per manshift to 4.21 tons – over 70%.[100] Enthusiasts for machinery quoted higher figures still, even in thicker seams – for

Table 4.2 *Net cost saving from using coal-cutters under certain assumptions (pence per ton)*

Percentage rise in productivity by machines	Labour cost of getting by hand 1s 6d	2s 0d	2s 6d	3s 0d	3s 6d
25	−2.4	−1.2	nil	1.2	2.4
33.33	−1.5	nil	1.5	3.0	4.5
50	nil	2.0	4.0	6.0	8.0
66.66	1.2	3.6	6.0	8.4	10.8
75	1.5	3.7	6.5	9.5	12.3
100	3.0	6.0	9.0	12.0	15.0

Sources: see text.

instance, W.E. Garforth's report of a rise from 3.25 tons to 6 tons per manshift in a four-foot seam.[101] But there were no doubt failures to be set against these high claims, and those did not get the same publicity.

So far as the second aspect is concerned – the cost of labour, or, to be more precise, the cost of the labour of getting and filling ready for the haulage workers to take over – it is important to understand its relationship to productivity. For every unit of labour cost, a given increase in productivity implies a certain saving in cost. To give two examples, for each shilling per ton of the original labour cost, a one-third increase in product per head implies a gross cost saving of 3d per ton; and a doubling of productivity implies 6d per ton gross cost saving.[102] Out of this saving has to come the costs of power, maintenance, repairs, and allowances for interest and depreciation, which together were put at around 6d per ton by the author of the contemporary textbook which paid most attention to economics.[103] Deduct this from the gross cost saving, and one can get the figures shown in table 4.2, indicating the net cost saving (i.e. the profit or loss) produced by mechanical cutters according to the rise in productivity which they achieved and the labour cost per ton of getting and filling by hand. However, note that any increase in the average value of output is not taken into account in this table.

This table brings out quite clearly the conditions necessary for profitable use of mechanical cutting if there had been no change in the average value of output. With a productivity increase of a third, which was what one might expect in seams of over three feet thick, the hand-hewing cost had to be above 2s per ton for there to be any net saving.[104] Any profit below this level depended on a more valuable product being achieved. Now before 1914, and particularly before 1907, the labour cost of handwork was rarely above 2s per ton in any seam of over three and a half feet thick; and in these seams, the proportionate change in large coal pro-

Table 4.3 *Annual profit per machine in various conditions (in £)*

Annual output per machine (tons)	Net saving per ton 3d	6d	9d	1s 0d	1s 6d	2s 0d
5,000	67.5	124	192.5	250	375	500
10,000	125	250	375	500	750	1,000
20,000	250	500	750	1,000	1,500	2,000
30,000	375	750	1,125	1,500	2,250	3,000

Sources: see text.

duction was small, and became smaller the thicker the seam. Moreover, the increase in productivity effected by machines was reckoned to be smaller the thicker the seam. It is easy to appreciate, therefore, why mechanical cutting was seldom employed in seams of over four feet thick, and virtually never in those of over four and a half feet. Only an exceptionally hard seam of four feet plus could justify the use of machinery.

It seems, then, that economic considerations probably ruled out the use of mechanical cutters so far as a majority of British coalowners was concerned, since well over half of them in 1914 were still working seams of four feet thick or more. This included most of the pits in South Yorkshire, North Nottinghamshire, Warwickshire, Cannock Chase, and South Wales outside the anthracite district. It also included a large proportion in Fifeshire, Northumberland and Durham, Lancashire, Derbyshire, and Leicestershire. In fact, the only districts where most collieries were mining seams of less than four feet thick were West Yorkshire and the Scottish coalfields other than Fifeshire – plus the Welsh anthracite district, in which, of course, there was no extra price to be got for large coal.

It is possible to say, then, that it was reasonable for most coalowners in the thicker seams to stick to hand mining. But there was a substantial minority of thin-seam operators, and by no means all of these used machinery. How reasonable were they? This is difficult to answer. It is impossible to translate the representative figures of savings per ton at all accurately into terms of returns on capital; but it can be done roughly on certain assumptions, and this has some value in showing what was needed in order to make investment in machinery worthwhile. Table 4.3 does this by showing the annual profit per machine on various assumptions about output and saving. Now the average annual output per cutter throughout the country in 1907 was only a little over 10,000 tons, and in 1913 it was barely 500 tons more. The highest figure we have seen quoted is 25,000 tons, in a seam nearly four feet thick. Barber, Walker & Co. achieved 20,000 tons per machine per year in the last few years before 1914 in a seam between three and three and a half feet thick.[105] On the other hand,

Stringer & Jagger, a West Yorkshire firm, were only averaging a little over 5,000 tons a year in the early 1900s in a seam one foot ten inches thick, which is as low a thickness as was ever worked. It would seem likely, therefore, that in the sort of medium-thick seams where 20,000 tons could be got per machine per year, one should expect the annual profit to be a maximum of £500, and more likely well below that.[106] At the opposite extreme from the medium-thick seams, a really thin-seam pit, like those of Stringer & Jagger, might achieve very large net savings per ton, most of it coming from the higher proportion of large coal.[107] But the annual output per machine in these thin seams could not be expected to be much more than 5,000 tons. So, once again, a figure of about £500 annual profit emerges as a maximum – or, at any rate, as something which was only likely to be exceeded in unusually favourable circumstances.

How does this sort of figure translate into terms of return on capital? The answer to this must vary, since the necessary investment also varied a good deal according to individual circumstances. It was reckoned that each cutting machine plus its share of the necessary power plant, transmission lines, etc., cost between £1,000 and £1,250 in the mid-1900s. This was for a battery of six cutters, up to which point there were economies of scale in plant, transmission, etc., but beyond which these rapidly dwindled. But this was only the beginning, at any rate in a colliery which was already operating on handwork. There had to be a reorganisation of the faces and underground roadways consequent on the longer working face required by machines; there had to be a concentration of the workings to economise on power transmission; and if large productivity increases were achieved there would need to be an increase in haulage facilities. The total effect might be to double the investment needed for the plant itself, sometimes even more. For Barber, Walker & Co. the capital outlay worked out at £2,500 per machine; whilst the manager of Wharncliffe Silkstone Colliery, an enthusiast for cutters, reckoned on £2,000 per machine.[108] These figures indicate the order of magnitude involved. They suggest that, leaving aside the really exceptional cases (of which Wharncliffe Silkstone seems to have been one), a maximum return on capital of 20% to 25% could be expected from the innovation in seams of less than four feet thick, and perhaps half these figures should be regarded as more typical, probably less for seams between three and four feet thick.

These rates of return are undoubtedly very respectable ones for an innovation which no longer involved pioneering; and they must be regarded as the inducement which had attracted the 122 known users of cutters in 1904 out of a total of about 250 owners of pits working seams of three feet thick or less who also employed at least 100 men (excluding South Wales, where geological conditions were almost uniformly unfavourable). The 130 or so who did not use cutters (plus an unknown number working slightly thicker seams) would seem on the surface to have been turning down the change of up to 25% return. Why should this be so? There is no single answer, but a number may be suggested. First, some of them must surely have been right in thinking that their particular conditions were unsuitable for machinery. Others had mines approaching exhaustion, where costly installation costs would not have been worthwhile. A few did not have the finance available because they were scarcely making any profits to plough back and were scarcely in a position to attract external funds.[109] Yet for all these arguments, there must also have been some entrepreneurs who were viewing the apparent likelihood of a 10% return or more with an unduly sceptical eye. However, this is surely nothing unusual at any time and in any industry. At all times there are leaders and followers in technological innovation, and the followers tend to be those who view the prospect of risk and uncertainty with a more jaundiced eye. It is, surely, quite a normal situation that some entrepreneurs should require a larger risk premium than is implied by a yield of 10% or 12%. Certainly this is the implication of Carter and Williams's studies in the 1950s.[110] These entrepreneurs may be, in some sense, inefficient, through being excessively cautious risk-averters rather than risk-takers. But they are a phenomenon which is always with us, and the proportion which they seem to have constituted of all the entrepreneurs in the coal industry before 1914 does not seem to be unusually large.

Finally, before leaving the subject of mechanical cutters, let us look briefly at a factor which some of the entrepreneurs put forward in defence of their high assessment of the uncertainty of the return on investment in these machines. This is the hostility of the workforce. It is extremely difficult to judge this, but hostility did certainly exist. The *Colliery Guardian* wrote in 1901 of profit being 'more than swallowed up in the increased cost of filling after the coal was cut, the fillers wilfully making the cost heavier in the now well-known manner'.[111] The *Colliery Guardian*, however, was not generally well-disposed towards miners, and it certainly cannot be accepted as unbiased evidence about the normal and continuing reaction of the workers. That there was often difficulty on first introduction of machines to a pit or seam is undoubted, and scarcely surprising.[112] That it continued for very long after the initial period of adjustment is unlikely. Barber, Walker & Co.'s experience was probably typical. It had a short strike on the introduction of cutters, followed by a period of a few months when productivity was somewhat reduced, before it picked up again and surpassed its previous level. The Coal Conservation Committee, during the First World War, may have been too optimistic when it wrote: 'Reference is sometimes made to the hostility of labour to changes in methods of mining and more particularly to the use of mechanical appliances underground. We do not regard this impression as well-founded . . . Investigation of particular cases where difficulties appear to have arisen rather suggests that in these cases the human factor has not been sufficiently considered in the changes introduced.'[113] Perhaps this is too easily ignoring an element of Luddism amongst miners; but it seems likely that it is much closer to the truth than the *Colliery Guardian*.

When people in Britain talked about the mechanisation of mining, either before or after the First World War, they did so almost entirely in terms of the coal-cutter, or holing machine as it was sometimes called, which undercut the coal in much the same way as the hewer did with his pick. The mechanical pick, which in both America and Europe was at least as much used as the cutter, was rather uncommon in Britain, except for driving headings, and to a limited extent after 1900 in connection with pillar and stall working. The main reason for this was the hardness of British seams outside South Wales. For the most part they had to be undercut, or else an excessive proportion of small coal would be produced, and even a protagonist of the mechanical pick, like W. Blakemore, had to say that 'it must be admitted that the longwall machine is more economical where it can be used'.[114] The only exception to this was in steeply inclined seams, where the weight of the pick was on the coal and did not have to be supported. The only important examples of such seams in Britain were in South Wales, Lancashire, and, above all, the small Potteries coalfield; and mechanical picks were used to some extent in the last two of these.[115] In some parts of Europe, notably the Ruhr, conditions tended to favour the mechanical pick rather than the coal-cutter, and this was the main reason for its widespread use there.[116] The American-style pick on wheels, however, met no acceptance either here or in the rest of Europe. The main reason was probably the fact that it could only be used in fairly thick seams where there was room for it, and handwork was still the more economical proposition in those conditions given British wage rates. A subsidiary reason may have been the uncompromising and unrelenting hostility of the miners, which, for once, was sympathetically received by the owners. Undoubtedly it did shake its operators severely.[117] It also probably required more room than was generally available in the deep seams typical of Britain.

The introduction of mechanical cutters in the longwall system necessitated a concentration of working faces, and that, together with the increased productivity, meant that a greater quantity of coal was produced from a given area in a given time. This presented a problem of haulage, notably of auxiliary haulage, which was aggravated by the need to keep the faces clear if the cutters were to work efficiently. The first answer to this problem was the ripping of a great number of roadways; but this involved a considerable increase in costs, especially in the thin seams which were the very ones where mechanical cutters were most used.[118] In addition to this rise in the cost of ripping and maintaining roadways, there was an increase in costs through the greater number of haulage workers which had to be employed. As a result of all this, managers of pits where thin seams were worked were especially anxious to find a better solution to the haulage problem than the multiplication of roadways. The introduction of face conveyors in 1902 went a long way towards providing such a solution, and they spread quite rapidly compared with other innovations.

Face conveyors did not escape technical teething troubles, but these were of comparatively short duration. And though conveyors designed for one seam were not usually adaptable to other seams and circumstances without cost, a large variety of types was available soon after the original introduction of the apparatus. There does not appear to have been the same reluctance to experiment strongly in the use of conveyors that there was with cutting machines. The reason probably was that, once the initiative had been taken by the application of cutters, the adoption of the next innovation followed almost as a matter of necessity.[119] Moreover, it was easier to do this both psychologically, because the ice of managerial or labour conservatism was broken, and physically, because the power plant and transmission lines were there, and cost less to extend than to instal. Finally, the saving of costs needed in order to make the installation of a face conveyor profitable did not seem formidably large, and was a good deal lower than the savings suggested in the textbooks as feasible, or those recorded in the trade journals as having been achieved. A hundred-yard conveyor cost about £300.[120] The amount of coal it would require to handle in order to pay off this investment would depend on the net saving of cost per ton; but even if this was a mere penny per ton, it would only amount to 72,000 tons – which a hundred-yard thin seam might be expected to produce in from two to four years.[121] Net savings of 2½d per ton seem to have been achieved in one case where conditions were suitable and the management was good, and in another a *gross* saving of 11d per ton was reported.[122] And a leading textbook went so far as to suggest that gross cost savings should be as much as 1s per ton in a thin seam of hard coal, not counting an increase in receipts from more large coal.[123]

The connection between the use of conveyors and of mechanical cutters is fairly clear from the Inspectors' returns of the numbers of them employed in the different coalfields. In 1913, Scotland led with 125 conveyors, and Yorkshire and the East Midlands was second with 86, most of which were in the thin-seam pits of West Yorkshire. South Wales and the Northern Inspection District had 61 and 58 respectively, the former being the only coalfield where conveyors were employed with hand holing to any significant extent.[124] The other coalfields were distinctly less progressive in this respect. Lancashire and North Wales only had twenty-two conveyors, and the Midland and Southern Inspection District had a mere two, even though there was nothing in their conditions to account for this.

The last major technical innovation of the nineteenth century in the mining and bringing of coal to the surface with which we shall be concerned is the application of electricity to colliery purposes, which first occurred in 1882.[125] For some time it was chiefly used for lighting and pumping only, though its spread for these purposes in the late 1880s and early 90s was said to be rapid.[126] The larger collieries and the larger coalfields were the leaders in this, above all the East Midlands, which had the least risk of firedamp explosions. By the 1900s, though, the other coalfields had caught up, and when Special Rules under the Mines Regulation Act were first drawn up in 1905, very few collieries with 100 or more employees were not using electric lighting on the surface and at the shaft bottom, and usually in the main roadways. In Yorkshire, where the inspector noted this in drawing up the *List of Mines* for 1907, only three pits of this size did not have electricity.

By the beginning of the 1890s, the use of electricity was said to have got beyond the lighting and pumping stage in a few places, and it was 'already successfully applied for underground haulage in some South Yorkshire pits . . . but this was a very exceptional novelty'.[127] The relatively rapid progress, however, was not kept up. The main reasons for this were technical – at least for a time. First, there were difficulties in transmitting the quantities needed for power purposes, and then there was the danger of sparks causing explosions.[128] When these had been overcome, by the later 90s, with armoured cable and alternating current motors, fears remaining from earlier stages remained to reinforce the conservatism of some managers. A further hindrance to the spread of electricity for power purposes was probably the fact that, as the *Colliery Guardian* put it, 'the continuous use of air-compressing plant for a great number of years produced a large number of men . . . who were conversant with the working of the apparatus'.[129]

Despite the difficulties, real and fancied, several of the larger collieries, especially in Durham, South Wales, Yorkshire, and Nottinghamshire, used electricity for everything except winding by 1904,[130] although the majority everywhere preferred steam or compressed air for power purposes; and it is symptomatic of the slowness of its adoption in the previous decade that it could still be referred to at that time as 'this new agent'.[131] Figures are not available, but judging by various accounts in the trade journals of numerous individual collieries, we may doubt if much over half even of the relatively modern pits of the most progressive areas used electricity for any underground power purpose in the early

1900s. The main reason for this slow pace of adoption was the extremely low cost of power generation by steam so far as most collieries were concerned. In existing pits the steam plant was already installed, and the coal which it used was exceedingly cheap, and virtually costless in those seams which produced slack that was useless for coking purposes. In the light of this, it is not surprising that the most progress had been made in those coalfields with major outputs of coking coal.

In the decade after 1904 there was a much more rapid spread in the use of electricity for underground power. Two changes were principally responsible for this. In the first place the price of slack rose proportionately as much as that of large coal in the boom of 1899–1900, and it did not fall back to anything like the same level as before. In South Yorkshire, for instance, the price of best slack rose from 3s 6d per ton in 1895 to 10s 6d per ton in 1900, and fell back to 5s 3d in 1905, whilst the price of house-coal rose from 8s 6d to 16s, and then fell back to 9s.[132] Small coal thus became proportionately more valuable, raising the relative cost of steam power even to collieries. At the same time the technical innovation of the exhaust steam turbine for the generation of electricity was introduced in 1901, and economised very considerably on the use of coal in those cases where large-scale production was economic, i.e. where a large output could be used or sold.[133] In optimum-sized plant, costs per horsepower-hour were more than halved.

Just what this meant in terms of return on capital is impossible to assess accurately, because some of the results of using electricity are unmeasurable, and, in any case, there are no industry-wide statistics that can be useful. We have, in fact, got just two pieces of evidence from individual company records, and they tell roughly the same story. The first is from Trimdon Grange Colliery in Durham, which was fully electrified in 1911, with an apparent yield on the investment of 20 to 25%.[134] The second is from the Cannock Chase Colliery Company's pits, which were electrified a year later, and the manager made an estimate of the return in the first year as £3,946 on an investment of £21,259 – nearly 19%.[135]

In the light of this, there would seem to have been a clear case for the adoption of electric power from about the early 1900s onwards, and in fact there was a great extension of its use. By 1913, when statistics are available for all coalfields, well over half the pits employing fifty or more men underground were using electric power for some purpose underground, as can be seen from table 4.4. In the districts where output had been expanding relatively strongly in the previous decade the proportion was over two-thirds. In the other coalfields, where output growth had been sluggish, the proportions were lower – most notably so in the areas where peak production had been reached before the end of the nineteenth century and there had been little recent sinking.

Table 4.4 *Proportion of British coal mines using electricity for power, 1913*

	Percentage of all mines	Percentage of mines employing 50 men or more underground
Durham & Northumberland	55	71
Cumberland	42	60
Yorkshire	48	73
East Midlands	49	70
Leicestershire	63	79
Warwickshire	69	77
Lancashire & Cheshire	30	39
North Wales	34	48
North Staffordshire	35	68
South Staffordshire & Worcestershire	12	36
Shropshire	4	13
Gloucestershire & Somerset	29	54
South Wales	48	74
Scotland	59	75

Sources: *Reports* of HM Inspectors of Mines and *List of Mines*.

Whilst the proportion of collieries using electricity for power had reached quite substantial levels by 1914, the average amount used at each colliery was still fairly modest, as table 4.5 shows. There is a number of interesting differences between coalfields brought out by this table, some unexpected and not easily explicable. One of the rather surprising things is the very imperfect correlation between tables 4.4 and 4.5, at any rate so far as the inland coalfields are concerned. Particularly noticeable is the contrast between Leicestershire, where 79% of sizeable pits used electric power, yet only at the rate of 0.45 HP per thousand tons, and Warwickshire, where the figures were 77% and 1.50 HP respectively. There is no obvious reason why this should have been so, except that there were more large, recently sunk pits in Warwickshire, and these had probably been laid out to use electricity for haulage, whereas its use may have been largely confined to lighting in the older pits of Leicestershire. Another rather unexpected thing is the relatively small use of electric power in Yorkshire and the East Midlands, despite the early use of electricity for lighting there, and despite the high proportion of their output produced by recent large sinkings along the eastern edge of the coalfield. In fact it seems to have been just these new collieries which helped to reduce the proportionate importance of electric power, firstly because no coal-cutters were used in the thick seams they were exploiting, and secondly because steam-powered haulage was still the most economical where

Table 4.5 *Electric horsepower used per thousand tons of output, 1913*

	Underground	Total
Durham & Northumberland	1.27	2.33
Cumberland	0.90	1.82
Yorkshire	0.71	1.73
East Midlands	0.68	1.09
Leicestershire	0.39	0.45
Warwickshire	1.27	1.50
Lancashire, Cheshire & North Wales	0.90	1.25
North Staffordshire	1.62	2.41
South Staffordshire & Worcestershire	0.68	0.87
Shropshire	0.41	0.61
Gloucestershire & Somerset	2.13	2.31
South Wales	1.60	3.20
Scotland	2.20	2.82

Sources: as for table 4.4.

working could be laid out for the most up-to-date version.[136] The greater economy of electricity in working older mines is exemplified by the North-East and by the small coalfields of Gloucestershire and Somerset. It may also have had an influence on the relatively high figure of horsepower used in Scotland, but the greater progress of mechanical cutting there was probably the main factor.

The last technical development with which we shall be concerned is a rather different one from those looked at hitherto. These have been innovations which improved the methods of getting the coal to the pit bank. But as time went on, coalowners and managers came increasingly to concern themselves with what came to be known as coal preparation. Up to the 1880s, the preparation of coal for the market was a simple, not to say primitive, matter. Especially in the first half of the nineteenth century, it took place to a considerable extent underground, in that only large coal was brought out of many pits.[137] Where everything was wound, plain fixed screens were used for sorting the coal into a small number of grades. Part of the reason for this simplicity lay in the low level of demand for grades other than large coal and engine coal; part in the relative cleanliness of the first seams of coal to be exploited. As J.T. Forgie put it in his presidential address to the Mining Insitute of Scotland in 1898:

> The screening and cleaning of the round coal and sizing and washing of the dross by substantial, elaborate, and expensive machinery have enabled us to work seams which otherwise would have remained unworkable, and we thus see all

over Scotland the lives of the older collieries being extended far beyond expectation, and even the resurrection of some of the dead and dismantled pits.[138]

Washing and screening plant, however, was not initially developed either because the demand for grades of small coal went up, or because it was essential to the exploitation of certain seams. It was developed out of the combination of a competitive industrial structure and the excess capacity created by the great boom of the early 1870s, as a means of selling more coal at the expense of one's competitors, and so pushing one's output up nearer to capacity. In this it was successful (for the pioneers),[139] and once adopted at one colliery in a district, others were compelled to follow suit on pain of either losing customers or having to be content with a lower price. As a result, the innovation spread rapidly in the 1880s and 90s.[140] In passing, it is worth noting how well the two innovations, washing and screening plant and coal-cutting machinery, exemplify the saying that in times of high wages, labour-saving innovations are more rapidly adopted, whilst in times of low prices product-improving ones are more acceptable.

There may be no reason in the models of neo-classical economics why the saving of one particular kind of cost should ever be preferred to any other kind of cost, but there does seem warrant for the belief that it takes place in certain circumstances and Paul David has shown why this should be so.[141]

We began this chapter by quoting Arthur Taylor's remark that 'in coal-mining the role of the innovator is fundamental . . . the indispensable condition of the industry's continued existence', and perhaps it is well to end by re-emphasising this. The nineteenth-century British coal industry, as we have seen, expanded enormously, largely in response to the growth of demand. It was only able to do so because of very great and continuing improvements in technique. Such improvements had been going on in previous centuries; but as the rate of increase of demand stepped up, in the 1830s, so did the pace of innovation. However, apart from this surge in the 30s, which, of course, took more than that decade to work through, it is impossible to discern extended periods of especially rapid innovation, although there seems to have been a sustained quickening of change from the early 1870s onwards. Nor is this particularly surprising, for whilst the rate of growth of output slowed down from the 70s, there continued to be large increases in absolute terms right up to 1913. One cannot but agree with Bulman and Redmayne, who, writing in the 1900s, said: 'The chief impression left by an historical review of coal mining in the United Kingdom is the enormous progress made during the last two or three generations.'[142] Yet is it also true that at all times the majority of managers lagged behind the best practice of the leaders. One can see this in the slow spread of longwall mining from its birthplace in the West Midlands; in the persistence of corves and sledges long after the invention of rails and wheeled trams; in the time taken between the first underground haulage by steam engine and its general use; in mechanical ventilation, coal-cutting by machine, electric power, etc., etc. Some inventions were applied quite rapidly, because they were cheap (e.g. shotfiring) or offered obvious high returns (e.g. face conveyors where mechanical cutters were already used). But it is generally true to say that it took at least two decades after an invention had passed out of the experimental stage before it was in really general use. This is not peculiar to coalmining in nineteenth-century Britain, but there are a number of explanations for it which are specific. A major one was the fact that mines are subject to exhaustion. Apart from the obvious effect of this in inhibiting technical change in pits whose life was near its end, it also lay behind W.H. Chambers's dictum that it was 'generally a safe rule for managers to follow, that no additional capital expenditure should be incurred after the plant of a colliery was completed, except when a quick and large return on the investment was assured'.[143] Arthur Taylor believes that another reason for delay in the general spread of costly innovations lay in the capital structure of the industry. 'The predominance of small units both in an entrepreneurial and operative sense, undoubtedly militated against technical experiment and innovation', he says.[144] He is almost certainly right, though not, as he suggests, because small firms tended to distribute a greater proportion of profits than large ones. This was not the case, since distribution of the bulk of profits seems to have been practised by a majority of all kinds of firm.[145] The main reasons were first the existence of some, albeit limited, economies of scale in coalmining at the greater depths being reached by the last decade or so of the nineteenth century, and economies of scale in the use of certain sorts of power plant, which were also becoming increasingly noticeable at the same period, especially with the increased employment of power-using technologies. These economies of scale led to larger firms tending to have proportionately more profits for ploughing back and/or greater ability to attract outside finance. Second, large firms had greater ability to attract the more go-ahead managers – a special kind of scale economy. Coupled with these positive reasons, there was the negative one of the staying power of established but technically inefficient collieries, for, as Taylor points out, 'as long as a pit was profitable in the sense that weekly receipts covered weekly outgoings, it was in the owner's interest to keep it in operation: an abandoned mine and its machinery have little value as a realizable asset'.[146] Finally, no explanation of the slowness of general adoption of some technical changes would be complete without mentioning the notorious conservatism of many managers. So much has been made of this in subsequent discussion of the technical backwardness of the industry, especially in the interwar period when, for a time, that backwardness was most marked, that at times reactionary incompetence has seemed to be offered as its whole *raison d'être*. This clearly does not fit the facts. Probably only a few managers were technically conservative on principle, and all the rest were at least attempting to make rational decisions with costs, risks, and profits in mind. But, equally probably, most managers were inclined towards the side of caution, and the pioneers were therefore tempted to raise the cry of conservatism.[147]

5

LABOUR

The great expansion in the British coal industry in the nineteenth century required not only a large increase in the number of mines and in the capitalisation of collieries, and changes in the technology of the industry, but also a much bigger input of labour. By far the most important component of this increase was the growth in the number of men employed. There may have been some increases in the regularity of work, and this certainly varied very considerably in the short run, at both employer's and worker's initiative; there was a marked decrease in the average number of hours worked in a day by each miner in the last half-century of the period, though in part this may have been a mere offset to an earlier increase in hours above the average ruling at about 1800. There were other, less tangible influences on the input of labour, which have the nature of changes in its quality. But whatever the effects of all these things on the labour input – and each will be examined below – it is unquestionable that the prime factor in its increase was the simple growth in size of the colliery labour force.

Before going into questions of number and time, the ingredients of the economist's concept of the input of labour, it will be useful to look first at the nature of the work that was required.

No doubt all kinds of workmen in the mines were necessary to their efficient operation, but few will quarrel with the statement that the hewer was the most important man, the man whom all others were, in effect, employed to serve. Hewing was, without doubt, severe labour in the vast majority of British mines. Only in the soft seams of South Wales is there room for even the slightest qualification of the adjective severe – and there, though the undercutting of the coal was easier and wedging it down usually unnecessary, the hewer had to perform other heavy tasks, such as timbering, and do so in conditions which required even more vigilance than was necessary elsewhere.

The normal operation of the hewer in most seams worked by hand, is still as well described by Brand in his late eighteenth-century *History of Newcastle* as by anyone.[1] 'In the process of hewing the coal from its bed, the hewer first digs as far as he can into the bottom of the stratum; then he nooks or corners off the part measured out, and afterwards the great coals come away by a wedge or mallet.' When, and where, explosives came to be used, the wedging process was replaced by shotfiring, which eased the task whilst, for some time at any rate, increasing the danger. The hard labour of the hewer was compounded not only by danger but by discomfort, often in an extreme form. Tancred described it, a shade less dispassionately than Brand, to the Royal Commission of 1871 thus:

> The ordinary posture is sitting with one leg doubled beneath him and the other foot resting against the coal. Inclining his body to one side, so as often to nearly touch the ground with one shoulder, he digs his pick with both hands into the lower part of the coal, and haply into some softer material beneath the coal . . . When he has, after two or three hours, undermined enough for safety, he inserts iron wedges by means of a heavy hammer between the coal and the roof above it, by which means he detached the coal. The cramped posture, the closeness of the subterranean atmosphere leaded with coal dust and the smoke of his lamp, and sometimes the sulpherous exhalations, together with the bodily exertion cannot fail to be very exhausting. He is subject to severe accidents from unsound part of the roof or from masses of coal falling upon him, from the inflammable gas which may burst in upon him and envelop him in flames, or from the deadly black damp which may stifle him.[2]

The cramped posture was particularly exaggerated in the thin seams, which were predominant in two or three districts – northern Northumberland, western Yorkshire, Ayrshire, and some others. The danger from gas was greatest in some seams in Lancashire, Durham and Northumberland, South Yorkshire and South Wales. The risk of falls of coal or roof was worst in South Wales or in the Ten Yard seam of South Staffordshire, where the working 'chambers' were described by A.W. Grazebrook in 1906 as 'a vast and ugly-looking place, and none but men who are used to it will work in them.'[3]

In all districts the hewer was, to greater or lesser degree, a specialised workman. In Northumberland and Durham he did nothing but get the coals. Elsewhere he usually set the timber in his own working place, and, especially in Scotland and Wales, he might help 'fill' the coal – i.e. put it into corves or tubs – and even, sometimes, join in the drawing (or 'hurrying', or whatever the local word was) the tubs to the main haulage or, in the early part of the century, the pit bottom. This was most common in the first fifty or sixty years of the nineteenth century, when, in some coalfields, hewers generally employed their own 'hurriers', often their children, or even, in Scotland up to the 1840s, their wives. Other skilled specialised workers included those who made and repaired the underground roadways – timbermen, stonemen, rippers, brushers, etc. Partly at least, these were older men who had given up the severe work of the hewer. But some moved straight into the oncost jobs when they became adult, usually because they preferred the regular income of the day wage man to the fluctuations of the pieceworkers. There is some evidence, in fact, that these occupations were partly hereditary.[4] Then, of course, there were the NCO's of the management, also mainly consisting of former hewers – overmen, deputies, firemen, etc., who were responsible for supervising operations and for inspecting the workings to check for gas or other danger. It must be stressed, though, that for the most part these were not supervisors in the sense that factory foremen were. They might visit the working places once or twice a day, but the hewers were almost completely in charge of their own operations. The deputy's job was really to guard against explosions and falls, and to fire shots.[5]

In the inland coalfields, and to some extent elsewhere except in the North-East, a separate group of fillers was employed to load the tubs at the face. These mostly consisted of youths and young men who were waiting for a place as hewers. From the face, the full tubs were taken by the hurriers to the main haulage or, earlier, to the pit bottom. In Durham and Northumberland, the task of filling and hurrying was combined in that of the putter – again youths and young men waiting for strength or opportunity to become hewers. Hurrying was the occupation which included most of those females who were employed underground in certain coalfields[6] until they were barred, along with boys under ten years old, by the Act of 1842, for which the 7th Earl of Shaftesbury is perhaps best known. The work of the hurrier was not light, as may be gathered from the description given above,[7] and in thin seams it was accompanied by discomfort just about as acute as that of the hewer. From ten years old, which was the normal age of beginning (until the minimum age for going below ground was raised to twelve in 1860) up to about fourteen, the children usually worked in pairs, one pulling and the other pushing. Thereafter, the youth usually managed a tub on his own, until such time as he could graduate to facework, either directly as a hewer, or as a filler, or, in the North-East, as that half-man-half-lad, the putting hewer. Prior to 1842, children were taken into the pit to work at a very early age, though the six- and seven-year-olds referred to in an account of 1800 quoted by Ashton and Sykes[8] were uncommon by that time, according to the *Reports of the Children's Employment Commissioners*, when eight was usually the youngest age of entry, and many owners were at least nominally putting the limit at ten. These children under ten were mainly employed as trappers, whose job was to open and shut the doors as the hurriers and others passed to and fro, these doors being the means of keeping the ventilation currents moving in their proper course. None could call this a heavy job, but it carried greater responsibility for the safety of the pit than we should now regard such children as capable of bearing, it was exceedingly monotonous, certainly not without danger of injury from passing tubs, and, above all, one with just about the longest length of working day. Young children also stoked the furnaces.

The last major category of worker required at the collieries was those who worked on the surface. The most responsible, though not always the best-paid, and usually one with amongst the longest hours, was a small group, the winding enginemen. More numerous were the banksmen, who unloaded the tubs, and the screensmen, who sorted the coal. The latter job became more elaborate as time went on and screening machinery was developed. In the early part of our period it consisted simply of 'picking' – that is, picking out stones, and perhaps also selecting the larger coal in a few sizes – and was often done by women. This continued to be the case to quite a large extent until into the 1870s, if not later.[9]

There is no point in dwelling on the hardships of the miner's life, and on what were already, by 1840, widely regarded as the evils of women and young children working underground. To do so might make one wonder how the mines gathered the vastly expanded labour force they required without some element of compulsion. There were, however, compensations in the miner's lot, of which perhaps the two main ones were the relative freedom from supervision,[10] and the factor summed up in the Lancashire song quoted by Ashton and Sykes:[11]

> Collier lads gets gowd and silver,
> Factory lads gets nowt but brass.

We shall return to this theme in the next chapter.

The increase in the number of coal miners in Britain is reasonably well documented from 1841 onwards, and between then and 1914 it was, in absolute terms, larger than in any other occupation except domestic service. The course of this increase, as shown by the decennial censuses, was as shown in table 5.1. These figures, of course, relate only to one date in the year concerned, and they do not purport to cover only miners in employment at that time. They are, in fact, statistics of those people who described themselves to the census enumerators as coal miners (or, in coalmining communities, simply miners). It is likely that a proportion of the unskilled surface labour force about the collieries did not so describe themselves or were not so classified by the enumerators, especially prior to 1881. As against this, unemployed miners and retired miners under sixty-five years old are included, though it is impossible to say if this bias offsets the former one. However, for most practical purposes it does not seem unreasonable to take table 5.1 as a measure of the actual labour force in coalmining in the years in question.

Another measure of the labour force is available from 1851 onwards, and on an annual basis. This is the returns collected by HM Inspectors of Mines of the number of wage-earners 'ordinarily employed'. These were required by statute from 1873 onwards, having been collected previously on a voluntary basis. These statistics are shown in table 5.2. Like the

Table 5.1 *Coal and shale miners under sixty-five years old at censuses, 1841–1911*[12]

	1841[13]	1851	1861	1871	1881	1891	1901	1911
Males	141,392	211,517	276,471	351,426	434,362	594,253	747,911	1,025,634
Females	3,248	2,586	3,813	6,033	3,500	3,847	3,852	5,252

Table 5.2 *Numbers employed in coal mines in Britain according to the returns of HM Inspectors of Mines (in thousands)*

1851	216.2	1867	333.1	1883	514.9	1899	729.0
1852	222.8	1868	346.8	1884	520.4	1900	780.1
1853	229.5	1869	345.4	1885	520.6	1901	806.7
1854	236.1	1870	350.9	1886	520.0	1902	824.8
1855	242.7	1871	370.9	1887	526.3	1903	842.1
1856	249.3	1872	418.1	1888	549.2	1904	847.6
1857	256.0	1873	514.1	1889	581.8	1905	858.4
1858	262.6	1874	538.8	1890	632.4	1906	882.3
1859	269.2	1875	535.8	1891	668.0	1907	940.6
1860	275.8	1876	514.5	1892	683.6	1908	987.8
1861	282.5	1877	494.4	1893	683.0	1909	1,014.0
1862	291.0	1878	475.3	1894	705.2	1910	1,049.4
1863	299.0	1879	476.8	1895	700.3	1911	1,067.2
1864	307.5	1880	484.9	1896	692.7	1912	1,098.1
1865	315.5	1881	495.5	1897	695.2	1913	1,127.9
1866	320.7	1882	504.0	1898	706.9	1914	1,133.7

census figures, they do not provide the ideal measure of the size of the labour force, even after 1872, since, for one thing, they relate to all miners under the Coal Mines Acts, and so, like the census figures, include a small number of men working other minerals; and for another thing, no uniform definition was provided for the phrase 'ordinarily employed', so that employers may have interpreted it in different ways. For example, it would be possible for an employer to include all men whose names appeared in the pay book at any time during the year; or the number which appeared in the last entry he made before making the return to the inspector; or an average of the numbers receiving some payment in each month or fortnight; or an average of the numbers receiving other than small casual payments in each pay period; or an average of the number of men who entered the colliery for work each day. This does not exhaust the list of possible interpretations. Intuitively, the most likely seems to be that the employer would give the last figure he had before making the return. This is supported by the opinion of the only inspector who commented on the matter, James Willis of the South Durham District, who believed that most employers gave the numbers paid on the last pay-day of the year.[14]

On balance, then, it seems reasonable to accept the figures given in the Inspectors' Reports as being end-of-year figures, applying to a period when, *ceteris paribus*, employment was likely to be high – because the last pay-day included the pre-Christmas period, as well as because demand for house-coal was seasonally high. Moreover, the statutory returns from 1873 onwards can be taken as complete, even though the figures for 1873 itself had to be partly estimated by the inspectors. Whether the same can be said about the statistics for 1851–72, based on voluntary returns, is more doubtful. Morris and Williams say of South Wales that 'the inaccuracy of the inspectors' estimates of numbers employed in the 1860s became apparent when the 1871 census figures were published',[15] and they prefer to use the census figures. However, they appear not to have realised that the inspectors' figures for South Wales for that period did not include Monmouthshire, which was bracketed with Gloucestershire and Somerset. Certainly, so far as the aggregate figures for the country taken as a whole are concerned, there seems little warrant for their view. The inspectors' numbers for 1851, 1861, and 1871 are (rounded) 216,200, 282,500, and 370,900, which are respectively about 4,000 above, approximately the same, and about 9,000 above the adjusted census figures. These do not seem to be major discrepancies. However, some of the district figures are clearly suspect – notably those for Lancashire and South Staffordshire, as well as South Wales. Equally worrying is the leap from 370,900 to 538,800 between 1871 and 1874 – a 45% increase at a time when output only rose by 10%. Obviously some considerable part of this increase *was* real, in what was probably the biggest boom the coal industry has ever enjoyed, taking place at the same time as working schedules were having to be adapted and hours reduced, following the Mines Regulation Act of 1872, as well as by trade union pressure. But was it all? Unfortunately, there is no way of answering this question; but at the very least, one must regard the figures before 1873 with a good deal of caution.

With the exception of the single census figure for 1841, there is very little precise that can be said for the period prior to 1851 about the coalmining labour force of the country as a whole, though there is a certain amount about some of the individual coalfields in the mid-1840s, notably that of Durham and Northumberland. Perhaps the safest thing to do would be to leave it at that. However, the censuses prior to 1841 do allow us the possibility of making estimates; and though these are in the nature of informed guesses and should not be taken too solemnly, the possibility is worth following up. Two methods of using the census material suggest themselves. The first is to take the proportion of coalminers to the total British population in 1841 and simply extrapolate backwards, using that proportion as a constant. This method is extremely crude and unreliable. Given the fact that the percentage of coalminers rose at every census after 1841, it is hardly likely that it was constant earlier. However, if the rate of growth of the percentage in the period 1841–81 is projected backwards to 1801, one arrives at an estimate of the colliery labour force in the latter year that is below 40,000, which seems unlikely. The second method is to relate the number of miners to the population of those districts which were the mainly-coalmining areas of the country at each census. Again, there is an objection to using the proportion of miners in 1841 as a constant, though in this case the objection is not so strong, since the ratio only fell slowly after 1841, and perhaps had been doing likewise earlier, under the impact of the slow spread of other industries onto the coalfields. This movement, of course, only applied to certain coalfields; and, since we shall in any case want to make estimates of the labour force for individual coalfields (both for their own interest and because hours of work, regularity of employment, etc., varied between coalfields), it seems best to use this method to build up estimates of the aggregate national labour force district by district, adjusting the ratio only where this seems necessary. Of course, this method depends on subjective judgment, both as to what constitute the 'mainly-coalmining areas' at each census and as to whether there is a clear case for altering the 1841 ratio of miners to population in a particular district at a particular time. But provided the judgment is sound, it ought to provide more plausible estimates than the rough-and-ready first method which suggested itself. In the hope that the judgment is trustworthy, and stressing once again that these are no more than informed guesses, the estimates by the second method are given in table 5.3, together with figures derived from the censuses for 1841 and 1871. For counties which contain more than one coalfield it has been necessary to estimate the division of the county total amongst the various coalfields by reference to the occupation statistics given only for males over the age of twenty by registration districts in 1851 to 1871.

By way of postscript to this table, it is worth mentioning those figures of the labour force in the Durham and Northumberland coalfield given by authoritative contemporaries. The first is a conjecture by Matthias Dunn, a well-known viewer who became a Mines Inspector, that around 1810 the North-East had 9,700 pitmen – presumably meaning underground workers.[16] This fits well into our estimate of total employment. The second is the figure 21,000 for 1829, made by the famous John Buddle to the Select Committee on the Coal Trade,[17] which matches our estimate for 1831. The third is 33,898 for February 1844, given in the *Report of the Commissioners to Enquire into the Operation of 5 & 6 Vict. c. 99, and into the State of the Population in the Mining Districts*,[18] and apparently based on an exhaustive enquiry. This figure suggests that the expansion

Table 5.3 *Numbers employed in coal mines in British coalfields: estimates 1801–31, census returns 1841–71 (in thousands)*

	1801	1811	1821	1831	1841	1851	1861	1871
Durham & Northumberland	14.2	16.6	18.8	21.0	24.6	38.0	47.7	59.6
Cumberland	1.4	1.8	2.0	2.0	2.1	3.5	4.2	4.9
Yorkshire	7.1	8.2	9.0	10.1	12.2	20.5	31.1	38.6
North Derbyshire & Nottinghamshire	3.0	3.5	4.0	4.6	5.4	8.3	16.0	23.0
South Derbyshire & Leicestershire	0.6	0.7	0.9	1.0	1.2	2.3	3.6	4.9
Warwickshire	0.6	0.6	0.8	0.9	1.0	1.4	1.8	2.2
Lancashire & Cheshire	9.6	11.6	13.7	16.0	18.7	31.0	43.9	52.3
North Wales	2.7	3.0	3.8	4.1	4.8	5.1	6.4	10.2
North Staffordshire	1.6	2.1	2.5	3.0	4.0	6.9	8.8	12.0
Shropshire	3.4	4.0	4.1	4.6	5.4	4.7	3.8	4.5
South Staffordshire & Worcestershire	5.4	7.0	8.7	10.9	14.1	20.9	26.2	27.7
Somerset & Bristol	2.6	2.9	3.9	4.6	5.2	6.0	6.4	7.9
Forest of Dean	1.2	1.4	1.6	1.9	2.4	2.2	2.9	3.2
South Wales	7.2	10.7	13.0	16.8	25.8	30.9	37.9	50.6
Scotland	8.9	11.0	12.4	13.6	17.5	32.6	35.4	50.6
Great Britain	69.6	84.0	99.2	115.1	141.5	214.1	276.5	351.4

Note: for the sources and methods of estimation see the text. The composition of the 'mainly-coalmining areas' referred to there is too lengthy to be given here, but details may be had from the author. The following adjustments were made to the 1841 ratio of miners to population: Durham & Northumberland +1% in 1831, +3% further in 1821, +1% further in 1811; South Wales +1% in 1821, +1% further in 1811; Scotland +1% in 1821; West Yorkshire +0.5% in 1821. As in table 5.1, miners aged 65 and over have been excluded in 1851, 1861, and 1871. The figures for 1861 and 1871 relate to males only. Columns do not necessarily sum to the total because of rounding.

between 1841 and 1851 was most rapid in the early part of the period, prior to the great strike of 1844.

For the period for which statutory returns of numbers employed in each district are available, these can be treated as being as reliable as those for the country as a whole. But for the years prior to 1873, when the inspectors made estimates on the basis of voluntary returns, it is clear that some of the district figures are inaccurate, and none can be taken as completely reliable. The district statistics are accordingly presented here in table 5.4 only for the period of the statutory returns.

Perhaps the most obvious thing that stands out from tables 5.3 and 5.4 is the fact that, whilst the labour force of the country as a whole grew by

Table 5.4 *Numbers employed in coal mines, by coalfields, 1873–1914 (in thousands)*

	Durham & North-umberland	Cumb-erland	York-shire	N.Derby-shire & Notting-hamshire	S.Derby-shire & Leicester-shire	Warwick-shire	Lancashire & Cheshire	N.Wales	N.Stafford-shire	Shrop-shire	S.Stafford-shire & Worcester-shire	Somerset & Bristol	Forest of Dean	S.Wales	Scot-land
1873	99.8	5.7	57.5	36.0	6.4	4.6	61.2	13.0	23.0	7.6	36.8	...	...	...	58.6
1874	99.1	5.7	62.5	39.9	7.7	4.8	65.2	14.1	20.8	8.1	31.8	8.0	5.1	73.3	76.1
1875	102.1	6.5	62.2	39.9	8.0	4.6	67.3	13.6	19.6	7.7	35.7	8.0	4.7	72.6	71.7
1876	101.1	6.0	61.0	40.4	7.4	4.6	63.2	12.0	18.6	6.6	32.8	7.6	4.4	57.5	69.9
1877	98.6	5.8	60.6	38.6	7.6	4.0	61.3	10.9	18.0	5.8	28.8	7.3	4.1	65.3	67.8
1878	92.5	5.8	59.8	38.2	6.9	4.2	61.6	9.6	16.9	5.2	25.1	7.1	4.0	63.7	66.4
1879	89.6	5.8	60.1	39.3	7.3	4.3	62.5	10.1	16.0	4.8	23.6	7.2	4.3	66.5	67.0
1880	94.8	5.9	60.5	38.2	7.0	4.2	62.7	10.2	15.6	4.8	23.5	7.2	3.8	69.2	68.2
1881	96.0	6.2	60.5	38.8	6.9	4.2	63.7	10.2	17.3	4.6	23.8	7.2	4.4	72.9	69.3
1882	98.9	6.1	61.5	39.1	6.8	4.0	64.7	10.1	18.3	4.4	26.7	7.2	4.1	74.9	67.9
1883	99.5	6.4	63.2	40.4	7.5	4.3	66.5	10.1	18.0	4.3	23.8	7.4	4.2	80.7	68.4
1884	102.4	6.5	64.0	41.6	7.4	4.3	66.1	10.2	16.7	4.1	23.8	7.6	4.2	83.9	69.4
1885	102.0	6.8	63.6	43.0	7.8	4.4	66.3	9.3	15.5	4.0	23.1	7.8	4.2	85.2	70.0
1886	101.8	6.7	65.7	43.4	7.4	4.4	67.1	9.3	16.7	3.7	22.0	7.7	4.2	85.1	63.1
1887	101.6	6.9	66.9	44.0	7.3	4.6	67.5	9.8	15.5	3.4	21.7	7.7	4.4	86.9	71.2
1888	104.0	6.5	67.0	45.8	7.3	4.9	68.1	10.2	16.4	3.7	22.0	7.5	4.3	91.4	70.5
1889	109.0	6.4	70.1	48.2	7.8	5.2	70.6	10.8	17.0	3.6	23.2	7.7	4.5	99.4	74.1
1890	116.8	7.1	76.9	52.5	8.4	5.6	76.5	12.3	18.3	3.6	25.3	8.0	4.9	109.9	81.1
1891	123.4	7.8	82.2	56.5	8.9	6.1	80.9	12.8	18.0	3.7	26.3	8.3	5.0	116.6	85.1
1892	124.1	7.6	86.7	59.2	9.4	6.1	84.0	12.7	18.3	3.7	26.4	8.0	4.8	116.7	89.0
1893	130.7	7.8	88.7	60.1	9.8	6.1	86.7	12.6	19.0	3.8	27.5	7.0	5.4	118.0	90.8
1894	136.3	8.3	91.2	63.1	10.0	6.4	88.1	12.6	19.3	3.8	27.9	7.0	5.3	124.7	93.8
1895	135.6	7.7	88.7	62.9	9.5	7.6	87.3	12.2	18.2	4.0	26.7	7.1	5.2	126.1	94.1
1896	137.1	7.8	89.6	62.5	9.2	7.8	83.8	11.5	18.5	3.8	26.9	7.1	4.9	125.2	88.8
1897	138.1	7.8	89.6	63.1	9.1	8.1	84.4	11.7	18.2	3.7	26.0	6.9	4.7	126.8	89.0
1898	140.1	7.8	91.2	64.6	9.1	8.6	83.4	11.5	18.9	3.9	26.0	7.4	5.2	128.8	92.0
1899	145.4	7.9	94.9	66.4	9.5	8.9	83.4	12.0	20.1	3.8	27.3	7.2	5.0	132.7	95.9
1900	152.6	8.5	100.8	70.5	10.4	9.5	88.0	12.6	21.9	3.7	28.6	7.5	5.6	147.7	103.8
1901	155.1	8.9	105.3	74.0	11.2	10.6	92.8	13.4	23.2	3.9	28.9	7.5	5.9	150.4	108.0
1902	158.9	9.0	107.7	76.7	11.5	11.4	93.4	13.3	25.3	3.9	28.8	7.9	5.5	154.6	109.0
1903	160.5	9.3	109.8	78.3	11.5	11.8	95.3	13.2	25.3	3.9	28.1	7.7	5.3	159.2	111.1
1904	165.4	9.0	112.0	79.2	11.6	11.6	94.0	12.9	23.9	3.8	26.9	7.5	5.2	163.0	112.8
1905	170.0	9.0	112.6	79.4	11.8	12.2	94.0	12.7	23.8	3.8	27.2	7.9	5.2	165.6	114.3
1906	177.0	9.1	115.5	81.6	11.9	13.2	92.4	13.0	24.6	3.5	27.5	8.0	5.4	174.7	115.7
1907	186.4	9.5	123.9	84.9	13.1	14.6	95.5	14.4	25.9	3.6	28.1	8.3	5.9	190.3	124.8
1908	194.1	9.6	135.2	88.3	13.5	14.7	100.6	14.9	26.8	3.5	28.5	8.4	6.0	201.8	132.1
1909	200.5	10.0	141.0	90.1	14.0	15.3	103.8	14.9	27.8	3.5	29.4	8.8	6.1	205.0	133.6
1910	212.4	10.5	146.9	91.0	14.6	15.7	104.6	15.2	28.9	3.7	29.3	8.7	6.1	213.3	137.9
1911	216.7	10.6	148.0	93.2	14.2	16.9	104.0	15.4	29.4	3.7	29.4	8.9	6.3	220.9	138.4
1912	218.9	10.7	152.8	93.6	14.7	17.6	105.8	15.7	30.4	3.7	30.5	8.7	6.4	225.5	143.3
1913	226.8	11.0	161.2	96.0	15.4	19.2	108.0	15.9	31.9	3.7	31.2	8.9	6.7	233.1	147.5
1914	227.6	11.4	164.3	96.4	15.1	19.1	107.7	16.3	32.4	3.7	32.3	9.0	6.7	234.1	146.2

something between ten and twenty times between 1801 and 1914,[19] there were great variations in the rate of increase between different coalfields at different times. However, since, by and large, these matters have been discussed in chapter 2, or will be discussed in chapter 10, we shall not be concerned with them much at this point. Like long-term movements in production, those in the labour force of individual coalfields were related primarily to their differing abilities, for geological and geographical reasons, to meet the expansion in the demand for coal which was so marked in our period. In so far as output and workforce did not move together, there were changes in productivity which will be considered later.

Year-to-year variations in employment were mainly connected with temporary fluctuations in the demand for coal, and, in some degree, with the changes in wages dependent on them. To begin with, a period of strong demand led to more regular work for existing miners, with little engagement of additional men. Wage rates, at any rate by the 1850s, rose early in the boom,[20] but this was rather the result of custom and of fear of labour troubles than of the need to attract extra workers. However, if, as it generally did, the boom continued long enough for new investment to increase the capacity of the industry, there was a rise in the number of men needed, and this additional labour was attracted by the rising wages. The increase in numbers, however, lagged behind the rise in wages, partly because it takes time for men to change jobs, and partly because the greatest expansion of capacity normally came after the peak of the boom was passed. In those districts where the long-term expansion of capacity came to an end, the fall in demand, prices, and then wages which followed a boom led to some drift of workers out of the industry – partly through the failure of recent recruits to stay, and partly through the closing down of pits which had become economically unviable. These pits tended to have a more than normally elderly workforce, and some of these men failed to find new jobs, whilst others were temporarily unemployed or moved into other occupations, at any rate until the next boom. In the majority of districts, however, there was never any marked fall in the labour force, except for a short time after the dizzy heights of the early 1870s. The normal case was for the cyclical downswing to be accompanied by stagnation in the numbers employed. Instead of men drifting away, it was rather a case of the industry failing to attract its usual quota of recruits. The direct means by which the long-term trend in these expanding districts made itself felt in the year-to-year variations was through the tendency for owners to retain men even when they could not provide them with regular employment at those seasons when demand was at its highest.

Having established, so far as possible, the size of the colliery labour force during our period, it is of interest to ask whence it came. The traditional answer to this question has been that it came largely from the sons of miners, who had an extraordinary high birth rate and who invariably followed their fathers into the pits. Using data from the census from 1841 to 1911, A.K. Cairncross[21] provided some suggestive statistical backing for this view when he found that, over the period as a whole, five-sixths of the increased population of the main colliery districts came from their natural increase, and only a sixth from net immigration. The birth rate in the colliery districts was not notably higher than in northern urban areas, but it was significantly above that of the South and of the northern rural areas. In terms of long-run comparisons with other times and other countries, its level was relatively high, except when compared with newly settled lands.[22] It climbed to a peak of 41.9 per thousand in the decade 1871–81, which was 42% above the English average, and higher than the average of any contemporary European country except Hungary, Russia, and Serbia; and it continued at an historically high level until after 1900. Colliery district death rates, meanwhile, remained unchanged for long periods – at a little above 23 per thousand from 1841 to 1880, and at between 19 and 20 per thousand from 1881 to 1910. Cairncross is surely correct, therefore, when he says that 'the colliery districts depended for their growth . . . , in fact, on a high birth-rate'.[23] Rather cursory inspection of the age-structure of the population of colliery districts suggests that the age-specific birth rates might be rather closer to the national average, and age-specific death rates might be a little higher still than the national average, but it does not indicate any need to alter the essential conclusion that the population of the colliery districts increased principally because of its own large families.

This conclusion needs only comparatively little modification to be applied to the various individual coalfields separately. Cairncross does not show the decade-to-decade or even the total increase in population for any individual colliery district, but he does give a table of the decennial effects on each of them of net migration, and the total increases can be worked out from the censuses. The results are given in table 5.5.

The most obvious long-term gainer from migration was South Wales, with a total net gain over seventy years of 372,000, an average per decade of 30% of the total increase, with a maximum percentage gain of 40 in the 1880s.[24] In terms of numbers, no other district approached the Glamorgan group's gain, though some other districts had high percentage rates of gain for short periods. Over a third of the Durham group's growth of population came from net immigration in the period 1841–70, and it continued at a lower rate until 1900; the Wigan group's increase averaged about a fifth from net inflow from 1841 to 1880; and the Yorkshire and East Midlands districts experienced net immigration amounting to

Table 5.5 *Population increase and gain or loss from net migration in selected colliery districts, 1841–1910 (thousands)*

	1841–50		1851–60		1861–70		1871–80		1881–90		1891–1900		1901–10	
	Pop. growth	Net migration	Pop. growth	Net migration	Pop. growth	Net migration	Pop. growth	Net migration	Pop. growth	Net migration	Pop. growth	Net migration	Pop. growth	Net migration
Durham group	114.1	+49.9	170.1	+63.4	236.5	+84.3	241.1	+31.6	231.4	+8.2	269.6	+23.4	513.2	−23.8
Glamorgan group	116.5	+52.1	105.0	39.6	98.2	+11.0	126.5	+12.2	217.9	+87.2	216.1	+40.3	362.8	+129.3
Wigan group	31.2	+8.2	46.8	+10.8	51.5	+7.3	83.7	+17.8	79.0	+4.8	69.2	−13.6	55.4	−28.9
Chesterfield group	15.1	−8.8	25.1	−5.4	38.5	+0.1	77.8	+15.9	71.3	−1.3	106.7	+22.2	127.9	+26.9
Barnsley group	8.1	−6.6	21.8	+2.0	24.5	+2.2	45.1	+11.2	35.8	−1.0	52.3	+5.7	65.1	+11.3
Doncaster	2.3	0	4.7	−0.3	5.8	+0.6	8.7	+1.8	9.6	+0.7	20.4	+9.0	37.7	+22.0
Wrexham	0.5	−0.2	7.9	+2.7	5.8	+2.0	6.3	−1.9	6.6	−1.1	9.9	−0.5	8.9	−2.9
Nuneaton	3.4	+0.4	1.0	−3.6	−3.4	−7.3	2.0	−2.0	2.8	+ 0	10.8	+7.3	12.3	+8.4
Cannock	0.8	−0.8	1.8	−0.3	5.0	+2.0	12.4	+5.7	4.4	−3.3	7.2	−1.3	7.0	−2.8
Ashby de la Zouch	1.7	−2.2	5.9	+1.7	3.0	−2.4	1.8	−4.0	2.8	−3.3	7.8	+0.4	6.2	−2.5

The groups are the same as in Cairncross (1953) with the addition of Doncaster Registration District, and with the exception that the Wigan group is composed slightly differently. It consists here of the Prescot, Wigan, Warrington, and Leigh Registration Districts.

between a fifth and a third of total population growth in the 1870s and from 1881 to 1910. The other, smaller districts were generally net losers by migration, but each experienced one or two intercensal periods when net inflow was a large proportion of the total increase. These periods corresponded to the great enlargement of previously small coalfields, or, in the case of Cannock Chase, its initial opening up.

Discussion so far has been in terms of the population of colliery districts; but though this is useful as a guide, provided we can ignore cross-migration, and tells us that the majority of the labour force's recruits came from their local communities, it does not say anything definite about the relative importance as a source of colliery labour of the mining community proper. The colliery districts as defined (necessarily) by Cairncross include the agricultural, and in some cases urban, areas surrounding the pits, and the proportion of non-miners in the adult male population of every single registration district was at all times substantial.[25] Is it possible to get a clearer picture of the inflow into the mining community from *all* outside sources, including these local non-miners? Certainly this cannot be done with any pretension to accuracy; but provided one can accept a number of assumptions and approximations, rough estimates of the flow can be made. The main assumptions which need to be defended are these: (1) that each male miner had three dependents on average; (2) that the birth rate for the mining communities equalled that of the colliery districts and can be assumed for earlier periods than 1841; (3) that the death rate of children before ten years of age was the same for the mining community as for the colliery districts; (4) that the annual wastage of the mining labour force can be inferred from the death rate for miners given, for certain periods, by the Registrar General, together with an allowance for men permanently incapacitated by injury. The first of these assumptions is the most crucial, and it is also, fortunately, the easiest to defend. The number of female miners was never other than small in proportionate terms and they can be ignored safely; and, since it must be approximately correct to say that each male miner had a female coeval, one can say that the mining community was equal to twice the number of male miners, plus dependents outside the active age-group. The age of entry into the labour force, of course, changed somewhat over our period; but we cannot be far wrong at any time if we regard all miners' sons who reach the age of ten as being the potential internal recruits, though it will be as well to bear in mind that this will tend to inflate the figure in the last forty years or so of our period, and especially in the last twenty. At the other end of life, there was no official retiring age, and some men went on working as miners, though probably not in very active jobs, into their seventies. Rather more were compelled by the state of their health to cease to be miners at an earlier age, sometimes much earlier. If we take sixty-five as a conventional figure for retirement we shall probably be erring on the high side; but since the numbers involved in the higher age-groups are small compared with the younger, the total of dependants will not be much exaggerated – certainly by no more than it is understated in taking ten rather than fifteen as the age of potential entry. If this reasoning is correct, then the fact that at every census from 1841 to 1911 the proportion of the population of the colliery districts which was under ten or over sixty-five years old was very stable at just over a half suggests a ratio of three dependents per adult miner.

The second assumption, namely that the birth rate for the mining community was equal to that of the more widely defined colliery districts, may result in some understating of the internal flow of recruits, but it is unlikely to be very considerable, for the highest rate of any individual registration district is only a little over 10% bigger than the average, and the results are not very sensitive to this sort of variation. The third assumption is also one in which divergence between the experience of the mining community and of the colliery districts is possible, but is unlikely to be so large as to affect the results. In this case, the error, if there is any, is likely to be in the opposite direction to the last case: the death rate for the children of the notoriously ill-housed miners may well have been above the average of their district.

The fourth main assumption is based on the average death rates for miners given in the *Registrar Generals' Reports* for the three years centred on each of 1861, 1871, 1881, and 1891. These rates have been used for the 1860s to the 1880s; and because Cairncross showed that the overall death rate in the colliery districts was stable, apart from one fairly sharp break between the 1870s and 80s, it seems reasonable to take the 1861 figure as applying to earlier periods and the 1891 figure as applying to later periods. It might be argued that one should posit a gradual decline in the death rate after 1891, in line with national experience, but the evidence of death rates for males aged fifteen to sixty-five in the colliery districts in 1901 and 1911 suggests otherwise. In addition to death, there is need to allow for wastage through injury, and this is much less easy to estimate. HM Inspectors of mines reported all deaths from accidents from 1851 onwards, but they provide only limited evidence on injuries. However, at various times some of the inspectors did give details of the length of time injured men were off work and of the proportion who never returned to work. These suggest a ratio of three incapacitating injuries to each accidental death, and this has been used to make estimates for the period after 1850. For the years prior to 1851, for which there are no inspectors' *Reports*, it is assumed that the ratio of persons employed to numbers incapacitated in 1851–60 was unchanged. This could perhaps be justified on the hypothesis that there was no change in the accident rate until the beginning of government inspection.

The last assumption, that all miners' sons followed in their fathers' footsteps, and that once a miner always a miner, is not completely accurate, of course; but it is near enough to make no material difference, certainly prior to the last twenty years or so of our period.[26] Prospects for the children of the vast majority of manual workers were limited by their lack of education, and within these limited prospects, coalmining afforded one of the best wage rates at most times. Moreover, it offered a degree of freedom from regimentation and of independence in controlling one's pace of working, which became decreasingly common in other manual occupations as the nineteenth century went on.[27] In any case, in most districts there was little alternative employment available to a boy within easy distance of his home. And, as Alan Griffin, the historian of the East Midlands coalfield, has said, 'even for boys living in places like Basford, where other work was available, underground pay, coupled with what was expected of them, took them to the pit'.[28]

There are various other less important, or at any rate less questionable, assumptions necessary to any calculation of the flows into and out of the colliery labour force, but these have been relegated to the notes to table 5.6, which shows the stages by which the final estimates for Britain's coalfields as a whole are arrived at. It should go without saying, but for the sake of the wary or the credulous must be stated emphatically, that these figures are only designed to show rough orders of magnitude, and they are clearly subject to very wide margins of error, especially prior to 1841, when so much depends on sheer guesswork. However, the picture which is suggested is one which fits in quite well with such literary evidence as we have, though going well beyond it in one important respect.

This latter is, in fact, one of the most striking things about the table. For it suggests that prior to the 1880s – or, to be more precise, from the 1840s to the 1870s inclusive – the traditional view that the colliery labour force was mostly self-generating is quite a long way from the truth. If the orders of magnitude shown are correct, not far short of 60% of the industry's recruits in this forty-year period came from outside the mining community. And even if the most crucial assumption – that about the number of dependents in the community – were wrong, and there was an average of 3.5 rather than 3 per male miner, well over a third of recruits was external. To bring the proportion down to as low as a sixth, the birth rate would also have to have been underestimated by 4 or 5 per thousand. This is not without the bounds of possibility, and is, indeed, more likely than that a dependency ratio of 3 is wrong. But it is highly improbable that both dependency ratio and birth rate are wrong; and if the birth rate on its own were to reduce the estimated external recruiting to a third, it would have

Table 5.6 *Estimates of contributions to the colliery labour force of those inside and outside the colliery community, 1801–1910 (in thousands except as otherwise indicated)*

	1[a]	2[b]	3[b]	4	5[c]	6[d]	7[b]	8	9[b]	10[e]	11[f]
	Number of male miners	Total mining community	Birth rate per 1,000	Male births to mining community	Deaths before age 10	Potential recruits from mining community	Miners' death rate per 1,000	Wastage through death	Wastage through injury	Potential net increase of labour force from mining community	Deficiency of labour supply from mining community
1801–10	77	308	35	54	18						
1811–20	91.5	366	35	64	21	36	16	15	11	10	5
1821–30	107	428	35	75	25	43	16	17	13	13	3
1831–40	128	512	35	90	30	50	16	20	16	14	12
1841–50	176.5	706	35.4	125	41	60	16	28	22	10	61
1851–60	244	976	39.0	190	59	84	16	39	30	15	49
1861–70	313.5	1,254	39.6	248	80	131	16	50	32	49	26
1871–80	423	1,692	41.9	354	97	168	14	59	34	75	69
1881–90	581.5	2,326	36.9	429	109	258	12	70	31	157	16
1891–1900	746.5	2,986	35.9	536	135	321	12	90	30	201	−44
1901–10	946					401	12	114	37	250	−8

[a]Means of beginning and end years.
[b]See text.
[c]Based on the proportion of death under age 10 to births in the colliery districts in each decade from 1851 to 1860. For earlier periods the deaths are assumed to have been a third of the births.
[d]Column 4 minus column 5, set forward in time by one decade.
[e]Column 6 minus columns 8 and 9.
[f]The increase shown in tables 5.1 and 5.2 minus column 10.
Sources: unless otherwise stated in the notes above, all statistics in this table are based on the census reports, on the annual *Registrar General's Reports*, or on tables 5.1 and 5.2. The figures in the table have been rounded, but the calculations were done before rounding.

to be over 50 per thousand – which, again, is extremely unlikely. There is, of course, a serious possibility of errors going in the opposite direction, especially in the death and injury rates.

All told, then, one can be fairly confident in saying that somewhere between a third and two-thirds of the inflow of labour between 1841 and 1880 (or, more likely, between 1843 and 1874) came from outside the mining community.

Prior to 1841 the estimates cannot be regarded as at all reliable, but the impression which they give coincides with what little comment there is in the literature – namely that the mining community was quite closed, and recruited comparatively little outside until around 1830.[29] Dr James Mitchell, the Children's Employment Commissioner for South Durham, for instance, noted that 'in 1831, when there was a general strike amongst the colliers, many hundreds from a distance were induced to come to Durham'.[30] However, his colleague for Northumberland and North Durham said that of the Welsh and other southern blacklegs imported in 1831, 'scarcely one of them is now to be found remaining'.[31] Perhaps the rapidly expanding pits of South Durham, greatly helped by the coming of the railways, had been able to absorb them without the same bitter hostility that the established Tyneside and Wearside collieries had shown.[32] Actually, this concentration of attention on the North-East is unjustified, for it would appear (see table 5.7) that the only coalfield to recruit labour from outside the mining community in any large quantity in the 1830s was South Wales. Little documentary evidence survives apart from the censuses themselves, but it is clear that the mid-30s were years of expansion in the metal trades of South Wales, and the first steam coal seams were tapped in the Aberdare Valley in 1837.[33] There is some evidence of Irish immigration, even before 1831, and one may legitimaitely infer that more would follow later; but the rural districts of South Wales counties would seem to have been the main source of outside recruitment, followed by similar areas in the neighbouring English counties.[34] In Scotland, which was the only other district to require more than a handful of entrants from other jobs or backgrounds before the 1840s (and then only in the 30s), the sources from which they came were probably similar to South Wales.[35]

In the 1840s, when there clearly was a large inflow of labour, the bulk of contemporary references is to the importation of blacklegs during major strikes in the middle of the decade. Two broad classes predominate amongst those reported as immigrants – the Irish, and workers from other coalfields, who, of course, only represent transfers *within* the mining community as we have defined it. Scotland and Durham are the two districts where there is most evidence for inflows of workers, and in both cases the Irish are singled out specifically. The *Report of the Com-*

missioner to Inquire into the Operation of 5 & 6 Vict. c. 99, and into the State of the Population in the Mining Districts noted in 1844 the evidence of an overman at Dundyvan Colliery in Central Scotland that every year new men were brought in because of strikes – 'mostly Irish' – and that 'since 1837 at least 4,000 new men have stuck to the trade',[36] and in the following year the manager of the Monkland Coal and Iron Company referred to Irish labourers coming in as drawers to replace men who had become hewers as a result of reductions in the darg.[37] The same source also noted Irish strike-breakers in Durham in 1844, about a third of whom remained.[38] In a recent study of aspects of the social history of Durham mining, Robert Moore writes that

> on the basis of reading newspapers – without intending to construct a history of the Irish – the following impressions emerge: Irish labour was used in the 1840s to build the railways in the county, there followed a period of unemployment during which a number of the Irish turned to highway robbery . . . With the sinking of the shafts in mid-Durham . . . new jobs were found for Irishmen, again the heavy labour involved in shaft-sinking. When the shafts were sunk the Irish were employed in labouring and on coke-ovens . . . Only slowly were the Irish able to gain skilled piece-rate jobs at the face.[39]

This takes the story on to the 1850s or 60s; but back in the mid-40s many other than the Irish were brought into the North-East. Welbourne noted the recruitment of Cornishmen, who remained,[40] and doubtless were a focus for those other Cornishmen who came to Durham and Northumberland in the 1864 strike, and who 'learned fast, and . . . had much to teach in return, in stone-work, and timbering, and the use of the heavy hand drills', but who were not accepted until at least the next generation.[41] An earlier historian of the miners of the North-East, Richard Fynes, thought the sources of immigrant blacklegs were even wider. They were sought 'all over England, Wales, and Scotland', he said, and 'large numbers of men arrived from Wales and other places, guarded by the police, and in some instances soldiers'.[42] When the strike was over, there was rioting against Welshmen at Seaton Delaval and against the Irish at East Holywell; but despite this and other hostile measures, many remained.

The concentration of the literary evidence on immigration in the 1840s on Scotland and the North-East, and above all on the Irish element in it, is probably somewhat misleading, though there was certainly a large movement of Irishmen into Scottish pits. Similar calculations to those made in table 5.6 have been made for the main individual coalfields, and the resulting estimates are given in table 5.7. Once again, the uncertainty of these figures and the large margins of error must be stressed. Even so, it is clear that *all* the major coalfields, with the partial exception of South Wales, required a substantial flow of recruits from outside the native

Table 5.7 *Estimates of contributions to the colliery labour force of those inside and outside the colliery community – main coalfields, 1811–1910 (in thousands)*

	North-East		S. Wales		Scotland		Yorkshire & E. Midlands		Lancashire & Cheshire	
	A	B	A	B	A	B	A	B	A	B
1811–20	2.4	−0.2	0.2	+2.1	1.9	−0.5	1.9	+0.6	1.4	+0.7
1821–30	3.0	−0.8	0.5	+3.3	1.4	−0.2	2.1	−0.4	1.7	+0.6
1831–40	3.2	+0.4	0	+9.0	2.2	+1.7	2.1	+0.8	2.1	+0.6
1841–50	2.6	+10.8	0.7	+4.4	1.6	+13.5	1.5	+9.7	1.4	+10.9
1851–60	1.2	+8.5	2.3	+4.7	3.8	−1.0	1.6	+16.7	1.9	+11.0
1861–70	11.4	+0.5	4.6	+8.1	8.5	+6.7	8.3	+6.2	6.5	+1.9
1871–80	14.6	+21.8	4.1	+18.2	11.8	+6.9	13.2	+24.5	11.2	−0.2
1881–90	30.5	−3.1	15.3	+28.4	21.9	−6.1	33.5	+5.9	21.2	−4.0
1891–1900	41.3	−9.6	24.1	+9.7	23.8	−0.9	44.2	−3.6	25.7	−13.8
1901–10	45.1	+16.5	42.1	+28.4	28.2	+2.2	56.4	+5.5	31.0	−19.8

Sources: as for table 5.6. Columns headed A are of the potential net increase from inside the mining community; columns headed B are the implicit flow from (or to, if the sign is negative) outside occupations or backgrounds.

mining community at this time. Perhaps the movement of the Irish attracted attention simply because they *were* Irish, and were more visible, not so easily absorbed into the community as English or Scottish farm labourers, textile operatives, or nailers. Though we cannot be sure of it, these latter groups almost certainly provided the great majority of the inflow to coalfields other than Durham and Lanarkshire, and probably a substantial part there too.[43] But at this disaggregated level, one should note the importance of the movement of miners between the different coalfields. As well as those movements into Durham which were noted above, the Lancashire coalfield proved particularly attractive to the Welsh, especially in 1845–6, when wages were better maintained than in the North-East.[44] Looked at from the other end of this movement, it would seem that the inflow of labour into South Wales in the 1830s had been temporarily overdone, for E.W. Evans says that it may well be that the labour supply was more than adequate there at various times in the 40s – 1842–3 and early 1844, and again by the end of 1847.[45]

The inflow of labour to the mining community, begun in a big way in the 1840s, continued in the 50s and 60s, though it would seem at a declining rate overall, if not in each individual coalfield. One cannot, however, be absolutely sure about this, for there was certainly some emigration of members of the mining community overseas, and it is possible that this was greater in the 1860s than earlier.[46] The main source of inflow almost certainly continued to be the rural migrant, though for the most part such people took surface jobs initially, and if they did in time go underground, they seldom graduated to facework unless they made the movement from farm to mine at an early age.[47] The Irish continued to come into coalmining, in South Wales at least, but the historians of that district do not regard them as a major source of colliery labour, and probably the same is true elsewhere. Certainly the rate of emigration from Ireland fell after the catastrophic outpouring of the 1840s, and the emigrants would appear to have gone mainly to overseas destinations.[48] Even Scotland, for which the Irish had been a more important source of labour than anywhere else, found the stream dried up. Writing in the early 1860s, the historian of Coatbridge, A. Miller, was able to say of newcomers to coalmining that 'the great majority . . . [were] Irish'.[49] But by the boom of 1866, the owners had had to turn elsewhere, to Cornishmen, hit by the decline of copper and tin mining. Youngson Brown reports the *Colliery Guardian* as saying 1,500 had been drawn into Scottish pits during the autumn of 1866, despite a notice sent by the Scots miners to the South-West saying that men were not needed, and 'the object of the employers is to bring a surplus population among us, so that they may drive us back to working 13 or 14 hours a day . . . and to reduce wages back to their favourite figure – 3s per day'. Youngson Brown goes on to say that 'the immigration seems, in fact, although on so large a scale and composed of such "foreigners", to have caused comparatively little trouble, once the initial difficulties of absorption were overcome'.[50] By the great boom of the early 1870s, which saw the influx from outside the mining community higher than ever before or since, the Irish had ceased to be of much significance,[51] and by far the main source of newcomers was agricultural labour, though the North-East was an exception.[52] There is an enormous amount of testimony to this;[53] but in that extraordinary boom it was not only farm labourers who were attracted by the great rise in wages which occurred,[54] it was practically every kind of wage-earner. The president of the Mining Association, Alfred Hewlett, who was the manager of the Wigan Coal and Iron Company, told the Select Committee of 1873 that 'during the last few months a great number of persons, railway servants, policemen, all classes of men, have gone to the pits rather than continue their ordinary avocations'.[55] The chairman of a Yorkshire colliery company, Robert Tennant, said that the 'high wages that are paid now are attracting labour from other pursuits – agricultural labour, and even tailors and shoemakers, and all sorts of people come to get employment'.[56] In Nottinghamshire, according to the Midland Inspector of Mines, T. Evans, the new collieries were employing a lot of former stocking weavers.[57] So great was the demand for labour that one firm, at least, Pease and Partners, opened a recruiting office in the East End of London.[58]

It is perhaps to be expected that at that time, when experienced miners could earn far more than ever before, and when labour disputes, though frequent, were short, there should be little or no mention of migration between coalfields, such as we have already referred to in the 1840s, and as was also experienced in South Wales in the 1860s.[59] However, one interesting, though in overall perspective not very important, migration of miners was mentioned by Alexander Macdonald, the President of the Miners' National Association. This was the return of at least 500 men, and probably more like 1,500, from the USA. He even referred to many commuting for the summer season to the United States collieries, and working in the winter at home.[60] This must have been a short-lived phenomenon!

It is entirely to the boom peaking in 1873 that the large inflow of workers from outside the colliery community in the period 1871–80 has to be attributed. In the rest of the decade, the numbers employed fell, and many of those who had come into the mines, went out again. There were also, clearly, some established miners amongst the emigrants to the USA in this period, as earlier, and especially perhaps in the 1880s.[61] The movement of miners as blacklegs between coalfields also resumed, as witness the Welsh at Dodworth in Yorkshire in 1877, and unemployed Black Country miners at Denaby Main in 1885.[62] However, it would be wrong to stress the strike-breaking element in internal migration too much. During the 1880s, in fact, there was only a modest inflow overall, and that was mainly the result of the continued expansion in South Wales. In most other districts, except Yorkshire and the East Midlands, the mining community was growing faster than the demand for miners. So internal migration seems to have been more important than recruitment from outside in this decade.[63] In the 1890s it became true of the mining community in the country as a whole that it was growing faster than demand, though South Wales continued to attract outsiders, as it did right through to 1914, on a scale not experienced elsewhere. Some of these certainly came from other coalfields, notably the Forest of Dean, the stagnant or declining Somerset and Bristol coalfields, and Lancashire, but also from Staffordshire and North Wales.[64] The surplus sons of the districts where expansion had slowed or stopped may often have emigrated overseas or left the industry altogether; but many moved to South Wales, and some to the new collieries of Yorkshire, Nottinghamshire, and Warwickshire. There they joined that other stream which, whilst intermittent and weak in the 1880s and early 90s, continued to spring again in times of brisk demand, despite the fact that the mining community could have provided, overall, all the miners needed on its own – the farm labourers and others, who took the lure of relatively high wages. H.S. Jevons, in his authoritative description of the British coal industry at the outbreak of the First World War, characterised them like this:

Large numbers of agricultural labourers from South and Mid Wales, also from Somerset and Devon, have flocked to the mines during recent years in South Wales . . . In the new pits, and in older pits where the price-lists are low relatively to their conditions of work, one finds a most extraordinary assortment of men of different occupations. Working with a few men who have been miners from their boyhood there is, perhaps, one who has just been a baker's van-driver, and one who was a grocer's assistant. Beside him is a late traveller for a firm of paint manufacturers; and asking another you will find he is a market gardener who has fallen on evil days. Yet again, there is a man brought up in a saddlery business of his father's, which after many years' struggle came to an end, and working with him is an intelligent young labourer who came from a Devonshire farm to the tin mines of Cornwall and then thought he would try his luck in South Wales; a postman from Pembrokeshire, and a railway signalman dismissed from service owing to a mistake which derailed a few trucks. All these, and many like them, have found in South Wales their El Dorado with steady work at high wages, if none too pleasant and somewhat dangerous. The same kind of thing is going on in South Yorkshire and in every growing coalfield.[65]

This is a picturesque, yet no doubt accurate, illustration of the heterogeneity of the sources of the labour force in some pits in the fast-expanding districts. But, of course, it must not be allowed to obscure the fact that there was no overall net inflow into the mining community of the country as a whole; and probably the majority of the net inflow even of South Wales, and certainly of Yorkshire and Nottinghamshire, was provided by migrant workers from other coalfields.[66] A remark of a Warwickshire manager to the Eight Hours Day Committee in 1907 may be allowed to stand as a corrective to any exaggerated idea of the importance, by the 1900s, of the outsider. W.G. Phillips said that in his district 'we have utilised the agricultural labourer to a very large extent, but we cannot make a very good collier of him as a rule', and he added that 90% of the faceworkers were colliers from boyhood.[67]

One can sum up this discussion of the sources of colliery labour by saying that the generally accepted view that the mining community itself was by some way the most important generator of new recruits holds good prior to the 1840s and after 1874; but in the interval, flows from outside were of roughly equal importance. At most times, these outside recruits came mainly from amongst the farm workers, with the Irish element being of major importance in the 1840s in Scotland, and perhaps in Lancashire and the North-East, and also contributing in South Wales in both the 40s and 50s. Incidentally, the Irish element retained its identity for long after its arrival. Reference has already been made to this in Durham (see above, p. 116 and n39), but it was probably even more the case in Scotland. As late as 1890, the secretary of the Fife and Clackmannan Coal Owners' Association, John Connel, told the Royal Commission on Labour that 'in the West of Scotland they have a larger infusion of Irish workers . . . which we have not, and I believe that we have a better type of men in Fife'.[68] The 'infusion', as we have seen, had been mainly thirty and more years before. Another quite significant source of outside recruiting was metal mining, especially in the 1860s and 70s, when Cornish copper and tin mining were in decline, but also to some extent earlier. Lead miners were also of some significance in the North-East and in Yorkshire and the East Midlands, both in the 1870s and again in the early 90s and around 1900, when Pennine mining was falling off.[69] Men from other trades *were* drawn into the pits, most notably during the boom of the early 70s, but also at other times of labour scarcity and high wages; but they were never a major inflow. Finally, from the 1850s onwards, internal migration of coalminers from one district to another became important at other times than when strike-breaking was required, and after the end of the great boom of the 70s this probably became the biggest source of external labour for the coalfields which were growing fast.

Before we leave the question of the size of the labour force, there is the important matter of its composition to discuss; for the proportion of underground to surface workers, and of faceworkers to so-called offhand men, of hewers to fillers, and so on, needs to be established, so far as it can be, since the wages and earnings of the various groups could differ substantially, and their proportions to one another had a major bearing on the course of productivity. As in so many things, it is impossible to say anything very precise in the aggregate about many of these matters until towards the end of our period, since the Departmental Committee on the Eight Hours Day in 1906 provides the first country-wide statistics for anything other than the surface/underground division. But the latter is probably the most important structural division, and there is some earlier material on it, from the early 1870s, and earlier for the North-East.

First, then, the division between those employed underground and at the surface. There is little evidence before the mines inspectors' *Reports* began to give firm annual data in 1873, but what there is suggests a very stable proportion before that date, just as there was afterwards. The earliest figure we have for a whole coalfield is for Durham and Northumberland in 1844, and was provided by T.J. Taylor, a North Country viewer, to the Select Committee on Accidents in Coal Mines in 1854.[70] He gave total colliery employment as 32,820, of which 25,383 were underground workers; that is 77%. He also gave figures for 1853, which came to the same proportion. By the 1870s the proportion in that district was in the range 79 to 81%. The 1880s began at 80%, but had fallen to 76% in 1889. It then rose to 78% early in the 90s, and gradually returned to 80% between 1901 and 1913.[71]

The earliest record of the underground ratio for any other coalfield is that given in his *Report* for 1853 by HM Inspector Joseph Dickinson for each of Lancashire, Cheshire, and North Wales. In each case the figure came to 82% of the labour force. Twenty years later, when the inspectors began to make regular returns, the proportion ranged from 76 to 78%, and apart from occasionally touching 79% in the 1880s and early 90s, it stayed there until 1914. For other districts the inspectors' *Reports* provide our earliest data. Again they show stability, with no trend during the whole of the rest of our period except in Nottinghamshire.[72] Fluctuations in the short term were very small, and the range between coalfields was from 73 to 75% in South Staffordshire to 84 to 86% in South Wales. The United Kingdom average hardly varied, being 80 or 81% in every year.

This stability is in some ways surprising. The amount of preparation to which the coal was treated at the colliery increased a very great deal, most particularly after the boom of the early 1870s. It is true that much of this washing, screening, sorting, and so on, was mechanised, and that the increase in the elaboration of the preparation was almost entirely associated with the application of more machinery. Nevertheless, none of the machinery was anything like fully automated. An influence in the other direction was the increase in the average size of pits, for there were certainly some economies of scale to be had in the tasks for which surface workers were needed; and the fact that South Staffordshire had the lowest proportion underground and South Wales the highest is itself suggestive of this. That these two influences practically offset each other, and left the proportions of the workforce above and below ground virtually unchanged, certainly from the 1870s on, but so far as one can tell throughout the entire century, may have had something to do with the efforts of managers to keep the different components of total costs steady. But it is more likely to have been largely coincidental.

The stability which characterised the division of the labour force between underground and surface workers seems also to have been found to some extent in the proportion which consisted of hewers, at least in Durham and Northumberland, which is the only coalfield for which there is evidence covering a long period. The first data there relate to 1844,[73] and consist of lists of 124 collieries with the total numbers they employed and the numbers of hewers. The totals were 12,804 hewers out of 33,898 workers, or 37.8%. The range of the individual collieries was very large, from a mere 23% at Springwell to 56% at newly opened Trimdon and Shield Row. The importance for this particular proportion of how long a colliery had been open is shown not just by these individual examples, but by the contrast between the Blyth and Tees districts, where hewers constituted 44% of workers, and the old-established Tyne district, where the proportion was 35%. In between, at 39%, came the Wear district, which contained, in the south, some of the newest and best-organised pits, as well as old ones near Sunderland. It is only to be expected that older pits should have greater haulage distances, *ceteris paribus*, and thus require a higher proportion both of haulage hands and of repairers. Moreover, newer pits should, on average, have more effective equipment. But it is rather surprising that age seems to have been the only major factor affecting the proportion of hewers. Size of pit made no apparent difference. In pits employing over 500 men, 37.3% of them were hewers, which scarcely differs from the overall average. Of course, there clearly are diseconomies as well as economies of scale in underground haulage. For a pit of a given age, the bigger it is the longer the haulage is likely to be. One can also say that the deeper the shaft, the more likely is a pit to be large, in order to spread the heavier investment cost over a bigger output, and the more likely is it to be worked to a great age in order to spread the amortisation cost. However, in a period of expanding output, when the early-got coal near the surface had been used, some at least, and probably a majority, of the deeper and bigger pits are also likely to be the newer ones, still working coal comparatively close to the shaft, and with up-to-date equipment, capable of affording economies of scale. It would be wrong to put much emphasis on the latter in the 1840s, however, for the main opportunity at that time was in winding, which contributed only a small part of total costs. The wide spread of mechanical haulage underground, which offered the best hope of economies of scale in the nineteenth century, did not come until the 1870s,[74] though pit ponies and big tubs, introduced in the 40s and spreading fairly quickly, must have had some such effect.[75]

The next evidence on the proportion of hewers in the labour force comes from the evidence of a well-known Northumberland manager and owner to the Select Committee of 1873.[76] G.B. Forster gave the following figures for thirty-seven collieries in Northumberland:

	Hewers	Total	Hewers (%)
1871	7,249	15,512	46.7
1872	7,435	16,282	45.7
Jan.–March 1873	7,784	17,825	43.7

It is clear on looking at the totals employed in Northumberland that Forster's 'total' relates to underground workers only. If we raise these by the North-East average of a fifth to allow for the surface workers, the proportions of hewers were 39% in 1871, 38% in 1872, and 36% in the first quarter of 1873. Compared with the 1844 figure for the North-East as a whole, these suggest no change; though it is perhaps better to make the comparison with the 1844 average of Tyne and Blyth districts alone, which was 35%. This does indicate that there probably had been some small shift in the structure of employment towards proportionately rather more hewers and rather fewer offhand workers during the 1850s and 60s. That this was desired by managers is quite clear from the evidence to the same Select Committee of the President of the Mining Association, Alfred Hewlett. He told the Committee that even since 1870 'large sums of money have been expended in order to reduce the number of hands necessary for the haulage of the coal, and the staff underground, other than the hewers of the coal, and so liberate more men and allow them to become hewers'.[77] In Northumberland, at least, this policy does not seem to have been successful in the short run of the great boom. No doubt other factors offset it, especially the reaction of hewers to very high piece rates.[78] But did it even work in the long run? Still confining ourselves for the moment to the Durham and Northumberland coalfield, the evidence suggests that it did not; or, rather, that it did so to only a limited extent for fifteen or twenty years, and then the offsetting factors began to reduce the percentage of hewers once again. A return in 1890 in connection with the eight hour day question, which the Miners' Federation had recently begun to agitate, gives the number of workers for each coalfield in four categories – getters, conveyors, others underground, and surface workers.[79] In the other coalfields there are problems connected with the definition of the word 'getters', which we shall come to later. But for the North-East, with its more elaborate division of labour, it is clear that the definition of getters, or hewers, is consistent both with earlier and later sources. The proportions in the 1890 return – 43% in Northumberland and 39% in Durham – can, therefore, be taken as showing a noticeable, though not very large, shift in structure since the early 70s. Incidentally, the change in the relative positions of the two counties since 1844 is further evidence that the age of pits was the major influence on the proportion of hewers, for with the closing of the old Tyne pits in the long depression after 1873 (if not earlier, in some cases), and the opening of new collieries north of Blyth, the average age of pits was now lower in Northumberland.[80]

Sixteen years later, according to information provided for the Eight Hours Day Committee, covering the vast majority of collieries in the North-East, the hewers numbered 16,774 and 46,519 in Northumberland and Durham respectively, out of total underground employees of 34,720 and 95,285.[81] If the surface workers are added in the proportions derived from the inspectors' *Reports*, the implied proportions of hewers are 38% in each county. In other words, the situation had returned to more or less what it had been at the beginning of the 1870s.

Before going on to discuss the changes which took place between 1906 and 1913, partly under the influence of the Eight Hours Day Act, we must try to see what can be said about the proportion of hewers in the other coalfields during the nineteenth century. Unfortunately, there is no evidence whatsoever about any of them until the 1890 return, and hardly any, even, about any individual colliery. Let us begin then with 1890 and 1905.

The return of 1890 specified three classes of underground workers – getters, conveyors, and others – whilst the Eight Hours Day Committee's information was for hewers, fillers, drawers, repairers, and others; with adult hewers separately shown in another appendix derived from a different original source. All these series are summarised for the main coalfields in table 5.8.

Two aspects of this table merit attention. First, comparisons between coalfields, and second, comparisons over time. It is convenient to take the latter first, since the question of the definitions of the categories arises most naturally in this context. Since it is clear from the whole tenor of discussion of the industry at the time that there was no really dramatic change in the structure of the labour force between 1890 and 1905, it would appear that the 'getters' of 1890 and the 'hewers' and 'fillers' of 1905 virtually correspond with each other, and might be described collectively as the faceworkers. If this is so, then even though it would be unwise to reckon on exact uniformity of definition at the two dates, there is a definite suggestion that the decline in the proportion of hewers which we found in the case of Durham and Northumberland over this period – and which is reflected here, even though the 1905 figures for that district were provided for the Eight Hours Day Committee by a different source – is to be found in all coalfields. The decline was small – so small that we cannot really be sure that it actually took place – in Yorkshire and the East Midlands. It was a little bigger in South Wales and Scotland, whilst in Lancashire and Staffordshire the decline was more than a tenth of the original figure, and though part of this may have been due to changes in definition, especially in the case of Scotland, it is hard to doubt the reality of some structural change. The principal reason for it can only have been the increasing distances which the coal had to be hauled underground, which was, in turn, mainly a consequence of the rising average age of pits. This, in its turn, chiefly reflected the greater average depth of working,[82] which had gone up dramatically during the decade or so following the end of the boom of the early 1870s, as the smaller mines working coal close to the surface had been exhausted.

The dramatic effect of the process of ageing on the proportion of faceworkers is well brought out in the case of an individual colliery by Anthony Slaven's study of Govan Colliery, Glasgow. This shows the proportion of faceworkers to all underground workers falling from 67% in the late 1870s (when the records of the concern first make this infor-

Table 5.8 *Proportions of labour force in various categories, 1890 and 1905*

	1890			1905					
	Getters	Conveyors	Others underground	Adult hewers	Hewers	Fillers	Drawers	Repairers	Others underground
Durham & Northumberland	40	20	20	36	38	1	19	17	7
Yorkshire	50	20	12	33	38	10	16	11	4
Midland Inspection Division	53	16	10	28	37	15	16	7	4
Lancashire & Cheshire	51	16	16	32	37	8	15	16	4
Staffordshire Inspection Division	48	20	11	25	29	11	17	10	5
South Wales	52	13	21	35	47	2	13	19	5
Scotland	58	16	11	35	49	4	12	13	6
UK	50	18	15	33	40	6	16	14	5

Sources: see text.

mation available) to 52% in the early 1890s, 46% in 1903–7, and finally to 39% from 1908–12.[83] The last sinking at Govan had been in the 1860s, and the two pits still working in 1912 averaged approximately fifty years old.

It is impossible to make valid comparisons between the two dates shown in table 5.8 for the categories other than the faceworkers, since the definitions are not uniform, and the catchall category, 'others underground', includes in 1905 some who would have been called 'conveyors' in 1890, such as onsetters and trappers; and in 1890 it includes others besides the repairers, who were separately shown in 1905. Given the decline in the faceworkers' proportion and the reason for it, it seems likely *a priori* that both transport workers and repairers would be needed in larger proportions.

Turning now to the second aspect of table 5.8, the comparisons between various coalfields, these show considerable differences in some cases, and can be ascribed to a mixture of different customs and different physical conditions. The most obvious peculiarity amongst faceworkers is the low proportion in Durham and Northumberland. This reflects both the high average age of pits in that coalfield[84] and the elaborate division of labour in which, for example, hewers neither fired shots nor did repair work. The Staffordshire Inspection Division, too, was affected by the ageing of mines in Shropshire and the Black Country, though North Staffordshire was about average in this respect, and Cannock Chase had been mainly opened up in the 1870s. The comparative youth of pits had a contrary effect in the Midland Division, South Wales, and to some extent Yorkshire, but the proportion was also affected by the custom of hewers setting their own timber and doing some of the repairs at the face. This custom also ruled in an extended form in Scotland, but the relatively high proportion of faceworkers was also partly accounted for by the comparatively stable geological conditions. The Midland Division, too, had firm roadways and little or no gas. By contrast, the soft nature of the South Wales bituminous seams not only led to faceworkers doing their own repairing as it was needed (and with some urgency), but also required a relatively large number of repairers on the main roadways. The great depth of many Lancashire mines, and the consequent liability of the roadways to 'heave' or 'creep', had a similar effect in that district.

Since the change in the proportion of hewers in the labour force which we found in Durham and Northumberland between 1890 and 1905/6 was also found, often to an even greater extent, in the other coalfields, it is tempting to assume that they had also experienced the same rise in that proportion in the period from the 1840s to the 1880s. This may well have been the case, for technical progress was at least as much applied elsewhere as in the North-East from the 1850s or so onwards. But we can do no more than speculate about this, except to note that the increase in

Table 5.9 *Percentages of labour force who were hewers, 1905, 1913, and 1914*

<table>
<tr><th></th><th>1905</th><th>Nov. 1913</th><th>June 1914</th></tr>
<tr><td>Durham & Northumberland</td><td>38</td><td>35</td><td>35</td></tr>
<tr><td>Yorkshire</td><td>38</td><td>35</td><td>37</td></tr>
<tr><td>Lancashire & Cheshire</td><td>37</td><td>21</td><td>30</td></tr>
<tr><td>Midland Inspection Division</td><td>37</td><td rowspan="2">34</td><td>41</td></tr>
<tr><td>Staffordshire Inspection Division</td><td>29</td><td>32</td></tr>
<tr><td>South Wales</td><td>47</td><td>40</td><td>32</td></tr>
<tr><td>Scotland</td><td>49</td><td>38</td><td>40</td></tr>
<tr><td>UK</td><td>40</td><td>33</td><td>36</td></tr>
</table>

Sources: CIC (*SP* 1919 XIII), appendix 28 (for Nov. 1913) and appendix 49 (for June 1914); *SP* 1907 XIV.

the proportion of pits fifteen or more years old between 1869 and 1889[85] is much the same in most coalfields as in Durham and Northumberland. The main exceptions are Lancashire and South Staffordshire, where the increase in older pits was significantly greater; the East Midlands, where the proportion was significantly smaller; and South Wales and North Staffordshire, where rather than any increase, there was a small fall. In the light of this we may speculate that there was possibly no improvement in the percentage of hewers in the labour force in the two former districts, at any rate after the early 70s; and that in the last two, or perhaps three, districts, the shift in the percentage was probably rather higher than the small rise achieved in the North-East. There is just one shred of evidence to support this so far as North Staffordshire is concerned. The inspector of mines for that district told the Select Committee of 1873 that hewers formed a quarter of the total employed there.[86] This implies that something like 35–40% were faceworkers. According to the 1890 return the proportion was 46% – sufficiently larger for us to be fairly certain that there really had been some improvement.

This is as far as we can get in any attempt to examine the change in the structure of the coalmining labour force in the nineteenth century, and it is time to turn to the last few years before the First World War. Unfortunately, since the Sankey Commission *Report* (on which we rely for our information) used different categories from those of the Eight Hours Day Committee, and, in particular, since it did not assign youths and boys to the adult categories, it is only possible to make the comparison for the proportions of hewers. This can be done for 1905 and November 1913 and June 1914. Even for this, there are some obvious disparities, and it would be unwise to take all the figures shown in table 5.9 at face value. The most glaring anomaly is in the Lancashire statistics, and in that case the Con-

Table 5.10 *Percentages of workers in pits thirty or more years old, 1905 and 1913*

	1905	1913
Durham & Northumberland	79	79
Yorkshire	61	64
Lancashire & Cheshire	73	78
Midland Inspection Division	68	67
Staffordshire Inspection Division	78	76
South Wales	53	60
Scotland	55	63

Source: HOLM.

troller of Coal Mines, who provided the 1913 figure, admitted that it had been influenced by his inability to distinguish getters from drawers, stonemen, etc., in some returns. It is probable that the same applied to some degree in Scotland and South Wales; and certainly something of this sort must have happened in South Wales in the June 1914 statistics, for whilst the fall from 1905 to November 1913 is just credible, that indicated in the following seven months is not. In contrast, it is the 1905 figures which are misleading in the case of the Midland Inspection Division, for clearly a good many daywage hewers – a common class in those coalfields alone – were included amongst the fillers.

Despite these reservations about the comparability of the figures at the different dates, it does seem reasonably certain that there was a quite considerable fall in the proportion of hewers in three major coalfields – Lancashire, South Wales, and Scotland – and a significant, though more modest, fall in Northumberland and Durham. In Yorkshire, and probably the East Midlands, there was virtually no change; whilst in the West Midlands there may even have been a rise in the hewers.

It is impossible to be sure to what extent these changes were associated with the effects of the Eight Hours Day Act, for the influence which had been at work since 1890 (or earlier) – the increasing age of the pits – continued to operate in most districts, and most especially in South Wales and Scotland. Table 5.10 shows the proportion of miners employed in pits that were thirty or more years old, which is as good a guide to this as we can take. It shows that in the two districts where it is possible that the proportion of hewers in the labour force actually rose between 1905 and 1914, there was a slight fall in the percentage of workers in elderly pits. In the North-East the latter percentage was stable; but everywhere else it rose, above all in South Wales and Scotland. Admittedly this does suggest that the falling proportion of hewers in Durham and Northumberland *was* largely the result of the Eight Hours Day Act. But the Act certainly had unusual effects there, because it affected the hours of the haulage workers very considerably and so disrupted their peculiar shift system. But it really tells us nothing about the other coalfields. In Yorkshire the Act affected relatively few workers, the ageing of the pits was only moderate, and there was little change in the proportion of hewers. In Scotland, South Wales, and Lancashire, on the other hand, all three changed quite markedly: the proportion of hewers went down drastically, the hours of many workers were affected by the Act, and the percentage of workers in elderly pits rose sharply. This coincidence makes it impossible to conclude for certain that either the Act or the continued ageing of mines was the more important cause of the change in the structure of the labour force. But given that this change was only quite moderate in the North-East, it is likely that the Act was not the more important influence.

Having discussed the size and structure of the labour force, we now come to the other major factor in assessing the input of labour into the industry – the length of time worked by the typical miner. This can be divided into two aspects. First, there is the number of hours normally worked per shift; and second, the regularity of work. These will be looked at in turn.

When we come to consider the question of the hours of work which constituted the normal working day, we encounter for the first time a problem which we shall meet again when wages and earnings are being examined. There was an immense range of experience even within each coalfield, especially prior to 1872, but by no means only then: the typical miner does not really exist. There is a further problem, and that is in deciding exactly what is meant by any particular reference to hours worked. From the point of view of the miner, one may say that his working day is from when he leaves home until he returns. But from the point of view of the economy of the industry, the time which may be regarded as the miner's input of labour, strictly speaking, is that which he spends on his job. For the faceworker this may exclude a considerable amount of time spent in travelling underground, and it certainly excludes any time allowed for meal breaks. This economic point of view is no doubt very close to the employer's way of looking at things; though since he could not usually keep a very strict check on travelling time, and since, in any case, the key class of labour, the hewers, were most commonly paid on piecework, employers came to regard the hours of their workers as being the same as those for which the pit was open – though in the last quarter of the nineteenth century the concept of 'bank-to-bank' generally replaced this. This could be defined as time spent below ground plus the time taken to wind all the men up *or* down; or, more strictly, plus the time between the middle man down and the middle man up. This concept was the one usually favoured by the men, and was the one in which they cast their demands in the late nineteenth and early twentieth centuries. Most employers came to use the concept because of this, but they were still anxious to steer negotiations and legislation into using the concept of time spent winding coal, and in the 1908 legislation they were successful in this. In attempting to make comparisons over a long period of time, these differences in concept must be borne in mind. So, too, must the extreme flexibility of attitudes of both many employers and workers towards the working day of those on piecework, up to 1873 at the earliest, and in some places until much later – even to the Eight Hours Day Act in a few cases. In many pits in the first half of the nineteenth century, and in some even into the twentieth, the hewers could come and go almost as they pleased, though this never applied to the larger and deeper mines, and was probably fairly unusual in Durham and Northumberland at any time in our period. Earlier, however, this flexibility had been common in the North-East too, and Welbourne could say of the mid-eighteenth century that 'except in the very deep pits they kept no regular hours, but came away when they were satisfied with their day's work'.[87]

A recent and valuable book by M.A. Bienefeld,[88] which devotes considerable attention to the coal industry, concludes that working hours in it rose in the first few decades of the nineteenth century. On the basis of rather casual statements by Adam Smith, Gabriel Jars, and Arthur Young, Bienefeld believes that hours in the eighteenth century 'seem to have been short'. Then, using mainly the evidence contained in the *Reports* of the Children's Employment Commissioners, he claims that:

> the condition of the hewer seems to have deteriorated from the time when he was able to enforce the working of the 'stint', limiting his labour to between 7 and 9 hours per day. The Commission did not find this system operating in any of the mining districts, except as a nominal unit of accounting for piecework. Worse still for the workman, the loss of the stint and the consequent extension of the working day had not been accompanied by the introduction of regular meal breaks in most cases. Thus while the industry adopted the day from 6 a.m. to 6 p.m. in conformity with general practice, most of the miners worked these hours with only one meal break of 20 minutes or less.[89]

There is much that is incorrect and misleading in this passage. In the first place, it is not certain that the stint system ever operated in all coalfields, though it certainly did in Scotland, Cumberland, the North-East, and South Staffordshire. In the last two, and maybe the last three, it had disappeared, except, as Bienefeld says, by 1842; but it was definitely still operating in a great many Lanarkshire pits at that time and much later.[90] But much more important than this, the idea that hours of work in the eighteenth century were short for *all* miners cannot be sustained. The specific references which Bienefeld cites are to 'colliers' (which is nearly always a synonym for hewers – the equivalent contemporary term for miners in general being 'pitmen', in the North-East at any rate); and they relate only to Scotland or the North-East of England; and one, at least, is to a two-shift system. Now, as Ashton and Sykes[91] said: 'In Scotland, Cumberland, and the Great Northern coalfield the hours of the adult coal-getter cannot be precisely determined', because of the stint system, which Bienefeld correctly described. Whatever they were, the putters, trappers, onsetters, etc., had to shift the coal that had been cut to the shaft bottom: 'they were always obliged to remain below after the hewers had left the pit, and their hours of labour were considerably longer than those of the adult pitmen'. Ashton and Sykes, like Bienefeld, cite Jars; but to show the hours of the haulage workers as well as those of the hewers – and they were fourteen, as opposed to six or seven. It is well to remember, though, that Jars's statements relate to just one colliery, Walker; and if one thing is clear from the extensive evidence of the 1842 Commissioners, it is the wide range of hours at different pits – that is, the wide range of different lengths of time for which different pits were open. Hewers' hours might vary even more. In the light of this, even Ashton and Sykes's conclusion that hours for both hewers and others rose between Jars's observations in 1765 and 1800 seems stronger than is warranted. Most of the rest of their scanty eighteenth-century evidence suggests that the typical pit – if such can be spoken of – was open for twelve hours a day, and non-hewers worked for that time. Hewers normally worked less, eight to ten hours being the common range. Walker Colliery seems, in fact, to have been untypical. It was, for that time, a large and modern colliery, and it may well be for that reason that Jars took note of it. It was certainly one of the first to operate on a two-shift system for hewers, and for that reason alone its hours cannot be taken as in any way representative of the great majority of collieries. Much more typical, surely, is Griff Colliery in Warwickshire, which Ashton and Sykes note had precisely the same hours in 1702 as 140 years later – namely twelve. Certainly there is nothing other than Jars, and the casual comments of Smith and Young, to support the hypothesis of a rise in hours in the period 1765–1800; whilst after that, there is a considerable amount of evidence to suggest that there was no general change in hours until after 1842. Bienefeld quotes various times of work from the 1842 Commissioners' *Reports* – Derbyshire (except Chesterfield area) 6 am to 8 pm; East Scotland 'over 12 hours much of the time'; Bradford and Leeds ten to twelve hours; Oldham nine, ten, or eleven hours; North Lancashire and Somerset eight to twelve hours; West Riding ten to eleven hours. But these are a mixed bag, referring some to lengths of time certain pits were open and some to hours worked by individual workers. Certainly they bring out the variations, both between and within coalfields; but the separate statements of the witnesses need to be considered with much more attention to exactly what they refer to, if we are to make a reasonable estimate of the 'typical' hours in each area. When that is done, what seems to emerge from the evidence of the older witnesses is that hours had not changed since at least the 1800s. Instances are reported of shorter hours than formerly, but the one cited by Ashton and Sykes, of a seventy-year-old woman who said that hours had been extended when bearing was replaced by putting (i.e. corves by trams), is the only case of an increase recorded.[92] However, the great majority who were in a position to speak to the matter either said nothing about it or reported no change. The one clear and important exception was that of the boys (mostly youths and young men actually) in the North-East, whose hours were fixed at twelve after the 1831 strike, a figure which some collieries had operated previously, but which at others had sometimes been considerably exceeded.[93]

Finally, Bienefeld's suggestion that in 1842 'most of the miners worked with only one meal break of 20 minutes or less' is contrary to the evidence of the Commissioners' *Reports*. In the mid-1830s in North Staffordshire – admittedly according to an owner – one and a half hours were allowed for meals in the twelve hours the pits were open.[94] In 1842 three-quarters-of-an-hour and one-hour meal breaks were reported from Derbyshire and Nottinghamshire, one hour in South Staffordshire, one to one and a half hours in North Wales, half an hour to an hour in the Forest of Dean and in some Lancashire pits, though others had no regular meal times; half an hour in Warwickshire and in the Bristol area; a quarter of an hour to half an hour in West Yorkshire; up to one and a half hours in some Leicestershire pits, but no regular allowance in others.[95] In other areas there were no regular meal times, but this did not mean that time was not taken. The piecework hewer could suit himself, and surely did, and those whose work was determined by the production of the hewer would usually find it possible to take such breaks as he took.

Whilst there had been little change in hours in the first four decades of the century, the position did begin to change a bit in the 1840s, though there were few overall formal changes until the 1870s. In those places where non-hewers were working twelve or more hours a day in 1842, the boom of the middle of that decade generally seems to have seen them brought down to twelve at first, and in most cases to eleven.[96] There was then little change in the 1850s, except for some reductions in West Lancashire, though there certainly must have been some drift downwards in the average hours worked by hewers in those places where the use of blasting spread.[97] The same was true for the 1860s, with Yorkshire in the middle of the decade as the main area of exception.

In explaining why there should have been such a long period of comparative stability, which then broke suddenly, and was again followed by an extended period of little change, Bienefeld's analysis seems very convincing. This is too complex to summarise adequately here, but perhaps its main thrust can be expressed in his own words:

> Our analysis . . . suggests that what actually took place was that in the short run each real wage-rate created a demand for leisure, but it was in the nature of the opposing interests being resolved under the collective bargain that this demand for leisure was repeatedly 'bought off' by the employers. This was possible because of the workman's continually shifting conception of his 'normal' or 'expected' income, which in a sense constantly made him revise the base line from which he measured the income increase to be compared to an increase in leisure. Thus, the fact that the thwarted desire for leisure cannot be treated as cumulative means that it may perpetually lose its unequal battle against the worker's own income preferences, and the employer's willingness to 'buy him off'.[98]

The one situation, according to Bienefeld, where this analysis breaks down is when 'a sufficiently rapid and sustained rise of money wage-rates weakens the income preferences of the employees until the slope of their indifference function approaches that of the employers and a reduction of hours becomes likely'.[99] The boom of the early 1870s produced a more rapid, and possibly more sustained, rise in money wage rates than at any other time in the nineteenth century. Hence, on Bienefeld's analysis, which we adopt here, the only sharp fall in hours which was spread over virtually all districts in the whole of our period.

That the fall in hours was experienced to a less extent by hewers than by others and that it was partly brought about by legislation, does not affect the analysis. The hewers, where they were pieceworkers, had already used their power of job control to set a combination of hours and earnings which suited themselves, at any rate during periods of high demand for labour. It was other grades of labour, who were more affected by formal hours of work, who were principally concerned to achieve a reduction. That they used union pressure to secure this in part by legislation affecting boys reflects firstly the efficacy of this method in the textile factories, and secondly the ease which the lack of supervision underground afforded to individual miners to defy union (as opposed to legal) attempts to impose restraints, and the consequent temptation to employers to lure, cajole, or threaten individual miners into undertaking such defiance.[100]

Table 5.11 is an attempt to represent in statistical form the discussion of the last few pages, an attempt to distil from Parliamentary Papers and from some trade union records figures which are as typical, or representative, as it is possible to be. Clearly there has to be a large element of subjective judgment in the interpretation of the evidence, especially prior to 1870 or so; but that is not the factor which induces the greatest need for caution. Rather this is the difficulty, if not the impossibility, of knowing to what extent such evidence as we have *is* reasonably representative. We have already suggested that Jars's observations on Walker Colliery in 1765 should not be so regarded, and it is difficult to argue with conviction that *any* of the scattered data prior to 1842 can be. The evidence to the Children's Employment Commissioners may well itself be regarded as biased, for it would be unfair not to recognise that most, if not all, of the Commissioners were out to make a case. However, there is so much of it, it comes from both owners and workers, as well as other residents in the colliery districts, and it is – sensitively interpreted – internally consistent to a large degree. The year 1842 may be taken, then, as a fairly firm beginning – and thanks to the retrospective reflections of the older witnesses, it can be taken as largely applicable to the beginning of our period. It should be remembered, especially when making comparisons with later figures, that table 5.11 relates to the hours the pits were working – i.e. from first man down to last man up, not bank-to-bank. Meal times, however, have been deducted as indicated in the evidence. Where there was no regular meal time, this has been assumed to be half an hour per day, except for the boys in the North-East.

Following the big reductions of the early 1870s, few changes took place until the Coal Mines Regulation Act of 1908, popularly known as the Eight Hours Day Act. The most important of these was the reduction in Lanarkshire in 1900, in the later stages of a boom which was second to that of the early 70s in vigour. In the previous boom the hours of Cumberland hewers were cut from about eight and three-quarters to eight, in 1891,[101] but no other change was reported as affecting a whole coalfield in either the early 90s or around 1900. Probably the nearest thing to a district-wide change was the gradual return after 1879 at many pits in the North-East to an eleven-hour weekday, or sixty-four hour week, for many of the boys over sixteen years old. This was reversed again in 1891.[102] Other changes affecting individual collieries or groups there certainly were, especially in Lancashire in the early and middle 90s,[103] but there is little sign of this in such data as are available for whole districts. As J.W.F. Rowe puts it, after mentioning the North-East and Lanarkshire, 'on educated guesswork it may be said that there was little or no change in the hours of work from 1888–1908, though the hewers in certain districts in the Midlands, and the transit hands in Somerset and the Forest of Dean, may have gained a small decrease'.[104]

The main miners' union outside Northumberland and Durham, the Miners' Federation of Great Britain, began agitating for legislation to limit the underground worker's day to eight hours from bank-to-bank almost from its foundation in 1888. Very early on in this agitation there came the Royal Commission on Labour, and much evidence was given to it on the hours of work of miners. Meanwhile there had also been a Board of Trade Enquiry, resulting in a Parliamentary Return,[105] and the MFGB

Table 5.11 *Typical weekly hours of colliery workers, 1842–73*

	Hewers	Boys	Others underground	Surface
Durham & Northumberland				
1842	60 (48/72)[a]	72	63	69
by 1850	54 (48/72)[a]	72	63	69
by 1870	45	72	63	66
by 1873	38	60	53	54
Yorkshire				
1842	60		66	66
by 1850	57		63	63
by 1870	51		54	57
by 1873	44		46	46
Cumberland, Lancashire & Cheshire				
1842	60		66	66
by 1850	60		66	63
by 1870	57		60	60
by 1873	48		50	50
North Wales				
1842	66		69	66
by 1850	66		69	66
by 1870	66		69	60
by 1873	50		53	50
East Midlands				
1842	66		66	66
by 1850	63		66	66
by 1870	60		60	60
by 1873	50		50	48
Shropshire & North Staffordshire				
1842	66		66	66
by 1850	66		66	66
by 1870	60		60	54
by 1873	50		50	46
South Staffordshire				
1842	60		66	66
by 1850	60		66	66
by 1870	54		60	60
by 1873	44		50	48
Gloucestershire & Somerset				
1842	54		60	57
by 1850	54		60	57
by 1870	54		60	57
by 1873	50		57	57

Table 5.11 (*cont.*)

	Hewers	Boys	Others underground	Surface
South Wales				
1842	69		69	66
by 1850	69		69	63
by 1870	69		69	60
by 1873	45		47	45
Ayrshire				
1842	50		65	...
by 1850	50		65	...
by 1870	50		50	...
by 1873	50		50	...
Lanarkshire				
1842	63		66	69
by 1850	63		66	66
by 1870	63		63	57
by 1873	50		50	50
Fifeshire				
1842	60		60	
by 1850	60		60	
by 1870	60		60	
by 1873	44		44	

A six-day week has been assumed everywhere up to 1870, and in Gloucestershire and Somerset in 1873; and a five-and-a-half-day week elsewhere in 1873.

[a]This represents the range from the pits with a double-shift of hewers to those working the single twelve-hour shift, though even in the latter few hewers worked as much as that, and an average of ten to eleven seems a reasonable estimate. In 1842, probably something like 10% to 20% only of hewers were on double-shift working. (Galloway (1898–1904) says five or six pits in a hundred had the double shift, but as they were the bigger ones, the proportion of miners concerned must have been two or three times as big.) It therefore seems fair to take sixty hours a week as the overall average in 1842. The rapid spread of the use of explosives was mainly what permitted the reduction of hewers' hours and the development of the North-East's peculiar shift system, and this was most speedy in the 1840s. The estimate of fifty-four hours a week as the overall average by 1850, besides being based on evidence in the sources, is also close to what would be arrived at on the assumption that practically two-thirds of hewers were on a double-shift system by then – an assumption which seems to fit the evidence about shotfiring.

Sources: evidence to the Children's Employment Commissioners (*SP* 1842 XV); SC 1873; RC on Trade Unions (*SP* 1867 XXXII); RC on Labour (*SP* 1892 XXXIV); returns in *SP* 1890 LXVIII and *SP* 1892 XXXIV, references in Evans (1961); Machin (1958); Welbourne (1923); and Williams (1962).

Table 5.12 *Hours of work, 1890 and 1905*

	Hewers		Others underground		All underground: MFGB	All underground: Home Office		Average meal time allowances: Hewers	Others underground	Hewers	Others underground
	1890	1905	1890	1905	1890	1890	1905	1890		1905	
Cumberland[a]	8.30	7.46	8.63	9.16	8.38	8.47	8.33	0.50	0.50	0.45	0.65
Northumberland	7.08		9.02		...	8.50		0.50	0.50		
Durham	7.22	6.82	9.40	9.59	...	8.91	8.30	0.47	0.68	0.43	0.62
Yorkshire	8.80	8.48	8.82	8.90	8.53	8.80	8.72	0.72	0.81	0.57	0.63
Midland Inspection Division	9.35	9.30	9.39	9.55	9.39	9.41	9.43	0.49	0.52	0.48	0.48
Lancashire & Cheshire	9.27	9.56	9.42	9.84	9.78	9.33	9.62	0.84	1.00	0.92	0.90
North Wales	8.86	9.10	9.10	9.62	9.38	9.08	9.36	0.65	0.64	0.62	0.58
North Staffordshire	8.95		8.92		9.48	8.94		0.62	0.62		
South Staffordshire & Worcestershire	7.82	9.13[b]	8.75	9.38[b]	8.85	8.44	9.30	0.83	0.83	0.72	0.70
Shropshire	8.48		8.98		9.22	8.83		0.67	0.70		
Gloucestershire & Somerset	8.56	8.53	9.37	8.76	8.82	9.11	8.73	—[c]	—[c]	0.53	0.53
South Wales	9.15	9.70	9.09	9.31	9.60	9.09	9.88	0.78[d]	0.66[d]	0.73	0.72
East Scotland	8.98	8.68	9.13	9.07	9.27	9.32	8.85	0.84	0.85	0.68	0.67
West Scotland	9.27	8.90	9.35	9.02			8.95	0.95	0.93	0.58	0.57
UK (weighted average)	8.71	8.60	9.37	9.47	...	8.91	9.05	0.70	0.72	0.65	0.65

had published the results of its own enquiry on hours.[106] All the data, however, are difficult to interpret and make comparable with later – or, for that matter, earlier – data, partly because of the ambiguities in the concept of working hours, which were referred to earlier. Both before and after the Royal Commission, bills were brought in in Parliament in practically every year, but got nowhere. One important reason for this was the firm opposition of the miners' unions of the North-East, where the dominant group, the hewers, feared that cutting the hours of the haulage workers might lead to total disruption of their shift system, and consequently to lower earnings, and perhaps longer hours, for themselves. Eventually, with the election of the Liberal government in 1906, a Departmental Committee was appointed to examine the matter, though, as Rowe puts it, 'with what seems to the statistician little short of criminal negligence, they, or the Board of Trade officials who actually conducted the matter, failed to insert any definition of what they wanted when they asked for "average hours bank-to-bank" '.[107] Nevertheless, comparisons can be made – with caution – between the statistics emanating from the enquiries of 1890 and 1905/6. These are shown in table 5.12.

The difference in the average hours of underground workers taken as a whole between the union figures and the official figures collected from the owners is of no account in many coalfields, but is enough to raise doubts in the case of Lancashire, all the West Midlands fields, South Wales, and, perhaps, North Wales and Gloucestershire and Somerset as well. Moreover, whilst the averages for the Midland Inspection Division are very close together, the Warwickshire figures on their own (which are not shown here) differ by more than any others,[108] and the Leicestershire and Derbyshire statistics also show significant divergences. In all cases of major difference, except Warwickshire and Leicestershire, the union found hours to be longer than the official figures claim, by as much as half an hour in Lancashire and South Wales. Which of these figures are we to believe? B. McCormick and J.E. Williams, in an article in 1959,[109] appear to prefer the union ones, though they admit that there is no way of confirming speculations about the reasons for the differences. Certainly they

Notes to Table 5.12

[a]The 1890 figures include iron miners.

[b]Averages in 1890 were as follows: hewers 8.32; others underground 8.84.

[c]The 1890 figures show no meal time allowance, but it may be assumed that at least half an hour was taken.

[d]The 1890 figures show no meal time allowance in Monmouthshire, etc. It has been assumed here that the average for the rest of South Wales applied to the whole coalfield.

Sources: *SP* 1890 LXVIII, appendix to first *Report*, table XI, and appendix 1 to second *Report* (for meal time allowances); and MFGB (1890).

are right in this, and probably the best way to proceed for the moment is to treat both series as evidence.

The reduction in hewers' hours in Durham is about what might be expected from the 1890 agreement that their true bank-to-bank hours should not exceed seven; and the reduction of the hours of others underground in Gloucestershire and Somerset also reflects changes which we know of from other evidence. The small changes in hewers' hours in Yorkshire, the Midland Inspection Division, and Gloucestershire and Somerset are also in line with what we should expect. But the reductions in Scotland are rather less than might be anticipated from the 1900 agreement in Lanarkshire, easily the biggest component, to reduce the coal-winding day to eight hours. More worrying still are the increases which are indicated in the official figures in the hours of both hewers and others underground in Lancashire, North Wales, the West Midlands, and South Wales, and of the non-hewers in the North-East, Yorkshire, and the rest of the Midlands. These were truly described by Rowe as 'very puzzling'.[110] His assertion that 'it may be taken for granted that there has been no increase in hours during this period', and his consequent suggestion that one or both sets of figures is false, must have a large measure of truth. If the union figures are compared with those for 1905, then most of the puzzles disappear. The Lancashire and North Wales hours would then seem to have changed but little, and those of the West Midlands would have fallen quite substantially, while the indicated increase in South Wales is reduced to more modest proportions. It would seem that on the grounds of plausibility there is more to be said for the union series than for the official figures in these cases. But the possibility that there really was some increase in hours in some districts, even though there was no formal change, cannot be wholly dismissed as Rowe does. There may have been a species of 'hours drift', caused by (a) some increase in winding times as more men were employed in the same pits; (b) perhaps also some increase in walking time which induced hewers to accept a small and gradual increase in their bank-to-bank hours; (c) some increase in the proportion of men employed at pits with hours above the original norm for the district; and (d) some increase in the proportion of haulage workers to repairers, etc., for the former usually worked longer hours than the latter. The first of these two suggestions may have applied to the districts where new sinkings were comparatively few, like Lancashire and parts of the West Midlands; the third may have occurred in South Wales, with the increasing relative importance of the collieries of big companies; the fourth appears to be a possibility only in Scotland.

Despite these possibilities, there remains the likelihood that there was some error in the statistics for a number of districts. As was pointed out earlier, neither in 1890 nor in 1906 was the concept of bank-to-bank defined precisely by the officials collecting the data, and though the union returns are less ambiguous in this respect, they still leave room for confusion over 'winding out' – that is, the time taken after the bulk of the men is up to finish winding out the last men. Certainly, it is clear from evidence to the Eight Hours Day Committee that mistakes about winding times did occur in both directions. In Scotland both winding times were included in some of the component areas, and this probably accounts for the rather low decrease in hours referred to above. In some parts of South Wales, on the other hand, both winding times were excluded, which, if corrected, would lead to an even larger than indicated increase in hours, unless the same mistake had been made in 1890, perhaps on an even bigger scale, which is what McCormick and Williams think, probably rightly.[111] This sort of mistake certainly did happen in some cases in 1890. In North Wales and the West Midlands there is evidence to suggest that *both* winding times were excluded then, and this may well have happened in South Wales too.[112] On balance, then, it is probably correct to add to the official figures for the different grades of labour the extra time suggested by the union figures for these four districts. What discrepancy remains after taking into account the probability of some 'hours drift', is more likely to result from errors in the 1890 statistics than in those for 1905, if only because the concept of bank-to-bank was more commonly employed and therefore likely to be better understood by the later date. It is, however, just possible that some of the 1905 figures may have been slightly magnified by the desire of employers to give an exaggerated impression of the likely loss of output resulting from a cut to eight hours bank-to-bank.

The upshot of this necessarily rather unsatisfactory discussion of unreliable, or partly unreliable, statistics is to confirm that, with a few exceptions known from other sources, the general enquiries of 1900 and 1905/6 suggest only small changes in hours worked. And in the one clear case of a district-wide reduction – Scotland in 1900 – it appears that a large part was offset by a reduction in the meal time allowance. It is time now to turn to the effects of the Eight Hours Day Act.

The Departmental Committee had considered the effects of an eight hours' day bank-to-bank, but, as eventually passed, the Act of 1908 excluded *both* winding times from the limitation of hours, so that it became in practice an Act to secure an eight-hour day for coal winding. It ensured eight hours bank-to-bank only to the man who could contrive to be both the last man down and the first man up – something which was hardly ever likely to happen in practice. In fact, the average bank-to-bank day for the country as a whole after the Act came into force was eight hours thirty-nine minutes.[113] This figure, computed for the Sankey Commission, is one of the surprisingly few that were provided for and by that body. In assessing the effects of the Eight Hours Day Act on the actual hours worked, therefore, one has recourse only to the scattered references in the evidences of witnesses before the Sankey Commission, aided by *a priori* reasoning about the necessary changes in the 1905 figures that would be produced by compliance with the law.

The one clear piece of statistical information which the Sankey Commission had was provided by the Monmouthshire and South Wales Coal Owners' Association, contrasting the hours worked before and after the Act in a large number of collieries in that coalfield.[114] This showed that coal winding had been reduced from nine to eight hours, and bank-to-bank hours from nine and a half to eight and a half. As a partial offset, the meal time allowance had been reduced by ten minutes. There seems to be no reason to question this assessment, although the reduction would have been nearer one and a half hours than an hour on four days in the week in those pits where the custom of the shorter Monday shift was not abandoned. Probably these latter were few. And in any case, the 1908 Act allowed sixty hours per year to be worked over and above the eight-hour limitation, and this was successfully insisted upon by the owners,[115] so that an average decrease of an hour per day is probably about right.

In Lancashire, except in the Burnley and Manchester areas where hours had been reduced earlier,[116] the Act cut at least an hour off the working day for all underground workers, probably more for haulage workers in the Wigan and St Helens areas. But in no other district was the impact of the Act likely to have been so widespread and so large as in South Wales. In North Wales and all parts of the Midlands except South Staffordshire,[117] hewers' hours were cut by quite a substantial amount – by nearly an hour in the Midland Inspection Division and in Shropshire, and by half an hour in the Potteries and North Wales. Elsewhere the Act apparently made little difference to the hewers. Other underground workers, however, gained something practically everywhere, though not very much in Yorkshire, Somerset, the Forest of Dean, and two of the biggest Scottish areas, Fifeshire and Lanarkshire. In all of these coal winding hours had already been reduced to eight per day, and the main beneficiaries of the Act in these districts were the few oncostmen who worked at repairs on night shift. In all Midlands districts, however, the haulage and oncost workers' hours were cut by an hour or a little more. Finally, in Durham and Northumberland the hours of haulage workers, which had still been ten before the Act except in a few three-shift pits, were necessarily reduced, but those of hewers and most of the remaining workers were unaffected. Almost universally in Durham, and widely in Northumberland, the pits which had operated the normal two-shift system – that is, two of hewers and one long one of putters – went on to a three-shift system, with the putters divided into two shifts working eight hours each.

Whilst most of the discussion of hours before the 1880s or 90s was in terms of the time the pits were open, and most of it subsequently related to hours bank-to-bank (however imperfectly this was defined at times), the effective supply of labour of the faceworkers depended on the number of hours spent at the coalface, and on the meal time allowances, which have already been noted. This varied from pit to pit and from district to district according to the time taken to raise and lower the men, and the time taken travelling underground. The former depended on the complement of the pit, the capacity of the winding equipment, and the general efficiency of the organisation. At the largest mines winding time could be as much as three-quarters of an hour at the beginning and end of each shift, despite the larger cages in use. At small pits it might be as little as ten or fifteen minutes, and in a drift mine it was nil. Other things being equal, it was longer at older mines, for the equipment was not so speedy. And whilst there was probably a gradual improvement in the efficiency of organisation, there were two periods when it was marked enough to attract attention. The first was following the reduction in hours in the early 1870s, partially associated with the legislative restriction on the hours of boys under sixteen.[118] The second was after the Eight Hours Day Act, which Sir Richard Redmayne, the Chief Inspector of Mines, told the Sankey Commission, in some cases 'led to a higher state of efficiency in point of the time of lowering and raising the men'.[119]

As a result of all these things, there was a certain degree of variation in the average winding time, both over time and between districts. But the latter variations were really negligible except for high figures in Somerset and Gloucestershire towards the end of our period; and the former seem to have been quite small, for the different influences of improving organisation and equipment appear to have offset those of deeper mines with bigger complements. The conventional figure, mentioned by nearly all witnesses as enquiries from 1873 to 1919, was half an hour each way. There is no earlier evidence, but also no reason to suppose that things had been very different, except in small and shallow pits, at any earlier date.

Apart from isolated tales of long distances and times spent going along narrow underground passageways to the working place, a good many of which are to be found in the *Reports* of the Children's Employment Commissioners, and, one suspects, lost nothing in the telling, there is little reference to the time spent in travelling underground before the enquiries of the Royal Commission on Labour. When hours were long and there was no question of restricting the time the pits were open, it was not a matter of much concern to anyone, except those unfortunate hewers who

Table 5.13 *Time travelling underground and time available for work at the face, 1890 and 1905 (in hours)*

	Time travelling underground 1890	1905	Time available at face 1890	1905
Cumberland	0.70	0.92	7.10	6.09
Northumberland	0.51		6.07	
Durham	0.93	1.27	5.82	5.12
Yorkshire	0.58	0.98	7.50	6.95
Lancashire & Cheshire	0.52	1.07	7.91	7.57
North Wales	0.59	1.02	7.62	7.46
Midland Inspection Division	0.87	0.95	7.99	7.87
North Staffordshire	0.70	1.07	7.63	7.34
South Staffordshire	0.50		7.32	
Shropshire	0.23		7.58	
Gloucestershire & Somerset	0.85	1.21	7.18	6.79
South Wales	0.59	0.96	7.78	8.01
East Scotland	0.48	0.73	7.66	7.57
West Scotland	0.53	0.85	7.79	7.47
UK (weighted average)	0.60	1.00	7.41	6.95

Sources: see text.

had to go distances to their coalface which were well in excess of the normal on which piece rates had been reckoned. So long as the average age of pits was not rising much there was probably little change in the average travelling time, either on account of longer winding or of longer walking underground. This suggests that there had been little change right up to 1890. At that time, the average travelling time for each district can be deduced from the data collected for the Royal Commission on Labour, though it suffers from the ambiguities connected with the inclusion or exclusion of winding times, which were referred to above. The figures, along with those officially provided for the Eight Hours Day Committee, are shown in table 5.13, together with the implicit time available for work at the face, after deduction of the average meal time allowances shown in table 5.12.

Ignoring for the moment North and South Wales and the West Midlands, where we are suspicious of the accuracy of the 1890 figures, it seems certain from these statistics that travelling time had risen in all coalfields, and in most of them quite markedly. In the Midland Inspection Division, the increase was only five minutes, but in most of the other coalfields it was around twenty minutes, and in Lancashire it was over half an hour. In South Wales, if we are correct in supposing that the so-called bank-to-bank hours in 1890 did not in all cases include one winding time, there may have been either an increase or a decrease in travelling time, but it was only a few minutes in any case – as we should expect in a still fast-growing coalfield, where the expansion was not brought about solely by opening very large new pits as in Yorkshire. The same applies to North Wales, though there is no *a priori* reason in that case for expecting it. In the West Midlands, however, even if our suspicions about the accuracy of the 1890 data are correct, there was still some increase in travelling time by 1905.

The consequence of the increase in time spent travelling, whether in the shaft or underground, was a fall in the time available for work at the face, even where there was no reduction in the bank-to-bank day, and despite the widespread paring down of the time allowed for meals. This applied everywhere except in South Wales; and, even there, the increase in time available may be more apparent than real, for it depends on our accepting that there was a genuine increase in bank-to-bank hours between 1890 and 1905. Whatever the merits of this, however, it is clear that there was little change in the time available for work in South Wales. The same applied in the Midland Inspection Division and in North Wales; but elsewhere the decrease was as much as three-quarters of an hour in Durham, where it represented a 12% reduction, half an hour in Yorkshire, and around twenty minutes in West Scotland and Lancashire – more in the latter, if the indicated increase in bank-to-bank hours in table 5.12 is not genuine.

So far in our discussion of working hours only average figures have been mentioned, other than casually, though the great flexibility and variation of hours, especially before 1873, was stressed earlier. This use of averages is desirable for most purposes. Nevertheless, it would be incomplete without an indication of the spread of the figures from which the averages are derived. This emerged naturally in the course of the discussion for the period prior to 1890. For the last few decades of our period it is possible to give quite a detailed breakdown from the information collected for the Eight Hours Day Committee. This is done in table 5.14.

The considerable amount of spread in the actual number of hours worked within each coalfield even at this late date was partly the consequence of differences in the hours of the various classes of worker, most importantly so in Durham and Northumberland. Other causes were (1) differences between pits, of long standing but with no obvious reason; (2) the existence of a double coal-getting shift at a few collieries outside the North-East; (3) differences at single-shift collieries between men employed on the main (or goal-getting) shift and those on 'back' shifts, when repairing was concentrated and the tubs were filled.

The final factors which must concern us in this review of the input of labour to the coal industry in the nineteenth century are the length of the

Table 5.14 *Percentages of all underground workers according to bank-to-bank hours, 1905*

	7.5 or less	7.5–8	8–8.5	8.5–9	9–9.5	9.5–10	10 or more
Cumberland & Northumberland	41	34	4	2	0.4	16	2
Durham	51	23	0.1	0.1	0.01	24	2
Yorkshire	3	22	31	33	4	6	1
Lancashire & Cheshire[a]	0.3	1	4	15	20	42	18
North Wales	—	4	2	11	69	10	4
Midland Inspection Division	0.01	1	3	19	34	32	10
Staffordshire Inspection Division	0.02	1	3	32	40	17	7
Gloucestershire & Somerset	0.1	0.2	0.04	5	38	38	19
South Wales	2	29	5	32	10	22	0.1
East Scotland	1	11	29	46	7	5	1
West Scotland	0.3	9	14	58	9	10	0.3
UK	12	11	8	18	19	24	8

[a]Lancashire and Cheshire includes Ireland.
Source: DC 1907, first *Report*, Part II, appendix C, table X.

working week, which was six days universally until after 1870 but was varied subsequently at different times and in different places, and the regularity of employment. This latter was controlled in part by the miners themselves, through the level of 'voluntary absenteeism', as it came to be called by those who deplored it – in plain language, missing a shift or two. Another influence was the temporary closing of the pits by the employers, usually because of slackness in the demand for coal – a regular seasonal phenomenon in some areas and seams, but also to be found happening in the depths of a depression. Yet another influence was the occasions when pits were closed owing to accidents, and there were often short periods of closure for repairs to winding, pumping, or ventilating machinery. In addition, in the days of ventilation by furnace, certain weather conditions prevented the effective operation of the furnace, and so led to closure of the pit – not always before some greater or lesser disaster had occurred. Finally, there was a growing practice in the later part of the nineteenth century of closing the pits for holiday periods. Easter and Christmas (or New Year in Scotland) had always been observed, usually by a two-day break, but from the late 1860s or early 70s, a summer break, usually of a week, connected with a traditional local fair or feast, became more common.[120]

The concept of the normal working week, in the sense of the number of days on which the pits were open for work rather than the average, or normal, number of shifts actually worked, only begins to have any significance during the boom of the early 1870s. Prior to that time, it was the customary practice everywhere for the pits to be open for a full six days each week, and no one seems to have thought of anything else – perhaps because there was no compulsion to work the full week, and most hewers did not in fact do twelve shifts in a fortnight, though oncost workers did.[121] The first change in the routine six-day week came in 1872, when the boom took off into unprecedented heights of prices and wage rates. But even then, it was only in the North-East (and a few scattered collieries elsewhere) that there was any formal change in the working week. In Northumberland, where it is possible that alternate six- and five-day weeks had come a little earlier,[122] the larger collieries, at least, went onto a five-day week in 1873, though there was a return to the eleven-day fortnight in 1877–8.[123] Durham seems to have gone over to having alternate Saturdays off, with a few exceptions, in 1872, and not to have had the five-day week interlude which Northumberland had.[124] The working Saturday was two hours shorter than the normal day in Durham, though not always in Northumberland.[125] This eleven-day fortnight, with one day slightly shorter than the other ten, continued to be the commonest system in both Durham and Northumberland at the time of the Eight Hours Day Committee, though there were substantial minorities operating six days a week, with both Saturdays being short (see table 5.15 below). In Scotland, it was only in Fifeshire that the six-day week was widely dropped in the early 70s in favour of the eleven-day fortnight. Likewise, in the Midlands, it was only in Warwickshire that this happened.[126] In Lancashire, though there was an unusual amount of variety in the practice of different collieries, the inspector for North and East Lancashire thought that by 1873 ten and a half days a fortnight was 'a fair average' of how long the pits were open for work.[127] In Yorkshire, the East Midlands, South Wales, and most of the smaller coalfields, however, as in most of Scotland, the six-day week persisted generally, though the Saturday shift was often shortened. In South Wales this happened to the Monday shift too.[128]

The situation which was established in the early 70s did not undergo many major changes in the next forty years. The twelve-day fortnight with four short days was still common in South Wales in 1905, though a variant in which the Saturday shift was even more curtailed whilst Monday became a normal day was also quite common. Yorkshire, much of eastern Scotland, Ayrshire, most of the East Midlands, the West Midlands, and the smaller coalfields were still mostly working six days a week, usually with short shifts on all or, in Yorkshire, alternate Saturdays. Lancashire still had a great variety of practices, though there had been a change for a substantial minority of its miners in the early 1890s to a straight five-day week.[129] The Lean Valley, in Nottinghamshire, had also been on a five-day week 'for years' in 1906,[130] though the rest of the East Midlands

Table 5.15 *Numbers employed at collieries according to their full normal working week, 1905*

	12-day fortnight				11-day fortnight		10-day fortnight	
	No short day	One short day	Two short days	Three or more short days	No short day	One short day	No short day	Other
Cumberland & Northumberland	301	5,473	5,338	—	9,112	21,057	42	6
Durham	—	1,218	21,084	—	—	70,891	—	—
Yorkshire	285	73,724	16	—	18	—	1,958	407
Lancashire & Cheshire[a]	226	2,105	46,730	230	—	2,884	12,244	911
North Wales	638	—	2,776	3,539	—	—	—	—
Midland Inspection Division	537	7	54,357	394	—	—	5,951	1,023
Staffordshire Inspection Division	4,517	—	20,738	7,723	567	—	610	153
Gloucestershire & Somerset	7,404	—	5,217	805	682	—	43	—
South Wales	1,124	—	32,237	87,019	—	—	13	22
East Scotland	3,528	221	12,799	487	13,614	2,318	6,146	378
West Scotland	351	1,051	4,521	—	2,369	2,167	20,361	591
UK	15,677	10,075	278,623	103,613	26,301	99,335	47,368	3,491

[a]Includes Ireland.

Source: DC 1907, first *Report*, appendix 1.

apparently still worked the short Saturday shift. Elsewhere, only Lanarkshire had experienced a major change, when, in 1901, it moved from six full days a week to five.[131] The position in the various districts in 1905 is summarised in table 5.15.

Whatever the length of the normal working week, one can say with a good deal of confidence that at no time anywhere was it worked by all, or even a majority, of miners. One reason for this was the action of the coalowners in closing the pits, either for necessary maintenance and repairs, or because of inability to sell enough coal to make it worth while getting it. The latter happened with seasonal regularity in some places right up to 1914, but it was almost certainly much more common before the spread of the railway network than it became later. The main reason for this seasonality was the fact that household demand for coal was at its height in the winter months, and there was insufficient willingness on the part of householders, merchants, or coalowners, for one reason or another, to stock coal during the summer for regular working to be maintained.[132] The only districts which still seem to have been seriously affected by this seasonal irregularity as late as the twentieth century, except in the very depths of depressions, were the Forest of Dean, Cannock Chase, Leicestershire, Warwickshire, and the East Midlands.[133] Elsewhere, seasonal irregularity pure and simple, unalloyed with cyclical influences, had probably not been of very great significance since the 1840s or even earlier. In South Wales there were still differences in wage rates in the sale coal (as opposed to the ironworks') pits between high season and low season in the 1840s, but not, apparently, later.[134] But there is no evidence of this sort of seasonality in any of the other coalfields where there was substantial development of manufacturing industry and/or shipment of coal by sea, and it is perhaps only in the East Midlands that it occurred to any considerable degree even before the spread of railways. This is not to deny, of course, that there was a small seasonal element in the regularity of working practically everywhere. The growth of the practice of taking holidays in summer has already been referred to, and it was common to have the biggest routine maintenance done in early summer; but there was probably a bit more to it than these things. For instance, in 1889, a boom year, a colliery in North Staffordshire missed four possible working days in the first quarter of the year, five possible days in the last quarter of the year (counting Christmas as possible), but an average of ten possible days in each of the middle two quarters – surely more than could be accounted for by routine closures for maintenance, or for national or local holidays.[135]

Closing of pits by the owners because of cyclical shortage of adequate demand was much more widespread than for seasonal reasons, though not so drastic as it surely would have been had wages been less flexible.

Nevertheless, there are plenty of references to irregular working of the pits in the eighteenth and early nineteenth centuries,[136] as well as in the enquiries of the mid-century period,[137] and it is certain that the following passage from the first article of the *Labour Gazette* on the state of colliery employment in 1894–5[138] was merely enunciating a long-established truth:

> Coal mining affords a good example of an industry in which the state of employment is best gauged, not by the proportion of workpeople entirely unemployed, but by the average number of days per week on which work is available. This results from the fact that, except in times of great depression or expansion of trade, fluctuations in demand are met rather by working more or fewer days per week, than by engagement of more or fewer men.

Two references from the Children's Employments Commission in the deeply depressed year of 1842 point to the prevalence of cyclical irregularity then. The first is the statement that in Durham only seven or eight days a fortnight were being worked at times owing to the inability of the employer to provide more work. The second refers to the opposite condition, and is in the report on Yorkshire: 'When there is an excess of demand, the men often work what is virtually overtime.'[139] We may be sure that the former practice recurred, to a greater or lesser extent, in each trade cycle, though the latter seems to have died out before the great boom of 1873 – to judge by the widespread talk of absenteeism at that time, which was scarcely, if ever, denied, though sometimes mitigated.[140] The first statistical evidence, other than for individual collieries, relates only to the North-East, and begins only in the early 70s. Anthony Slaven, however, has attempted to construct a series of average earnings at the coalface for Scotland, beginning in 1856, and this involved him in making estimates of the number of shifts worked per week. These estimates, of course, include the effects of absenteeism and deliberate restriction of output by the men, as well as closure of the pits by employers, and are described by Slaven himself as 'dubious'.[141] Nevertheless, as an indicator of the type of fluctuations which occurred, they are worth giving for the period up to 1889, whereafter a conventional figure of four and a half shifts is used up to 1898, and five shifts subsequently. They are shown in table 5.16. Very little reliance can be placed in the figures themselves, but they do make clear the presence of substantial fluctuations in work available – that is, the existence of quite severe short-time working, whether through the actions of the owners or through trade union restriction[142] in every decade prior to that for which country-wide statistical evidence is first available. But before that there are also series provided by the coal owners' associations of Durham and of Northumberland, which were given in evidence to the Select Committee of 1873, the Royal Commission on Depression of Trade and Industry of 1886, and the Royal Commission

Table 5.16 *Estimated average shifts worked per week in Scottish coalfields, 1856–89*

1856	4½	1862	4	1868	3	1874	4½	1880	4	1885	4½
1857	4½	1863	4	1869	4	1875	4	1881	4	1886	4
1858	5	1864	4	1870	4½	1876	4	1882	4	1887	4
1859	4	1865	4	1871	4	1877	3	1883	4½	1888	4
1860	4	1866	5	1872	4½	1878	3½	1884	4½	1889	4½
1861	4	1867	4½	1873	4½	1879	3½				

Source: Slaven (1966), appendix A, table 20.

on Labour. These are shown as Part I of table 5.17, the second part of which gives the *Labour Gazette* series beginning in 1894/5. These statistics are even more illuminating if expressed in terms of the average number of weekdays in the year on which the pits were closed, as has been done in table 5.18. Two things stand out from this table: the wide variation between coalfields, and the clear cyclical pattern. The differences between coalfields result mainly from two things: (1) variations in the length of the normal working week, which was itself complicated by the spread of multiple shift systems, especially after 1908; and (2) the seasonal factors already referred to. A third factor seems to have had some significance in Scotland, namely occasional shortage of wagons in wintertime, leading to a clogging of the transport system.[143]

The cyclical pattern, whilst apparent throughout the period, varied somewhat in intensity both between districts and over time, and seems to undergo a particularly large change in this respect in the last few years before 1914. The amplitude of the fluctuations for the period from the mid-1890s onwards is considerably greater for the inland coalfields, and even to some extent for South Wales (though at a much lower absolute level of days lost) than it is for Scotland, and most especially for Durham and Northumberland. The change from one year to the next only exceeded four days in the North-East on three or four occasions, and the average change of between four and five days per year was greatly influenced by 1895–7. Moreover, apart from that early period, the range of fluctuation between peak and trough was around 25 to 33%. In West Scotland the average change from year to year was five days, in the Lothians six days, and in Fife seven days, with a range between peak and trough of about 50 to 60%. In the East Midlands and Staffordshire, at the other extreme, the average change from year to year was between thirteen and sixteen days, and the range of fluctuation up to 1907 was about 200 to 300%. Cumberland, Gloucestershire and Somerset, Yorkshire, Lancashire, and North and South Wales came somewhere in between, though nearer to the stability of the North-East than to the volatility of

Table 5.17 *Average number of days on which coal was wound, 1871–1913*

Part I *Days per fortnight*

	Northumberland	Durham		Northumberland	Durham		Northumberland	Durham		Northumberland	Durham
1871	10.3 (1871–2)	9.2	1876	8.34	…	1881	9.30	10.33	1886	…	9.85
1872		9.1	1877	7.35	…	1882	9.64	10.61	1887	…	10.14
1873	9.58	…	1878	…	…	1883	9.84	10.63	1888	…	10.39
1874	8.89	…	1879	8.79	…	1884	9.33	10.12	1889	…	10.73
1875	8.67	…	1880	8.96	…	1885	8.66	9.83	1890	…	10.56

Part II *Days per week*

	Cumberland	Northumberland	Durham	Yorkshire	Lancashire	North Wales	Derbyshire	Nottinghamshire & Leicestershire	Staffordshire	Shropshire, Worcestershire & Warwickshire	Gloucestershire & Somerset	South Wales	Fife	Lothians	West Scotland
1894				5.03	5.10										
1895	5.18	4.75[a]	5.12	4.50	4.68	4.35	4.28	4.11	4.28	4.53	4.38	5.25	5.08	5.32	5.06
1896	5.12	4.79	5.23	4.78	4.90	4.75	4.38	3.95	4.65	4.83	4.51	5.32	4.92	5.41	5.21
1897	5.40	5.09	5.36	5.12	5.03	5.12	4.81	4.11	4.96	5.09	4.88	5.47	5.20	5.60	5.37
1898	5.66	5.18	5.45	5.26	5.23	5.52	5.07	4.52	5.04	5.31	5.47	…	5.25	5.55	5.32
1899	5.67	5.32	5.53	5.55	5.43	5.59	5.39	4.84	5.39	5.46	5.55	5.75	5.25	5.55	5.23
1900	5.58	5.35	5.53	5.57	5.42	5.59	5.53	5.04	5.39	5.46	5.58	5.61	5.29	5.63	5.33
1901	5.30	5.20	5.33	4.99	4.95	5.30	5.04	4.60	4.84	5.18	5.27	5.31	5.05	5.45	5.14
1902	5.42	5.25	5.38	5.00	5.02	5.30	5.12	4.77	5.07	5.18	5.22	5.64	5.22	5.49	5.17
1903	5.34	5.21	5.35	4.95	4.80	5.29	4.71	4.39	4.77	4.91	4.97	5.58	5.22	5.45	5.16
1904	5.20	5.20	5.29	4.91	4.77	4.99	4.52	4.28	4.78	4.98	4.99	5.61	5.05	5.54	5.05
1905	4.96	5.14	5.34	4.97	4.78	4.88	4.52	4.19	4.57	4.81	4.75	5.36	5.14	5.43	5.07
1906	5.39	5.33	5.43	5.32	4.97	5.40	4.74	4.52	4.84	4.88	4.91	5.63	5.40	5.58	5.21
1907	5.56	5.39	5.49	5.64	5.40	5.67	5.37	5.12	5.47	5.45	5.40	5.80	5.35	5.45	5.18
1908	5.24	5.29	5.35	5.26	4.84	5.17	4.99	4.65	5.05	5.38	4.98	5.58	5.05	5.04	4.87
1909	5.23	5.17	5.30	5.14	4.85	5.27	4.82	4.58	4.94	4.99	4.93	5.54	4.73	4.85	4.89
1910	5.44	5.21	5.29	5.31	4.86	5.39	5.00	4.76	5.16	5.12	4.89	5.72	4.84	5.01	4.99
1911	5.35	5.26	5.33	5.36	5.07	5.52	4.92	4.67	5.02	4.93	4.88	5.57	5.14	5.21	5.08
1912[b]	5.69	5.33	5.39	5.67	5.43	5.75	5.39	5.15	5.45	5.30	5.23	5.82	5.20	5.33	5.10
1913	5.70	5.40	5.48	5.75	5.42	5.80	5.55	5.29	5.49	5.52	5.37	5.92	5.29	5.26	5.09

the Midlands only in the case of the first and the last. The average change from year to year was nine days in Cumberland and South Wales, ten in Yorkshire and Lancashire, and eleven in Gloucestershire and Somerset and in North Wales. And whilst the range of fluctuation from peak to trough up to 1907 in the inland coalfields in this group was almost as high as in the Midlands (actually being higher in Gloucestershire and Somerset), in South Wales it was only above 100% in the movement to 1907 itself, and in Cumberland it did not much exceed that figure.

The partial series for Northumberland and Durham for the 1870s and 80s throw a little light on the period of the so-called Great Depression.[144] They suggest that the slumps of the late 70s and the middle 80s both had far greater effect on the working of the pits than did any after 1900 – a suggestion which the figures for various inland coalfields for 1894–6 indicate can be extended to include the slump of the middle 90s. In effect, there was a good deal more short-time working in these earlier depressions than later, so that whatever the defects and ambiguities of labels like 'Great Depression' and 'Edwardian Recovery', one can see how they might have come to be accepted by at any rate some of the working class. An examination of the files of the *Colliery Guardian* confirms this, for references to short-time working are abundant between 1876 and 1880, 1884 and 1886, and 1894 and 1896, but much fewer subsequently.

Whilst the number of days lost in many coalfields in the period after 1907 behaved in much the same fashion as it had done since the 1890s, in some there seems to have been a change in the pattern, which is probably best ascribed to the development of forms of multiple shift working in the wake of the Eight Hours Day Act. The level of days lost in South Wales in 1908–11 was about 20% less than in the period 1901–5, which was roughly the same stage of the previous trade cycle; and in the peak year of 1913 the almost incredible figure of only five days lost was achieved. This is so low as to imply seven-day week working in a sizeable proportion of collieries. The lowest previous figure had been eleven days in 1907; and

Notes to Table 5.17

[a]The inclusion of more collieries as from December probably reduced the figures for subsequent years compared with earlier ones. The reduction in December was from 4.82 to 4.55

[b]The strike period is excluded.

Sources: 1871–2 – SC 1873, X, Q. 3034 and 3401–6; Northumberland 1873–85 – *SP* 1886 XXIII, Q. 11, 688; Durham 1881–90 – *SP* 1892 XXXVI, Part III, p. 1004; 1894–1913– *Labour Gazette*.

The figures were arrived at by taking a yearly mean of monthly averages. Except for the very early years, these monthly figures were arrived at by taking the number of days on which coal was wound in a selected fortnight in each month, and dividing by two. In the first few years – it is not clear when exactly the method was changed, but it was certainly before 1906 – the actual number of days on which coal was wound in any month was divided by four.

The number of miners covered by the returns on which these figures are based varied from about a half to two-thirds. The *Labour Gazette* commented on this in September 1897 (p. 260): 'The gradual extension of the area covered by the returns will, no doubt, in some districts affect the value of the comparison of the average days worked, but for the larger districts and the country as a whole it has been ascertained that the value of the comparison has not been materially affected.'

Table 5.18 *Average number of days lost through closure of pits, 1871–1913 (assuming six days as normal week)*

	Northumberland	Durham		Northumberland	Durham		Northumberland	Durham		Northumberland	Durham
1871	45 (1871–72)	71	1876	97	…	1881	71	44	1886	…	57
1872		76	1877	122	…	1882	62	37	1887	…	49
1873	64	…	1878	…	…	1883	57	37	1888	…	44
1874	82	…	1879	84	…	1884	71	51	1889	…	34
1875	88	…	1880	81	…	1885	88	57	1890	…	38

	Cumber-land	Northum-berland	Durham	Yorkshire	Lanca-shire	North Wales	Derby-shire	Notting-hamshire & Leicester-shire	Stafford-shire	Shropshire, Worcestershire & Warwickshire	Glouces-tershire & Somerset	South Wales	Fife	Lothians	West Scotland
1894				51	48										
1895	44	66[a]	47	79	72	87	90	99	90	77	85	40	49	36	50
1896	48	65	42	65	59	67	86	109	72	63	79	37	58	33	43
1897	25	48	34	47	51	47	63	99	55	48	59	31	43	22	34
1898	19	44	30	39	41	26	49	78	51	37	29	…	40	24	36
1899	18	36	25	24	31	22	33	61	33	29	9	24	40	24	41
1900	24	36	26	24	32	23	26	52	34	30	9	22	39	20	37
1901	37	43	36	54	56	37	51	74	61	44	25	37	50	30	46
1902	31	40	33	53	52	37	47	65	49	44	27	20	42	28	44
1903	35	42	35	56	63	38	68	85	65	58	41	23	42	30	45
1904	44	44	39	59	66	55	79	91	65	55	41	22	50	26	50
1905	55	46	35	45	64	59	78	105	75	63	53	34	46	31	49
1906	33	36	31	36	45	32	66	78	61	59	44	20	32	23	42
1907	24	33	28	20	32	18	34	47	29	30	32	11	35	30	44
1908	42	38	36	40	62	45	55	72	51	34	55	24	50	51	61
1909	41	44	37	46	61	39	62	75	56	54	57	25	67	61	59
1910	30	42	38	37	60	33	53	65	45	47	59	16	61	52	54
1911	35	39	36	34	49	26	57	70	52	57	59	23	46	42	49
1912	17	37	34	19	32	15	34	46	31	38	42	11	44	37	49
1913	17	32	28	14	31	11	24	38	28	26	32	5	38	31	48

Major strike losses are excluded.
[a]See note a to table 5.17.

at the previous peak it had been practically double that. In Cumberland, Yorkshire, the East Midlands, and North Wales there was a similar, if less steep, progression from cyclical peak to peak, and a similar, or even shallower, trough in 1908–11 compared with 1901–5. In other coalfields, however, the slump of the late 1900s, though not perhaps as deep as that of the early part of the decade, had a roughly similar impact on the number of days lost.

The old double-shift system of the North-East, developed mainly in the 1830s to 1850s with the coming of shotfiring, as we have seen, was not disrupted by the Eight Hours Day Act nearly so much as had been feared. Instead, it was developed into a three-shift system for hewers, with two shifts of haulage workers, and by 1914, according to Rowe, 'very few pits in Durham were not working three shifts, but the tendency never operated to quite the same extent in Northumberland'.[145] However, the eleven-day fortnight seems to have continued to be the general practice, and there does not seem to have been any Sunday working, maintenance men apart. And since, within these constraints, there had been virtual full-time working at practically all times since the middle 1890s – the only significant exceptions being the landsale collieries of Northumberland – there was little scope for improvement after 1908. Elsewhere, double-shift systems had not been at all common prior to the Eight Hours Day Act, except perhaps in Yorkshire, though they were to be found at a few collieries in most coalfields. The usual system throughout the inland coalfields, and in South Wales and Scotland, did, in fact, involve two or more shifts, and some coal was normally got during more than one of them. But this was not a genuine double-shift system, for there was only one coal winding shift, the others, beginning late in the afternoon or at night, being used mainly for repair work and ripping, with enough coal being got to fill the tubs so that winding could begin promptly next day.[146] Just how big the balance in favour of only one major coal-getting shift was in 1905 is brought out in table 5.19. Unfortunately, similar data do not exist for later in our period, but it is clear that there was little change in many districts, Lancashire and Scotland being the most important of these. In others, however, owners began pressing for more and more double-shift working, though trade unions were generally hostile.[147] This pressure was quite often successful, and a South Yorkshire owner F.P. Rhodes's statement in 1915 that the Eight Hours Act had driven most South Yorkshire collieries into double-shift working,[148] applies with only a little less force to the large collieries in the East Midlands. There was also some such move in Cumberland and North Wales. In South Wales, the double-shift system proper was not much extended,[149] but some forms of overlapping shifts and Saturday afternoon shifts were widely adopted instead. Though the trade union leaders, still chiefly influenced by W. Abraham

Table 5.19 *Percentages of underground workers according to shift systems, 1905*

	Percentage at collieries			Percentage at all collieries		
	With one coal-getting shift	With two coal-getting shifts	With three or more coal-getting shifts	On 1st shift	On 2nd shift	On 3rd shift
Cumberland & Northumberland	2	36	62	65	23	12
Durham	2	34	65	60	26	14
Yorkshire	7	28	64	69	20	11
Lancashire & Cheshire	16	73	11	80	19	1
North Wales	45	41	14	76	17	7
Midland Inspection Division	42	39	19	86	10	3
Staffordshire Inspection Division	38	33	28	82	14	4
Gloucestershire & Somerset	45	49	5	84	15	—
South Wales	35	56	1	83	16	—
East Scotland	35	44	21	84	13	2
West Scotland	43	49	8	88	10	1

Source: based on appendix 1 to first *Report* of DC 1907.

(known almost universally as 'Mabon'), accepted that these were the best terms possible if earnings were to be kept up in the wake of the Eight Hours Day Act, there was much resentment amongst the general body of miners, and this contributed to the severe deterioration in labour relations in South Wales collieries at this time.[150]

It only remains to take account of the phenomenon which came to be known as absenteeism, a word 'which often carries a certain delinquent connotation'[151] – equally often quite wrongly. What it means in neutral language is the absence of miners from shifts which they could, in theory, have worked. This can be measured in different ways, but the most satisfactory one from the point of view of measuring labour input is to take the difference between the number of possible shifts that could have been worked, assuming all men on the books were present when the colliery was open, and the actual number of shifts which were worked. This measure includes both voluntary and involuntary absences – that is, absence through sickness, etc., and through unauthorised holidays – though one must point out that the use of the word 'unauthorised' is scarcely accurate, since it was never part of the accepted area of the owner's authority that he should be able to compel attendance. A different measure of absenteeism, which contrasted the average actual number of shifts worked in a period with the maximum ever actually worked during the same time of the year, was sometimes used, in an endeavour to eliminate involuntary absenteeism, or at least the greater part of it.

A good deal of information on absenteeism – we shall continue to use the word, though emphatically not implying any moral meaning thereby – was collected from the time of the Royal Commission on Labour onwards, and there is quite a lot of scattered evidence for earlier periods. It is possible, therefore, to establish a reasonably close approximation to the facts, at least for the half-century before 1914. However, it is well to be aware of the difficulties of explaining what is a very complex phenomenon. Indeed, F.D.K. Liddell and others, who conducted a lengthy enquiry into colliery absenteeism in the years soon after the Second World War, had to admit in the end that 'because of extreme difficulties inherent in the data, many of our enquiries did not produce positive results; those that were achieved were not on the whole surprising'.[152]

Perhaps the first thing to make clear is that, despite the puritanical views of a few employers, and of such men as Seymour Tremenheere, the first Chief Inspector of Mines, who said in 1850 that 'colliers cannot be induced to do a full week's work, or to abstain from their frequent habit of not working at all on the most frivolous pretences',[153] and in spite also of the connotations of delinquency which absenteeism acquired (though that did not become so marked until the First World War), coalowners in general did not expect their men to work every possible shift. This does not mean that they did not try to induce more regular working by their men – the bonds of the North-East are evidence that they did in that coalfield, and there were fines for absence in other districts too.[154] And, from a later period, there is the expressive complaint of an elderly coalowner from Lancashire, George Caldwell, who had been forty-four years in the industry in 1890, when he gave evidence to the Royal Commission on Labour: 'I can remember when colliers were the same as the employés in other trades; if they did not attend to their work they were told to go about their business; but we cannot enforce that discipline now.'[155] Not true, of course, but indicative of an attitude, and suggestive that there were some times when owners could and did enforce sanctions against absenteeism. However, even at those times, it is unlikely that many employers viewed one or two shifts missed per fortnight with much disfavour. Some would not even have been able to accommodate a 100% attendance of their workforce.[156] Perhaps the position was put best by J.S. Dixon, of the famous West of Scotland iron and coal family, in his evidence to the Eight Hours Day Committee, when he said: 'There is no hard and fast rule that a miner must work so many days a week. A miner takes an idle day whenever he wishes, but, of course, if he persistently does that, the manager may dismiss him.'[157] Notice that the phrase is that 'the manager *may* dismiss him'. Whether or not he did would depend mainly on the state of trade and the availability of men to replace him. In periods when demand was brisk and labour relatively scarce, managers felt that they had to tolerate more irregular working, and there is ample evidence to suggest that these were the times when absenteeism reached its highest levels. But the standard explanation for this is less the enforced tolerance of managers than the higher wage rates of such boom periods, and the greater consequent ease of earning an accustomed real income in a shorter time at work. Dixon, again, put this view – pretty well universal amongst employers – very effectively: 'in bad times a man loses part of his wages through enforced idle days when the pits are not working, and then he is anxious to get to work again, and in good times he knows he can get work any day, and he is more anxious to take holidays'.[158] There is, of course, a lot of truth in this explanation, which was not held only by coalowners but also by inspectors of mines, and even by trade union leaders, such as John Normansell of South Yorkshire and Keir Hardie.[159] But, though a supply schedule for existing labour that was inelastic with respect to earnings may be important in explaining fluctuations in absenteeism, it is surely not the only cause; and, in any case, there is merit in trying to go behind it a little.

The reasons which could lead a man to miss a shift are many and various. One of the most obvious is illness or the effects of injury, but this could be expected to show little variation over time, even in the long run. Its importance was variously estimated, but the records of the Lancashire and Cheshire Miners' Permanent Relief Society for 1913/14 give what is probably the most reliable figure, namely 1.15% of shifts over the year from July to June.[160] There were other causes of absence resulting from family affairs which were accepted as unavoidable, but the majority resulted from the miner's choice. Mining, as was pointed out at the beginning of this chapter, was very severe labour for most workers in the days before machinery was used for much beyond the main roadways. It is not surprising, therefore, that most miners felt that a fair week's work was something less than six full shifts, or even five. Some men, particularly older family men, paced themselves to work regularly for the full week, though most hewers probably reckoned on taking off one shift per fortnight at the least. But others, probably without consciously thinking about it, worked more rapidly than they could sustain for an indefinite number of consecutive shifts, and when they both felt like it and could afford it, took a break. One of the Mines Inspectors said quite firmly that 'a pikeman [the Black Country name for a hewer] could not go every day and work at his usual speed; it would be an impossibility'.[161] Maybe he exaggerated a bit; but perhaps the majority of the younger hewers behaved in this latter fashion, and some never gave it up as they became older. Amongst these were those who became involved in what Robert Moore has called 'the vicious circle of drink, gaming, debt, domestic disorder, and poor health',[162] which reinforced irregular attendance and eventually weakened the ability to make up for it by bursts of rapid working. In Anthony Slaven's study of the Dixon enterprises, there is an illuminating analysis of part of the Govan Colliery paybook for 1857, showing the attendance record over the twenty-six fortnightly pay periods of the ten hewers working in the Mossdale seam west side at Number Two pit.[163] Of 300 possible shifts, the number missed by each individual ranged from twelve to eighty-six, the mean being thirty-seven. Two men missed not a single shift in nineteen out of the twenty-six pay periods, but the worst man in respect of attendance had only one absence-free fortnight, and two others had only three or four such fortnights. The ten men fall quite clearly into two groups, with one odd man out. One group of six each missed less than a tenth of possible shifts, seldom more than one per fortnight. Amongst these six are the five top earners during the year, but the sixth had lower earnings than all but one man from the other group. That other group of three missed between 20 and 30% of possible shifts, and on one or more occasions missed at least ten shifts in a fortnight. One of these three missed forty-seven out of forty-eight shifts between mid-May and mid-July, having previously had a reasonably good attendance record, to which he had returned by November. Probably he was ill or suffered an injury. The same may have happened to one of the others, who missed twenty-three out of twenty-four shifts in February and March. But the third man had a much more regular record of irregularity, normally working eight or nine shifts a fortnight, occasionally less or more; yet despite this, his annual earnings were higher than those of four other men. The proportions of the different kinds of absence behaviour shown in Slaven's study are almost certainly unrepresentative of the industry as a whole, for Govan was an old-established colliery with a stable workforce and a tradition of good industrial relations, which was produced and reinforced by a sort of mutual attraction between the 'good' employer and the 'good' worker. Nevertheless, even at Govan, in this one small working face, a fair range of behaviour is displayed.

The ability of the responsible family man to earn what he regarded as a reasonable living in a shorter time than usual, when wage rates went up in a boom, may have led to some decrease in regularity of attendance by that class of worker. As Baron Duckham wrote about the Scottish miners in the eighteenth century: 'High wages did not so much make the pitman "idle" (as contemporary owners insisted); they presented him with a wider margin of choice between the rewards of money and the rewards of leisure. Mining was sufficiently unpleasant and the collective psychology of the pit village sufficiently strong for leisure (or "idleness") to be fre-

Table 5.20 *Percentages of possible shifts lost by hewers through absences, 1873–1914*

	Northumber-land	Durham		Northumber-land	Durham		Northumber-land	Durham
1873	10.4	...	1887	...	7.2	1901	...	11.6
1874	9.9	...	1888	...	7.6	1902	...	10.1
1875	10.1	...	1889	...	8.9	1903	11.0	9.5
1876	6.8	...	1890	10.6	11.5	1904	10.4	8.6
1877	...	...	1891	...	11.4	1905	10.0	8.3
1878	...	...	1892	...	8.7	1906	11.2	9.1
1879	5.2	4.0	1893	...	7.6	1907	...	10.2
1880	5.5	11.5	1894	...	7.8	1908	...	8.4
1881	8.9	8.1	1895	...	7.8	1909	...	7.2
1882	7.0	9.0	1896	...	7.3	1910	...	6.9
1883	7.4	8.8	1897	...	7.7	1911	...	6.8
1884	7.2	8.7	1898	...	8.8	1912	...	5.9
1885	6.2	7.8	1899	...	9.9	1913	...	7.2
1886	...	6.8	1900	12.8	11.7	1914	...	6.7

Sources: CIC, appendix 23; *SP* 1886 XXIV, Q. 11, 688; and DC 1907, Q. 10, 650.

quently chosen.'[164] But there is probably truth in the commonly held view that those who were normally irregular attenders were responsible for as much or more of the increase in absenteeism which accompanied each cyclical upswing.[165] That this occurred on virtually all occasions is clear from a variety of sources, and cannot be an illusion based on the prejudice of owners and others. The Durham Coal Owners' Association kept records of the proportions of possible shifts missed from 1879 onwards, based on the total number of men on the collieries' books, which were provided for the Sankey Commission in 1919. The Northumberland owners gave a similar series, though their's included men off sick, whilst the Durham figures did not, for the period 1873–85 to the Royal Commission on Depression in 1886, and some later figures to the Eight Hours Day Committee. These reveal the fluctuations very clearly.

There is no similar lengthy series for any other district, but what scattered statistical information there is – much of which is of little use because it confuses days lost through pit closure and through absence by the men – does not indicate the likelihood of any different pattern.[166] There were, however, quite considerable differences between the levels of absenteeism in different coalfields at all times for which comparable data are available. These are shown in table 5.21, which must be interpreted with great caution. Comparisons between 1899 and 1905 can be made, because the data were collected on the same basis, namely that the greatest number working on any one day in certain specified weeks was taken to be the maximum available for work in practice, and the difference between this and the average numbers actually at work was taken as the absenteeism percentage.[167] However, direct comparisons between

Table 5.21 *Absentee percentages, 1899, 1905, and 1913/14*

	Hewers		All underground		All workers
	1899	1905	1899	1905	July 1913–June 1914
Northumberland, North Durham, Cumberland	6.8	6.3	5.2	4.6	10.2[a]
South Durham	5.4	4.9	4.3	3.9	10.6[b]
Yorkshire	9.1	10.3	8.3	9.5	13.6
Lancashire	10.1	14.5	9.0	13.0	10.8 (Lancashire and North Wales)
North Wales	6.4	14.2	5.3	11.3	
East Midlands[c]	7.3	8.0	6.7	7.5	10.3
West Midlands[c]	11.1	10.1	9.3	9.1	12.2
Rest of England[d]			3.9	7.1	9.0
South Wales	5.4	5.5	5.5	5.6	10.8
East Scotland	6.1	6.7	5.0	5.8	6.8
West Scotland	5.6	7.3	4.7	6.2	6.1
UK (weighted average)	7.0	8.0	6.1	7.1	10.6

[a]Northumberland only.
[b]The whole of Durham.
[b]Warwickshire is included with the East Midlands in 1899 and 1905, and with the West Midlands in 1913/14.
[d]Includes Cumberland in 1913/14.
Sources: appendices to first *Report* of DC 1907 and to CIC.

these two years and 1913/14 are illegitimate, because the Sankey Commission's statistics were based on the total numbers on colliery books. (Quite apart from this, the districts are differently constituted in several cases.) The 1913/14 data do, nevertheless, suggest that, compared with earlier years, there was much closer similarity between the various coalfields, with only Scotland standing out as different. The lower rate of absenteeism there is probably to be associated with the existence of the five-day working week in Lanarkshire after 1900. In 1899 and 1905 there had been wider differences between coalfields, with Durham and Northumberland and South Wales, as well as Scotland, being below average, and Lancashire, Yorkshire, and the West Midlands being districts with relatively high absenteeism, as later. The explanation usually offered for the higher absence rates of these areas was that many collieries there were much more closely integrated into normal town life than in the other coalfields, and that absence was lower in rural than in urban collieries[168] – partly, it was often said, because where there were job opportunities for women, 'the men waste more shifts when they are not entirely dependent on their own earnings'.[169]

The changes in the absence rates between 1899 and 1905 are a shade puzzling in some cases, as is the movement upwards of the overall United Kingdom average, for 1899 was a year of boom in the coal trade, with wages rising, whilst 1905 was the worst year of the admittedly not very deep depression of the early 1900s. One would certainly have expected the behaviour of absenteeism in Scotland and the North-East to have been typical of the difference between the two years. However, 1905 showed a higher percentage rather than a lower one in every other district except the West Midlands, though the change was small everywhere except in Lancashire, North Wales, and the small coalfields of Gloucestershire and Somerset. Perhaps the explanation is to be found in changing attitudes to the relatively long hours still worked in these areas, though they were at least as long in South Wales, where the absence ratio was stable. Some support is lent to this argument by the fact that, despite difficulties of comparison between 1905 and 1913/14, it is clear that absenteeism fell quite markedly in Lancashire and North Wales, the districts which, along with South Wales, had their hours most reduced by the Eight Hours Day Act. That there was not a similar fall in the absenteeism rate in South Wales may perhaps be put down to the greater deterioration in industrial relations there, and possibly also to the greater impact of the Minimum Wages Act.

Having completed our analysis of the quantity of labour employed in the coal industry and of its regularity of employment, we now turn briefly to the quality of the men employed. One aspect of this can be dealt with quickly, for it would be unreasonable to expect that one generation was inherently more skilled and fit for its job than another, unless there is clear evidence of a change in overall standards of health, or in the standards of skill or of receptivity to the acquisition of skills. There is certainly nothing to suggest much change in health standards, at any rate before the last few years of our period, when adult death rates in the colliery districts did just begin to go down. Standards of skill required by the technology of the day were little changed from beginning to end of the period so far as the majority of miners is concerned, for the pick and shovel remained the basic tools for most, even if mechanisation was applied to much – though far from all – of the conveyance of the coal. As to receptivity to the acquisition of skills, there is no doubt that literacy rates went up rapidly in the last decades of the nineteenth century; but one may doubt whether this had any significance at all for the skills required by the great majority of miners. Possibly deputies and overmen were marginally more effective when they were literate, though one may suppose that most of that class had been so throughout the period.[170] But it is hard to see what impact literacy – or even the greater flexibility of mind which is supposed to come with literacy – could have on the ability to undercut, nick, wedge, fill, or putt. It is possible that where coal-cutting machinery was introduced, there was some temporary fall in skill, for it would take each set of men some time to acquire their old level of skill with the new tool. It is also possible that this technology did require some rather higher level of book-learning than previously. But such changes, as well as being unmeasurable, must have been of such minor influence on output and productivity as to be completely negligible.

On the other hand, it is at least possible that there was some change over our period in the efficiency with which the miner exercised his skill. It could be argued, for instance, that the decline of the old-time habits of hard drinking, which took place in the last quarter of the nineteenth century, improved the miner's efficiency.[171] So, possibly, did an improvement in nutritional standards which may have occurred at the same period. But probably far more important in this respect were certain technical inventions mentioned in the last chapter, notably the electric light and mechanical ventilation. The effect of these inventions in improving the personal efficiency of miners cannot be distinguished from any other effects they had on productivity, but it is likely that it was the most important one.

The question of the intensity with which miners worked received rather more contemporary notice than did their skill or efficiency. It was common for coalowners – and others – to argue that the modern workman did not work so hard as his forebears, despite his shorter hours of work. This is, of course, a perennial cry of all kinds of employers everywhere, often backed up by older workers. The argument was most often heard during periods of boom, and it usually indicated a confusion between absenteeism and the actual intensity of work down the pit.[172] When this distinction is clearly made, it appears more likely that the intensity of work was at least maintained as the hours fell. One of the mines inspectors, L. Brough of the South-Western Division, was surely intending to say this when he told the Select Committee of 1873: 'Men do not work worse underground, there is an *esprit de corps* with men, a sort of rivalry, and they will do their work well. I cannot believe in statements that the men are doing their work badly, but they are doing it in less degree.'[173] A Yorkshire owner, Robert Tennant, was more explicit: 'I believe that they have worked better during the time that they have worked, but that they have worked less hours.' The trade union leader from his district, J. Dixon, agreed with him.[174] His South Yorkshire colleague, J. Normansell, seemed rather to take the other side, however. He noted that

> it is said that the colliers do not work so hard as they used to do; and I do not think they intend doing that. They do not work as hard as they used to do, but they work, in my opinion, quite hard enough. They worked more like fiends, in my opinion, in former days, and they intend taking it a little easier.[175]

Perhaps he was referring to hours and absenteeism, rather than to intensity of work. Even if he was not, and if the intensity of work had gone down during the boom, there is little doubt that it picked up again in the later 1870s and the 80s, when wage rates fell, short-time working was more frequent, and even unemployment was not unknown as the size of the labour force declined from 1874 or 1875 to 1879. This period of anxiety and need for exceptionally hard work to maintain earnings lasted with little let-up until 1889, and it may be that it was the reaction from it that enabled a mining engineer to speak in the early 1890s of 'the natural and growing tendency on the part of the men to do less work, and that of a less laborious nature, than was done by their father'.[176] If there is, indeed, any truth in this statement, it is likely to apply only to the recent past of the late 70s and 80s, not to longer ago.[177]

6

WAGE BARGAINING

Before we take up the question of the cost of the labour input – that is, wage rates and earnings – it is necessary to look at the process of wage bargaining and at the institutional changes which took place during our period.

In the eighteenth century, except in those parts of Scotland where serfdom still persisted, the most usual form of wages agreement was the yearly hiring at a rate fixed in advance and normally unchanged from year to year. Sometimes slightly shorter periods than a year were adopted in order to prevent outsiders gaining a settlement for Poor Law purposes, but the yearly bond was certainly the most common form of hiring. That this was true at the end of the eighteenth century may have been largely due to the predominance of the Durham and Northumberland coalfield, where the yearly bond was still almost universal. But earlier, as Ashton and Sykes show, it was also normal in other coalfields.[1] By the beginning of the nineteenth century, however, although the bond had expanded into Scotland to take the place of serfdom, it was in decline or had disappeared south of the Tees. Even in the North-East there were signs that it was being eroded.[2] Nevertheless, it remained the main basis of hiring there until 1844, and it persisted in Durham for nearly three decades more. It also continued in Scotland until some time in the 1830s or 40s.[3]

In those areas where there were other industries close at hand, such as Yorkshire, Lancashire, and the West Midlands, there is no evidence that the yearly bond had any significance by as early as 1800.[4] And the latest reference to it which can be found for South Wales is in 1817.[5] Moreover, the East Midlands, where sub-contracting was common, as in the West Midlands, shows no sign of annual hirings in the nineteenth century. It seems fair to conclude, therefore, that it was only in the North-East and parts of Scotland that any considerable importance attached to the long hiring period in the nineteenth century.

Whilst the wage bargain, whether for long or short period, was in theory a wholly individual one between employer, or his agent – hereafter called the manager – and worker until a long way through the nineteenth century, in practice negotiations took place between the manager and groups of workers; or, when the groups were too large for face-to-face

BENTINCK COLLIERY.

PRICE LIST FOR GETTING THE DEEP SOFT COAL.

1.—Coal getting, including making gate, 3 ft. ripping, 20 cwts. per ton, 3/1.87.

2.—Coal to be filled with $1\frac{1}{4}$ in. forks, clean and free from bats and bind. An allowance of 8 lbs. per cwt. to be made for slack ; if this is exceeded, twice the weight of the excess to be forfeited.

3.—Pit bank machine weight to be paid for, except deductions as per Clause 2.

4.—One ton of House Coal to be allowed to every underground workman (being a married householder) or to a widow with one son or with two lodgers working for the Company, for every 24 shifts made in winter and 28 in summer. Men to have choice of leading. Coal to be supplied as it comes out of pit at 2/6 per ton.

5.—Main Gates for inclines and levels, which are made by stallmen, 10 ft. wide by 7 ft. high, extra per yard $4/7\frac{1}{2}$. Ripping main cross-gates going "awn" coal, putting in or cutting off stalls, 8/3 per yard.

6.—Ordinary stall-gates to be maintained 6 yards from face, 9 ft. wide and 3 ft. ripping, or if by reason of topholing, to be made $6\frac{1}{2}$ ft. clear in height.

7.—Leading or short stalls with fast ends, if under 24 yards (unless so made by stallmen) to be paid 4/2.55 per yard on gate.

8.—Stalls broken down through no fault of the men, side loose heading allowed 3/- per yard.

9.—New stall crossings put in by stallmen (brattice built by Co.). 9/3 each.

10.—Stallmen required to work on main roads 7/6 per day.

11.—Coal heading, with coal at current prices, 5/3 per yard. This does not apply to side loose heading or short heads in stalls where the coal has to be filled with screens in the usual way.

12.—Stalls with faults, throws, or steps, to be allowed $1\frac{1}{2}$d. per inch per yard, and where fault is full thickness of seam, heading price to be paid as per Clause 11, and in case of bad coal or difficulties Clause 18 to apply.

13.—Allowance for dinting and getting up floor in stalls where necessary, at the rate of per foot thick per yard 1/6, and stalls opening out and getting up floor for the first time to be paid also. This only applies when ordered by the Officials, and not until 6 inches in thickness is reached.

14.—Packing out airways, if left clean and packed both sides, per yd. 3/9, one side only $1/10\frac{1}{2}$. This only applies to gob and not coal side packs.

15.—Opening out new stalls with fast ends, if under 20 yards in length, to be paid for extra cutting and packing 3/- per yard.

16.—Stalls with deepening face and gate and dipping $2\frac{1}{2}$ in. to the yard to be paid extra $1\frac{1}{2}$d. per ton.

17.—Setting bars in gate roads :—6 to 7 ft. in length, 2/- 7 to 8 ft. in length, 2/3 ; 8 ft. and upwards, 2/9.

18.—In case of difficulties arising from bunkies, bad coal, bad roof, water, or stripping boundaries, and stallman fail to earn a day's wage, the Company to pay 7/6 per day.

19.—Making refuge holes 4 ft. x 4 ft. to be paid 7/-.

20.—The Company's bye-workmen not to use or fetch contractors' tools out of the stalls.

21.—Lamps $1\frac{1}{2}$d. per ton extra.

22.—Where stallmen tram their wagons in gates or heads 30 yards to the turnout $\frac{3}{4}$d. per ton to be allowed, and for every additional 30 yards $\frac{3}{4}$d. per ton.

23.—Packing both sides and ripping new gates, and leaving clean 9 ft. wide by $6\frac{1}{2}$ ft. high 13/6 per yard.

24.—Stallmen, emptying or filling dirt, $4\frac{1}{2}$d. per tram.

25.—The above to be the basis prices on which the percentage ruling in the district shall be paid, and subject to any advance or reduction that may occur.

Revision of DEEP SOFT PRICE LIST in accordance with the new 1911 basis rates.

The alteration in the case of Clauses 10 and 18 are hereby agreed.

Dated 6th January, 1916.

Signed on behalf of the Owners :

P. F. DAY, Manager.
W. SHARPE, Under-Manager.

Signed on behalf of the Workmen :

W. G. HANCOCK, Checkweighman.
A. BEARDSMORE, Stallman.
R. BUTLER, Stallman.
S. COTTERILL, Stallman.
C. BUNFIELD,
Secretary, Notts. Miners Association.

bargaining involving all members, with *ad hoc* deputations. Later, these representatives came to be elected in more or less formal fashion, and eventually they came to include officials from the district trade union. This last stage was not reached anywhere before the 1870s, however, and probably did not cover the majority of workers in every coalfield much before 1900, if then. Certainly the negotiation of the basis rate, the colliery 'price list', was nowhere in the hands of permanent union officials before 1914, for, as David Douglass, a one-time miner, has pointed out with feeling: 'at the same time as relying upon full-time representatives more than other working men at the time' – he is speaking of Durham from the 1870s onwards, but it soon applied to all coalminers – 'miners also distrusted them deeply', and 'the lodges hated the idea of leaving negotiations in the hands of full-time officials'.[6] It was in Durham and Northumberland, however, that union officials had the greatest influence on basis rates, for there the Joint Committee, composed of owners' representatives and union officials, had the power to vary basis rates from time to time so as to bring average earnings at a colliery (or in a particular seam) to within 5% of the county average.

The negotiation of a new price list – or, for that matter, the revision of an old one – was, then, an essentially local matter throughout our period. The initial negotiation, typically, took a matter of weeks, if not months, and, if applying to a major new colliery, could be, as described by H.S. Jevons, 'indeed a most dramatic event in the life of a mining town, the prosperity and character of the town hanging in the balance on the "give and take" of 1d. or ½d. per ton'.[7] He went on to outline the process of negotiation, and we cannot do better than quote him, though remembering that, for dramatic effect, he is alluding to the case of a big new colliery, and that for a small new winning in an existing colliery the negotiations would be altogether a more humble and expeditious affair, though similar in essence.

> The usual practice is for a number of working places to be tried by men paid by the day in order to test the productiveness off the seam; and this system is continued for many months. During the negotiations there are usually at various stages accusations and counter-accusations, the men asserting that the management have opened up only the better places so as to get a high average yield of coal per day per man, whilst the management assert that the men are purposely slacking to keep down the output. If the manager is a sporting man, he will do a day's work in a stall, and show that he can cut twice as much coal from the seam as the average man is sending out; to which the men retort that he has probably chosen an easy place, and in any case, if he liked to exhaust himself any skilled man could double his output for one day.

In establishing a price list:

> the ultimate result depends upon the relative skill in bargaining of the two parties,

and also on their potential staying power in case of a rupture . . . It will be understood, therefore, that very different results have been obtained in different collieries; so that there may be as much as 20 per cent difference in the cutting price of coal for the same seam, under practically identical conditions between one colliery and the next.[8]

Negotiations to fix the basis rates for hewers were the most dramatic and time-consuming of such affairs, but the rates for all other grades of labour had also to be agreed in the same way. Certainly this was true for other pieceworkers, such as putters or hurriers, though the negotiation was not so elaborate, and there was often resort to some conventional sort of relationship between the productivity of the seam (as established by the hewers), the nature and length of the roadways, and the basis rate – the 'price per score'. For day workers – both skilled oncost men and labourers – a conventional relationship to the basis rates of hewers was almost always accepted, though it was allowed to vary gradually over long periods.

The establishment of local basis rates was, thus, a somewhat tortuous business, and it is not surprising that, once fixed, price lists tended to be long-lasting. Once agreed, the list could become, as Jevons put it, 'precious almost beyond imagination to one side or the other', so that 'no alteration can be made without a long and bitter struggle'. Nevertheless, there were changes made in basis rates from time to time, either because one side or the other had 'got it badly wrong', or because conditions in the seam or pit had changed more or less dramatically. Reference has already been made to the powers of the Joint Committees in Durham and Northumberland to vary basis rates in accordance with the county average earnings. Such changes were not always accepted happily,[9] but they did take place more than just occasionally, and the number of local strikes over basis rates was very much lower there than elsewhere, especially than in South Wales, where the conditions in a seam were less uniform than in most other coalfields.[10] For all that, one must not give the impression that basis rates at any pit changed other than infrequently, even in South Wales: they did not. But there could be, and in some cases there clearly was, a change over time in the average of basis rates in districts taken as a whole. This could occur even though there was no change in any established price list, through the closing of some pits and seams and the opening of others. We shall meet examples of this in the next chapter.

We have, so far, been discussing the settling of the wage bargain in terms which assume that it took the form of an agreement, whether in the shape of a formal bond or merely in verbal terms, between the manager and the individual miner or his representative. This is entirely appropriate to the North-East of England, probably to Cumberland, and to many other collieries elsewhere throughout our period. In some districts, however, there prevailed, sometimes right through to 1914, systems in which a middleman, or sub-contractor, intervened between the coalowner and the great majority of miners. There were, broadly speaking, two types of sub-contracting, known by various names. For convenience these will be referred to here as the full butty and the little butty system.

The essential characteristic of the full butty system was that all men employed below ground, with the exceptions only of the manager and under-manager, were employed by the sub-contractor, and were not on the owner's books. The sub-contractor, whose formal title in those districts where the system was most common was the charter-master, but who was more colloquially known as the butty, might employ any number of men from two to two hundred, but from thirty to fifty was the usual range.[11] He might work the whole of a small or medium-sized pit, or only part. He had to have a certain amount of capital, though this was normally to be reckoned in hundreds, not thousands, of pounds, for he had to provide tools, tubs, and (where used) horses, and in some cases do preparatory work, such as making the underground roads and laying the rails. The owner, in these cases, only provided the shaft, drainage, winding gear, and so on, and had to cope with unforeseen difficulties, such as faults in the seams or irruptions of water. This was what usually happened in the West Midlands. In Warwickshire and the East Midlands, and probably farther north, the owners financed everything preparatory to the actual getting of coal.[12]

For most of the first three-quarters of the nineteenth century, the payment of wages was frequently entirely a matter for the butty to settle with the men he employed. But the growth of trade unions and of district-wide negotiations led, after 1870 or so, to wages being settled between the owners and the trade unions, and the agreed rates were then written into the butties' contracts. This was not always the case, however, and some owners took an active interest in wages quite early on in the century. The manager of Earl Granville's North Staffordshire collieries, for example, told the Select Committee on Accidents in 1835 that 'I regulate the price per ton that they shall give them . . . so that I see that the charter masters do not impose upon the workmen',[13] and it would seem that this was generally the case in Shropshire by about 1830.[14]

Since the butty received a price from the owners for each ton of coal delivered at the pit head, it was to his advantage to get as much coal as possible for as small a wage bill as he could achieve. To the owner faced with this problem in a pit which he was not sub-contracting, incentive wages were generally the answer, for the hewers at any rate. But the butty, being able to exercise close personal supervision, along with his 'doggie' (a sort of sub-butty, often receiving a proportion of the contract price) in the larger groups, mostly preferred to pay time rates, and achieve great intensity of work by bullying and the fear of not being offered work. Not unnaturally, the system was intensely unpopular with the great majority of men who could not aspire to becoming butties themselves.

In the little butty system all the workers were on the owners' books, though most of them were actually engaged by the sub-contractor (most often called a stallman). Once engaged, however, a man could not be dismissed without the consent of the manager. In addition to this fundamental difference between the full and the little butty systems, there were others of importance. In the latter, only the men at the coalface were employed by the stallman, except in those pits where haulage was also sub-contracted by the owners, as was often the case in Nottinghamshire and to a lesser extent in Derbyshire and the Potteries.[15] Even then, the haulage sub-contractor was always a distinct and separate figure. Moreover, the stallmen seldom employed more than five or six men, and often two stallmen, sharing a section of the coalface, only had six men under them.[16] In that there was a liability to abuse by slave-driving methods the two systems were rather alike where time rates were paid. But the little butty system was never hated with the same intensity that the full butty system inspired.[17] The main reason for this was probably the fact that the large number of sub-contractors involved made it possible for most hewers to reach that position eventually. And it also made it much easier to leave an unpopular stallman.

At the beginning of the nineteenth century the full butty system was prevalent throughout the Midlands, and was also to be found in Yorkshire, and both North and South Wales, and perhaps also in Lancashire, where it was certainly present in 1786. The system's development was probably connected with the evolution of the longwall method of working, and itself evolved out of a system of collective contracting by the 'pit's company'. 'The more or less democratic copartnery of earlier times, it would seem, had as its degenerate offspring the hated butty system of the nineteenth century.'[18] The system was probably not universal in any district, except perhaps Shropshire and the Black Country, after the failure of a manager brought in by the Earl of Dudley from the North of England, who had got rid of the butties in 1798.[19] But it was generally operated in the Midlands and perhaps South Wales, most especially in the smaller collieries owned by less well-to-do coalowners, who were relieved by it of the necessity to supply working capital. However, it was not confined to such collieries, and some big concerns, like the Blaina Company in Monmouthshire, were still using it in 1850,[20] though by then it was far from typical. In general, though, it is probably true to say that the system had neither spread nor receded by the 1840s.[21] By 1870, however, it had been replaced by the little butty system in the East Midlands, Leicestershire, and Warwickshire; the newly opened Cannock Chase coalfield was also using this system; there had been some changeover also in Shropshire and North Staffordshire, though the full butty system was still 'usual' in 1880, according to T. Wynne, the mines inspector for the district; and sub-contracting of any kind was no longer at all common in any other district, except in the form of the hewer paying his own hurrier, which survived in Lancashire and Yorkshire.[22]

One main reason for the change in the third quarter of the century was the hostility of the ordinary miners to the full butty system. The owners seldom had any direct economic reason to dislike the system, so long as it achieved the object of securing a large output per man; but strikes went a long way to preventing that, and, in addition, some owners came to feel that it was the butty who got most of the benefit from his slave-driving methods, and that they themselves would make a bigger profit from managers with a fixed remuneration.[23] The other main reason for change was the sinking of new deeper, and therefore bigger, collieries, with more expensive equipment. Not only could few, if any, butties have found the necessary capital to take on such mines, but the owners had too much invested underground not to want to exercise more direct control of the management.[24] By 1890, the full butty system was only found in the Black Country, and then mainly in the small, old mines, especially the 'broken' ones.[25] By that time it affected a few thousand miners at most. Elsewhere in the Midlands, the little butty system was used, and continued to be almost universal right up to 1914. In most other districts, however, sub-contracting had largely vanished by 1890. An exception, according to Keir Hardie, was Ayrshire, where he claimed that the little butty system was widespread.[26] There was also some sub-contracting of this kind in the eastern part of Fife, according to the local union secretary.[27] But these are the only references to the practice in Scotland.

Mention has been made incidentally, hitherto, of trade unions. The time has now come to examine their role in the wage bargaining process more fully, though it is not the intention here to embark on a history of the miners' unions: others have already done that task,[28] though it is undoubtedly true that, for the most part, they have done it in such a way as to produce 'a detailed and well-documented account of the past activities of a small and probably unrepresentative group of miners sitting in committee, but virtually no account at all of the miner in the pit, in his home, or in the community at large', as Christopher Storm-Clark put it.[29]

General historians of trade unions have tended to see unionism in the coal industry as an ephemeral and unstable thing until, at the earliest, the 1860s. The Webbs, for instance, say that up to 1844 'the miners, held in virtual serfage by the truck system and the custom of yearly hirings, had not got beyond ephemeral strike organisations'.[30] And though they pay a good deal of attention to Martin Jude's Miners' Association of Great Britain (founded in 1841, but not becoming really vigorous until 1844) and to the probably overpraised activities of W.P. Roberts, the so-called 'miners' attorney-general',[31] they accept what Alexander Macdonald told the conference of the National Union of Miners in 1873, namely that 'at the close of 1855, it might be said that union among the miners in the whole country had almost died out'.[32] And Clegg, Fox, and Thompson say that 'up to the sixties the development of trade unionism in the mines was marked by violent upheavals. Within two or three years unions could leap into prominence, win tens of thousands of members, and disappear.'[33] No doubt, they are, in a formal sense, correct. The first miners' trade union with a continuous formal organisation subsequent to its foundation was the South Yorkshire Miners' Association, formed in April 1858,[34] followed in Northumberland by what Welbourne called the third union, formed in December 1862 to fight an attempt to reintroduce the bond.[35] But to concentrate on the formal existence of a continuous institution is to miss much that is of significance in the way of collective action by miners. To be fair to Clegg, Fox, and Thompson, they are fully aware of this, and rightly note that there was 'a persistent rudimentary organization among the miners which provided the basis for rapid union growth', and that 'the calling of colliery meetings, the election of colliery delegates, and the holding either of county delegate meetings or of mass meetings of men from a county's collieries all preceded trade unionism proper'.[36] Ashton and Sykes may have been right when they concluded that 'for the student of labour organisation in the coal industry the eighteenth century belongs, indeed, to pre-history', but they themselves refer to several instances of collective action by miners at the colliery and even at the district level.[37] Indeed, whilst collective organisation covering more than one pit may have appeared only sporadically and been of a temporary nature until into the second half of the nineteenth century, it is clear that some form of collective bargaining within the pit (or seam) was general long before 1800, though it probably did not include everyone in the pit – members of the hewer's family employed by him were certainly excluded, and many repairing or ripping tasks may well have been the subject of individual bargain. However, this does not seem to have applied to the daywages of regular oncostmen, which were uniform in all the pay books we have examined.[38] That this well-established local collective action did not spill over into more frequent collective action on a wider scale at an earlier date probably owes a lot, outside Scotland and the North-East of England, to the small size of pits and the prevalence of the collective contract.[39] In Scotland, as Duckham has shown, serfdom, only finally ended in 1799, kept out 'new' labour and tended to lead to wage levels which, in contrast with other occupations, were higher than in England, and 'in practice many pressures, from desertions to strikes, could be exerted by the hewers to ensure the rate approximated to what they considered fair'. The whole system was described as 'a combination of theoretically strict work discipline and all-embracing paternalism'.[40]

Paternalism of the same kind, that of the aristocratic coalowners who, if they did not dominate were still highly influential, was also one of the reasons given by Welbourne in his explanation of the slow development, up to the late eighteenth century, of combinations of workers at different pits in Northumberland and Durham:

> The strong hereditary influence in the coal trade, the completeness of the monopoly which the Newcastle Hostmen enjoyed in the London market, the absence of internal competition assured by the 'Limitation of the Vend' . . . had made it the rule rather to accept labour costs as inevitable than to seek to diminish them except by engineering skill. At new collieries workmen's estimates fixed the rate of wages, and few mining matters were free from the rule of local custom. It was the influx of new capital, the sudden extension of the Wearside collieries, and the success of a less hidebound generation of viewers in working hitherto inaccessible seams, which persuaded the coal-owners to learn in the school of the Manchester manufacturers to treat labour as a commodity in a market subject to unalterable laws of supply and demand. Combination, strike and restriction were the only weapons ready to the hand of the pitman, weapons used with a vigour and success which kept the men, even in their later misery, from the worst evils of wage competition.[41]

The conclusion may require some modification, but the stress on paternalism and the usually successful operation of oligopolistic control of the chief market is surely justified. And the fact that there were district-wide strikes in the North-East in 1740, 1747, 1765, 1800 and 1810 is at the same time proof both that, though sporadic, there *was* collective action above pit level, and that its development *was* slow. That it grew in the 1820s into widespread attempts at restriction of output by the men, and that it flared up in a big strike in 1831 against some owners' practice of closing pits three days in the week and in favour of shorter working hours, shows that it was the factors which were bringing to an end the cosy world of very imperfect competition which were also making for more regular, if not as yet permanently organised, trade union activity.

This strike of 1831, and the lockout of the following year, are often seen as the results of the first strong (or even the first 'true') union in coalmining, and of the first genuinely aggressive move by a miners' combination.[42] This may be doing something of an injustice to earlier, but no more ephemeral, organisations about which less is known. Nevertheless, the union which was smashed in the summer of 1832 had had a formal existence since 1825, and is certainly the longest-lived miners' union known up to that time. Its main concern, as with most unions, was with earnings, though, for once, this was expressed not so much in efforts to manipulate wage rates as to change and strengthen the terms of the yearly bond. Its leader, Tommy Hepburn, an appealing figure as depicted by Welbourne, was a man of considerable shrewdness, and, according to Sweezy, he won success in the 1831 strike by taking the owners by surprise, and the gains of the reduction of boys' hours to a maximum of twelve per day and of the abolition of truck were permanent. (It is worth pointing out, though, that the former merely brought a minority of owners into line with the existing practice of the majority, whilst the latter was, as Welbourne pointed out, 'a reform made rather at the expense of the officials than the owners'.)[43] But the union's policy was not without its contradictions. Throughout the later 1820s, a period of rather dull trade, it had pursued a policy of restricting output, and it continued this after the victory of 1831. Yet one of the principal demands in that strike was for the protection of miners against an abuse of the clause in the bond which allowed viewers to lay the pits idle for up to three days in a week without paying the miners. This provision had been designed to meet the needs of major repairs or of an accident, but in slack times some viewers had taken to opening the pits only three days a week, and thus paying the men only for the coal cut on those days and avoiding meeting the (larger) minimum weekly payment stipulated in the bond. This abuse was certainly worth fighting; but the miners' own ca'canny policy offered the owners a justification for the counter-attack of 1832, when many of them – basically those whose pits were union strongholds – refused to re-bind their men if they would not leave the union and abandon restriction of output; that is, they locked them out. Local public opinion, which had largely sympathised with the men in 1831, was less favourable in 1832,[44] and was finally alienated by the murder of an elderly magistrate, Nicholas Foster, in June. The men's resistance, and the union, both collapsed before the end of the summer.

The formal union may have collapsed, but one may doubt whether informal colliery lodge combinations did, and one may be sure that the idea of union stayed alive. In fact, in a comparatively short time, in 1843, even a formal union was reconstituted, this time under the leadership of Martin Jude, for Hepburn had undertaken never to have anything to do with a union again as a condition of obtaining employment after 1832, and he kept his word. This time the idea of union reached out beyond even the confines of a single coalfield, and a union organisation was established at national level – the Miners' Association of Great Britain, set up in November of the deeply depressed year 1842.[45] The originators of the union were from Lancashire and Yorkshire, in both of which no district-wide formal union had yet been attempted, as well as from the North-

East, and they proseletysed in all the other major coalfields with at least a measure of success. Branches were established very widely, though their strength has been called in question so far as Derbyshire is concerned,[46] and the historians of the South Wales and Scottish miners make no mention of them. Some areas enjoyed a certain success in the cyclical recovery of the mid-40s, though more from local factors than from any help given by the National Association; but in what was still bound to be the key district, Durham and Northumberland, there was little but failure.[47] W.P. Roberts was engaged as the pitman's lawyer in 1842, and though he successfully resisted a number of individual cases of abuse of the law by owners and viewers, an effect of his activities was to induce the owners to abandon the bond and substitute monthly hirings. Many miners were alarmed at what they felt was a loss of security, and the strike of 1844 was at least partly for a return to a bond of at any rate six months. It failed, despite the beginnings of a cyclical upswing,[48] probably because of a mixture of determination on the part of employers, riled by Roberts's activities, and their ability to get large numbers of immigrants, mainly from Wales, Cornwall, and Ireland, to come as blacklegs. Once again, the formal union collapsed, though again, local collective action involving more than one pit was far from ceasing, and went on with unprecedented vigour and frequency in the 1850s.

It was not only in Durham and Northumberland that there was collective action through strikes and more or less organised restriction of output at district level in the first half of the nineteenth century. Reference has already been made to strikes in Scotland in the eighteenth century, but perhaps restriction of output was an even more characteristic collective practice in some, at least, of that country's coalfields. However, this practice of enforcing the 'darg' – the traditional day's stint, which was still the unit of payment, but which, where restriction was not practised, was generally exceeded on each shift by the majority of miners – was not operated formally on a district-wide scale, but in the individual pits, and by no means all of them, though in Lanarkshire it seems to have been effective at most.[49] It is hard to believe that it ever had more than merely local and temporary success, for, as Youngson Brown pointed out in connection with a later, more widely based return to the same policy by Macdonald's union in the later 1870s,

> the success of the restrictionist policy . . . depended on the demand for coal showing an elasticity of less than unity over the relevant range of output, on the absence of fresh inflows of labour should the policy show signs of succeeding, on the absence of any considerable stocks, and on the maintenance of a rigid discipline among the miners. There is no sign that McDonald ever considered any of these points except the last.[50]

The individual pit leaders in the first half of the century could not even do that much at a level which counted, except perhaps in a few isolated places on the periphery of the coalfield. Operating the darg did not mean that there were no strikes, though it seems likely that most were of a local nature. Page Arnot only cites one which affected a whole district prior to the founding of Macdonald's union in 1855 – Ayrshire in 1824 – though the Webbs imply a strike of 'Scottish miners' in 1830 or 31, which had been 'inadequately supported' by Owen's Grand National Consolidated Union.[51] There may well also have been something not very far from a district-wide strike in Lanarkshire in 1837 – or, more correctly, a lockout – for at least two big firms, Shotts Iron Works and Bairds, had a prolonged battle against the darg then, locking out their existing miners who refused to stop operating it and replacing them, partly at least, by labourers, often from Ireland.[52] It is likely that these two firms were not alone in their attempt to suppress the darg at this time.

In other districts, too, there is evidence of cooperation in collective action between men from different collieries prior to the Miners' Association of Great Britain in the early 40s, though seldom embracing the whole of a coalfield. Challinor says of the eighteenth century in Lancashire that 'it is possible to cite examples of violent struggles by workers, but they tended to be over a particular grievance. Once it was settled, workers would subside into apathy, leaving no lasting organization to commemorate it.'[53] However, by the nineteenth century there were more sustained efforts. In 1818, for example, there were strikes over wages in West Lancashire, when, in the St Helens area at least, the technique of having a succession of strikes at different pits, with support being given by those still (or back) at work was used.[54] Again, a union was involved in organising a strike in South-East Lancashire in 1831, though it failed in the face of a successful policy of using blacklegs from Shropshire and from amongst local textile workers.[55] In Yorkshire, too, there was joint action by miners employed by two or more concerns in 1819, when the big iron and coal firms of Bradford, the Bowling and Low Moor companies, were struck against. A few years later, in 1825, the miners of the Sheffield district won a temporary success by joint action, as did those of Wakefield early in 1833.[56] South Wales, also, had its experiences of collective action by miners at this period. Ness Edwards refers to successful resistance to a proposed wage cut in October 1816 through 'riotous assemblies . . . though violence was limited'.[57] This was followed by similar actions in the next twenty years. 'In these loose organisations', Edwards says, 'are found, in embryo, all the main characteristics of the miners' unions of today.'[58] A wider strike, in 1831, covering virtually the whole of the district, met with failure, and the union probably disappeared from formal existence for the next decade, though perhaps it was only from view. This was the period of the 'Scotch Cattle' episodes, the terrorising of blacklegs and others suspected of lack of solidarity with the union (or the 'general will', perhaps) by groups of miners who clearly were only too thoroughly imbued with the spirit of collective action. Whether or not any genuine union survived in the metaphorical as well as the physical underground, to reappear in 1843, it is clear that the impulse to joint action was there, and was translated into effect, if not into success, not all that infrequently. And whilst there was no continuous, minute-keeping committee, the idea of union was ever-present.[59]

It is evident from this brief survey that not only was informal collective bargaining within the relatively small group in the seam or the pit the general rule at the beginning of the nineteenth century, but there was, also, then or shortly afterwards, widespread willingness to combine with others in the neighbourhood in certain circumstances, in an attempt to alter (or maintain) the wage bargains of the whole group. This only took the form of a formal trade union on a comparatively few occasions, and no such formal union persisted for more than a few years. But it may well be asked whether the formal unions had any more influence than the more indestructible *idea* of union – that is, the existence of some sort of collective bargaining and a willingness to widen its scope at certain times. This brings up the whole question as to the effectiveness of any combination of workers in the nineteenth-century coal industry in changing wages from what they would have been in the absence of collective action; and this is better dealt with at a later stage.[60]

With the 1850s and 60s came the foundation of the first permanent – or, at any rate, long-lasting – formal trade unions, whose institutional histories have been well covered in the works cited in n28. A major characteristic of these unions up to the 1890s or beyond was a liability to extremely large fluctuations in their membership during the course of the trade cycle. In a sense, of course, this was simply a continuation of previous experience. Unions were founded, or waxed, in times of active trade and rising employment and prices; they collapsed, or waned, when prices and trade activity were falling, and stoppage of the pits was not to the employers' grave disadvantage, and they felt able, or even obliged for the sake of their business survival, to cut wages and fight the union. The pattern of expansion of 'social movements' in the upswing, which E.J. Hobsbawm[61] attributes, no doubt correctly in the general European context, to the period after 1850, had appeared rather earlier in the coal industry, with its proletarianised workforce and its, for that day, relatively capital-intensive organisation. The 1810 strike in Durham and Northumberland at the peak of a cycle,[62] is perhaps the first example of this pattern in the coal industry; but it was followed by similar examples of increased collective pressure on employers in every succeeding cyclical upswing, except maybe that of 1815, in which neither did the iron industry really participate nor did the shipments of coal from the Tyne and Wear grow.[63] Some of these examples have already been referred to – West Lancashire in 1818; Ayrshire in 1824; the Sheffield district union in 1825; Hepburn's union, founded in 1825, achieving its momentary success in the burst of activity in 1831; combinations in Lancashire and in South Wales in 1831 and in the Wakefield district in 1833; and finally, the widespread 'explosion' – to use Hobsbawm's phrase – of 1844. In all four quarters of these years, the Gayer–Rostow–Schwartz 'ranked relatives' series show an upward movement.

To insist that a 'modern' pattern of fluctuations in labour movement activity had come to the coal industry before the end of the Napoleonic Wars, and in most of the major component coalfields by the 1820s, is not to deny that the older pattern of 'explosion' during the simultaneous trough of a slump and rise in food prices (both connected with a bad harvest) continued alongside the modern one for a little while. The South Wales rioting of 1816 is a good example of this sort of incident, but there is none subsequently which is such a clear-cut one. Collective activity against employers, often involving some sort of union organisation, did occur in cyclical slumps; but it took a form more similar to the rearguard actions of the middle 1870s or 1890s than to the eighteenth-century corn riots described by Ashton and Sykes.[64] The four-month dispute in Lanarkshire in 1837 over the employers' attempt to break the darg, an attempt which was successful at least at the time, is of this nature. So, too, was the 1832 affair in Durham and Northumberland. Both were long, drawn-out, and bitter disputes – James Baird described that of 1837 as 'as determined and prolonged a strike as any we have ever had at Gartsherrie',[65] and 1832 was marked in the North-East by violence such as had not been experienced since the strikes and riots of 1740 and 1765.[66] This tendency for the defensive disputes associated with cyclical downswings to be relatively long and bitter – and mostly unsuccessful – was characteristic of the rest of the country, whether or not they were conducted by a trade union which remained in continuous existence. It is a tendency which is apt to mislead, by suggesting both that union activity was greatest in periods of depression, at least until it had been crushed for the time being, and also that unions were generally powerless in the face of determined employers. So far as the coal industry is concerned, the first of these suggestions is not true from very early in the nineteenth century; and, as we shall see, the second requires a great deal of qualification, for union activity in upswings had a considerable amount of apparent success.

The fluctuations of militancy of the first half of the nineteenth century continued, and were associated with wide fluctuations in union membership, in the second half, when continuous unions were established. The first of these to embrace a whole district, though not in either case a whole coalfield, were in South and West Yorkshire in 1858,[67] though Alexander Macdonald's Coal and Iron Miners' Association in Lanarkshire, founded in 1855, has some claim to priority; but it foundered in the vain strikes against reductions of wages which accompanied the aftermath of the boom of the early 1870s. Others followed in the North-East in the 1860s, in Northumberland in 1862,[68] and in Durham in 1869. But elsewhere unions remained more local and fragmented, and Challinor's statement that 'it was easier, say, to get the Hindley miners to agree to the principle of common action with miners throughout the rest of Britain than it was to get them to cooperate with the miners from neighbouring Wigan',[69] could apply equally to places in South Wales or in Scotland;[70] whilst in the Midlands there were only stirrings of unionism, helped from South Yorkshire, but not covering more than a few collieries.[71] In fact, the next move in these districts was into a national union – which proved, in the end, to be too ambitious a move for permanency. The revival of trade unions in Yorkshire, Northumberland, and parts of Scotland, had led, primarily under the influence of Alexander Macdonald, to a fresh attempt at a new national organisation, The Miners' National Association, founded (after one or two earlier conferences of delegates from small, scattered colliery lodges) in 1863. This union had different purposes, or at least different methods, from those of its predecessor of the 1840s. As Keith Burgess puts it, in a rather hostile fashion, 'leaders like McDonald and his ilk were able to steer it away from industrial activism and towards a course of legislative and parliamentary lobbying'.[72] Wages in the districts from which the MNU was mainly supported were above average, and were being handled by local collective action in a fashion which seems to have given reasonable satisfaction. To leaders like Macdonald and Thomas Burt, moreover, the disastrous failure of 1844 had discredited strike action as anything but a policy of last resort, and they saw more opportunities for worthwhile action for a national body in operating through national legislation. This course had a number of outstanding successes, beginning with the revised Coal Mines Regulation Act of 1860, which replaced the Act of 1855 which had been largely ineffective.[73] In addition to containing more and tighter safety rules, including the obligation to have two shafts per colliery by 1865, better provisions for enforcement, and the effective raising of the lowest age of permitted employment to twelve, this 1860 Act was the one which empowered miners at each pit to appoint a checkweigher from amongst their number. Though the Webbs admit that this provision was often rendered nugatory by the evasions of the owners, until it was strengthened in the 1872 and 1887 Acts, they nevertheless claim that it had the effect of producing 'a practically inexhaustible supply' of union leaders who were 'insensible to the bullying or blandishments of managers or employers.'[74] This position has been attacked by a recent critic, Keith Burgess, who has argued that

> the reality . . . was somewhat different. Although the position provided an ideal training for an aspiring trade union official, there were aspects of it which compromised checkweighmen as miners' leaders and differentiated them from the rank and file. If they did not keep on good terms with the employer, they were refused facilities to do their job . . . Grass roots discontent . . . might well halt production and jeopardize the checkweighman's comparatively privileged position. It was not uncommon, therefore, for the checkweighman to become an employer's man.[75]

He does not cite any evidence for this last statement, and we have been unable to find any, apart from the generalised hostility to full-time union officials (hardly the same thing as checkweighmen who were *branch* secretaries, in any case) apparent in David Douglass's *Pit Life in Co. Durham*, and echoed in some of the miners' statements recorded in the post-Second World War study by Norman Dennis *et al.*[76] These certainly reflect the attitudes of some latter-day miners, and were doubtless present in the nineteenth century too. But they appear to be based on the theory that all moderation by union leaders is a species of treachery,[77] not on any attempt to assess what was and what was not feasible or wise in certain historical situations. In any event, they do not call into question the Webbs' view about the significance of the 1860 Act for the leadership and continuity of miners' unions in the later nineteenth century.

Apart from the 1860 Mines Act, Macdonald and the other MNA leaders played a large part in getting its stronger successor in 1872, which helped bring down the working hours for many miners. They were also involved in the general trade union enquiry of the 60s, and their influence was felt in the beneficent legislation which followed in the next few years. To sneer at them for their involvement in this field is both to belittle genuine achievements on behalf of both miners and workers in general, and to assume without any warrant that the conditions for pursuit of a different policy, a policy of industrial militancy, were present.

Because they totally discounted the possibilities of nationwide militancy on the model of 1844, the leaders of the MNA were unwilling to let it get closely involved in supporting local strike actions, calls for which inevitably came mainly from branches in those districts where local organisation was most weak and fragmented. As a result, dissatisfaction in these districts, which came to a head in Lancashire first, led in 1869 to the foundation of what came to be, in some degree, a rival national body, though it was not originally intended as such, and did in fact cooperate in some ways with the MNA.[78] This was the Amalgamated Association of Miners. The MNA, and its major constituent districts, would not waste their resources on fostering unions in places like South Wales, where outside help and a tight, centralised structure were required. The AAM, as Challinor put it, 'can be viewed as an attempt, extending over six years, to apply the theory of centralized control of union wage bargaining'.[79]

The AAM had a mixed but, as one might expect, generally successful record during the boom of the early 70s, culminating in a peak membership of over 100,000, spread over Lancashire, Cumberland, the West Midlands, the small coalfields of the South-West, and both North and South Wales. The record of the MNA and its local constituents was very similar, as it was, too, in the downswing which followed. Both could take advantage of the price rises; neither could avert wage cuts after 1873, and the officials of both unions tried to keep the retreat orderly, and to prevent vain resistance *à l'outrance*. Only in the North-East was there even a modest measure of success in this, at any rate up to the middle or late 70s. Elsewhere there were many and bitter strikes and lockouts, notably in Scotland, South and West Yorkshire, and the Wigan area in 1874, and in Staffordshire and South Wales in 1875. This was the strike which broke the AAM in South Wales, and led in April to the absorption of its remaining district branches by the MNA in a renamed, but still federal and decentralised, Miners' National Union. In fact, in most of these districts, even after the major period of resistance was over, there were very many niggling, local disputes, involving one or two collieries at a time, throughout the rest of the decade, and in most of them no district-wide union survived into the 80s.

The two Yorkshire unions were exceptions to this latter; and the main exception to the rule of continuous petty disputes was South Wales. The main reason in this case, as also in Cannock Chase and in Somerset, was the adoption of a district sliding scale, relating wages to prices. This came in South Wales in the settlement following the 1875 strike, and it was greatly helped by the firm but conciliatory leadership of Mabon.[80]

The sliding scale was adopted two years later in Durham and Northumberland, in March 1877 in the former and in the following December, after an eight-week strike in the latter. But it is fair to say that the adoption of sliding scales in the North-East was but a formal expression of an attitude and policy which had been adopted for some time, certainly by the Northumberland leaders. It was of them particularly that the Webbs were thinking when they wrote that the achievement of recognition of the union by employers and the establishment of joint boards for negotiations (first set up in Northumberland in 1871):

> brought results which largely neutralised its advantages . . . The men gained their point at the cost of adopting the intellectual position of their opponents. When the representatives of the employers and the delegates of the men began to meet to discuss the future scale of wages, we see the sturdy leaders of many trade union battles gradually and insensibly accepting the capitalists' axiom that wages must

necessarily fluctuate according to capitalists' profits, and even with every variation of market prices.[81]

Whilst this probably does less than justice to the intellectual independence of Thomas Burt, who had apparently accepted that miners' wages necessarily varied with prices long before 1871, it is correct to say that the North-Eastern leaders did take this view, and were concerned to make the best of things for their men within what they saw as the constraints of the system, which they also accepted. That this brought them to accept sliding scales automatically relating wages to prices is not surprising after the violent fluctuations of the earlier 1870s, and the lost earnings that the miners themselves had suffered in the strikes and lockouts of the cyclical downswing, which had proved almost uniformly vain. Arbitration when the joint board failed to agree had been tried in 1874–6, but it was a source of expense and it had failed to avert trouble. The leaders, therefore, accepted a sliding scale, though agreeing that it 'had no pretence to be other than a practical solution of a permanent difficulty. They admitted that as a system of wage assessment it did not secure for the men full justice, but they were persuaded that it had the merit of averting strife and the poverty consequent on war'.[82] In view of the long time taken to make up for wages foregone in any strike lasting more than a few days, there is something to be said for their attitude, even if one admits that a certain amount of wages are foregone through industrial peace.[83] It was not, in any case, an attitude that was confined to Northumberland and Durham and South Wales in the 1870s and later. Cannock Chase had a sliding scale in 1874 and Somerset in 1876. Lanarkshire also adopted one in the mid-70s; and both parts of Yorkshire, which had followed Northumberland in achieving employers' recognition in the early 70s, accepted a sliding scale at the unions' insistence in 1879/80, though South Yorkshire failed to ratify theirs.[84] Most other districts probably failed to have district sliding scales at this time only because there was no district-wide trade union left to negotiate one; and wages unquestionably followed coal price movements very closely.[85] Still other districts came to sliding scales at a later date, after they had been abandoned in most of the inland coalfields. Fife, for example, began one in 1895, and Scotland as a whole had a *de facto* scale from 1899, whilst it is possible, too, that the Forest of Dean scale began in 1895, though it may have been earlier.[86] The great majority of mining trade union leaders who testified to the Royal Commission on Labour was in favour of sliding scales, including Keir Hardie;[87] and a number of economists who were sympathetic to labour were also in favour.[88]

With the trade revival of the early 1880s, albeit that it was a muted affair in the coal industry, there were the beginnings of revival in trade union organisation and activity in those districts where they had died out, and in some others in the inland coalfields there was a revulsion against the recently accepted sliding scales. South Yorkshire, as has been noted, failed to ratify theirs, and West Yorkshire's only lasted until 1881, when the first act of the newly combined Yorkshire Miners' Union was to end it. It is perhaps worth noting at this point, though, that the membership of the Yorkshire unions just before amalgamation was down to a mere 2,800.[89] Nevertheless, at about that time, representatives of several small local unions in Lancashire met those of Yorkshire, of the newly formed Derbyshire and Nottinghamshire unions, and of other small unions in North and South Staffordshire and in Shropshire. Together they agreed to demand a 15% increase in wages in the autumn, and to strike if necessary. In the event, however, the Lancashire unions were the only ones which actually came out on strike, and then 'they found their prospects of success jeopardised by the influx of coal into Lancashire markets from the very districts with which they had been banded together'.[90] Obviously, the time for a united movement was not ripe. But a major group of labour leaders in English coalfields was now definitely hostile to the sliding scale principle. Still, the upswing of the early 80s and the downswing to the particularly deep depression which followed came with little institutional change amongst the miners' unions, though Keir Hardie put together a loose federation in Scotland in 1886, which, however, failed to last two years before disintegrating into half a dozen or so component parts. Membership, outside the sliding scale areas, was down to very low levels by 1886 – 8,000 in Yorkshire, 7,000 in Lancashire and Cheshire, 2,500 in Derbyshire, and 500 in Nottinghamshire, and there were no reports from other districts, except for a union which had a sliding scale in Cannock Chase. This had 4,000 members.[91]

Major changes began to come again in the revival of trade which started in 1887. The old MNU had continued to survive, embracing only a few small scattered associations apart from the unions of Yorkshire and of the North-East. It consistently refused to become involved in wages questions, leaving those to the individual district unions. So when those unions which did not have sliding scales wanted to attempt to raise wages in the cyclical upswing, a separate conference of their district unions was called, including amongst them the Yorkshire union.[92] Indeed, the initiative for the meeting came from Ben Pickard, the Yorkshire leader. Out of this move there sprang the Miners' Federation of Great Britain. At first, this federation, which was not formally set up until November 1889, received the blessing of the MNU. There was, indeed, some attempt by the latter to rationalise the move as a division of functions – the federation for wages purposes where sliding scales did not rule, and the MNU for other trade union activities. But once there were two bodies in the field, rivalry was bound to develop; and it soon became clear that the new body would not confine itself to wages alone. Within a month of the formal foundation of the Federation, the MNU officials were objecting to its activities, and Pickard and his colleagues thereupon severed all connection with the old union.

It was not the question of a federation for wages purposes which split the coalmining unions, but the intervention of the new body in matters other than wages, and the first and most decisive of these interventions was the agitation for an eight-hour day. This was not only trespassing on the field of parliamentary agitation, which the MNU regarded as its own; it was completely opposed to the policy of the Durham and Northumberland leaders. The MNU still retained its position in the wider world of the Trade Union Congress, and one of its leaders, John Wilson of Durham, kept his place as the miners' representative on the Parliamentary Committee. The result was that the TUC failed to support the MFGB's agitation for the eight-hour day, and its influence was, if anything, thrown against the Eight Hours Bills in Parliament. The final cause of the breach was the failure of the North-Eastern unions to support the other English districts in the great lockout of 1893. Early in the previous year there was a similar strike (or lockout) in Durham over a 10% reduction in wages, which the men had entered into against their union leaders' advice. Just like the dispute of the following year, it was an attempt to establish the principle that a certain level of wages should be regarded as the minimum, whatever the state of trade and coal prices might be. Because of this, the MFGB gave generously to the Durham strike funds, though since Durham did not actually join the Federation until later in the year,[93] it did not operate Rule 20 and bring all its members out on strike. When, in February 1893, Durham (and Northumberland) accepted a further cut in wages without consulting the Federation they were, in effect, warned that if this happened again they would be expelled. When the lockout in the other English coalfields began in the summer, the Federation ordered the North-Eastern unions to join in by demanding the restoration of the February cuts. Durham refused, and, after a weak attempt, and a ballot which was narrowly against striking, so did Northumberland. So ended their connection with the MFGB for the time being. It may be doubted whether their leaders were other than thankful about this, for they continued to accept the necessity for wages to follow prices, preferred this to be effected peaceably by sliding scales, and, in addition, were by now divided by personal hostility from the Federation's leaders, whom they resented for having appealed over their heads to the Durham and Northumberland rank-and-file members.

One of the main principles on which the Federation had been founded was that wages should not be tied to prices, but that they should always be sufficiently high to provide a reasonable standard of life. The first test of this principle in the Federated Districts came in the summer of 1893, when the employers demanded a cut in wages, just as those in the more price-sensitive exporting coalfields had done in the previous year. A prolonged lockout ensued, 'by far the largest dispute the country had ever seen'.[94] It was, on the whole, a dispute without a great deal of accompanying violence, though the incident of troops firing on a disorderly crowd at Featherstone in September has achieved considerable subsequent notoriety – perhaps because of the later prominence of Asquith, who was involved peripherally as Home Secretary. The dispute was eventually ended in November, after government intervention and pressure on both owners and union, by reference to arbitration by the Foreign Secretary, Lord Rosebery, who said that the old rate should stand for three months whilst a conciliation board was set up. This board eventually negotiated a reduction of 10% in August 1894, and this rate was maintained without further cutting until the next upswing of the trade cycle.

The Federation leaders treated the outcome of the dispute as something of a triumph, and its official historian, R. Page Arnot, writes that 'it was a victory for a new principle, and for the new organization which had been built upon the assertion of that principle'.[95] This view has been called into question by J.E. Williams, who claims that 'the gains . . . were of doubtful value. The important principle of the minimum wage had been conceded by the employers, but only for two years . . . The artificially high wage rates which they had managed to maintain were more than counterbalanced by short time working.'[96] More recently, Griffin has attempted to defend the traditional view, by dismissing Williams's point about short-time working on the ground that the less efficient collieries alone had heavy underemployment, and they always did in depressions in any case.[97] This is irrelevant, however, for it does not meet the contention that there was more underemployment than there would otherwise have been, though Griffin does convincingly show that the coalfields with more flexible wage rates did not actually take over East Midlands markets through wage cuts.[98] Williams's case, whilst in the nature of things it cannot be proved, does appear more plausible; but it ignores one major way in which the outcome of the lockout did, indeed, prove to be a victory. Never before had wage rates declined by so little in a period of depression, and whilst this may not have applied to earnings, it is possible that increased leisure was regarded as some compensation and, more important, it is certain that the Federation's prestige was enormously enhanced. 'Hitherto unions that had opposed the downward trend of pay rates had been smashed. This time they emerged strengthened.'[99] This greatly helped the union to extend its control over the mining community and to end those violent fluctuations in membership which had characterised every previous trade cycle.[100] It was from this time that the union and the mining community began to become so closely interconnected that, by the twentieth century, they came to seem virtually synonymous, at least in areas where other industries were not much present.

One thing which the Rosebery Award did was to ensure that future movements of wages in most of the inland coalfields should be negotiated between the MFGB and the employers on the basis of the whole area, which soon became known as the Federated District. A series of conciliation boards succeeded that of 1894, with only one major gap between 1896 and 1898. All the percentage changes for these districts, which are shown in the next chapter, were negotiated by such boards. However, change in wages continued to be closely linked with price movements, though perhaps not with quite the same rigidity as when a sliding scale operated. This was, in effect, admitted by the union when, in consideration of an agreement by the employers not to cut wages in the autumn of 1909, the Federation leaders agreed that they should not subsequently be increased until the price of coal had reached a certain level – 7s 10.21d, in fact. This compromise of 1909, sensible though it no doubt was in many ways, certainly cut across the whole theory that the MFGB had been propounding for years, and was not significantly different from the ideas of Burt.

Whilst the inland areas which had previously been for the most part weak ones for formal union organisation were being taken over by the MFGB, the other big coalfields continued to go their own way. South Wales still had formal sliding scales until 1902, despite a big strike in 1898 before the last scale was established. Even when it was ended, the conciliation boards which followed operated on principles which were effectively little different from the scales.[101] Much the same can be said of the North-East.

It has become customary to say that one effect of the sliding scales was 'to suffocate trade unionism in the valleys of South Wales',[102] and that as a result South Wales miners got less than they would have done otherwise. We shall return to this last question later, but on the weakness of unions in South Wales there are two points to be made now. First, their weakness has been exaggerated sometimes. The men's representatives on the sliding scale committees were not generally feeble negotiators, and the changes in the successive sliding scale agreements reflected changes in relative bargaining power just as effectively as did the changes in wage rates elsewhere.[103] Second, there were other reasons which can be urged to explain the weakness of South Wales unions compared with those of the North-East in the nineteenth century as a whole, and compared with the Federated District in the 1890s. Evans seems to regard Welsh unwillingness to pay heavy union dues as the principal reason,[104] but that is only a superficial explanation, and others which he mentions are more fundamental. The attitude of the nonconformist churches was one thing which certainly influenced some miners against joining a union, but surely not many. The main thing seems to have been the lack, for most of the century, of any sort of cultural unity in a coalfield which had relied far more than the North-East on immigrants, and perhaps, on immigrants from a greater variety of sources. This is one of the main reasons which Page Arnot himself urges for the very similar tendency shown by Scottish unions, prior to 1899, to collapse onto their parochial units.[105] The biggest division in South Wales was one which continued right down to 1914 and beyond – that between the English- and Welsh-speaking miners. Many union historians in commenting on this have tended to blame the Welsh-speaking leaders, and Mabon in particular, for using this division as a means of getting their way, which was not the way of the MFGB leadership.[106] Undoubtedly there is some truth in this. But the division between the groups was being exacerbated for much of the 1890s from the other side too. As Evans pointed out, Mabon 'had laboured unceasingly to weld the weak District unions into a powerful organisation . . . The Miners' Federation had hindered, if not entirely frustrated, his efforts' by encouraging a rival English-speaking organisation, based in Monmouthshire and led by William Brace.[107] It was not until the two groups were reconciled, following the unusual disaster of a failed strike on a rising market in 1898, that a really strong union was possible in South Wales. But it is hardly fair to blame the earlier disunity on Mabon because he followed a policy which was widely accepted in other coalfields, which was, arguably, the best available in the circumstances, and which was, in any case, hardly different in anything other than outward form from that of Brace's group.[108] Certainly, to blame union weakness on sliding scales is to ignore the coincidence of scales and strong unions in other districts and in other industries.

As in South Wales, the conciliation boards, which succeeded the formal sliding scales of Northumberland and Durham in 1887 and 1889 respectively, worked in practice on principles little different from those of the scales. In Northumberland, indeed, there was a return to a formal sliding scale as late as 1914. In Scotland, too, similar principles operated, whether there was a formal scale or not. In Scotland, in fact, collective bargaining remained fragmented until 1899, despite the formation in 1894 of a new Scottish Miners' Federation which did at last manage to retain a continuous existence, though not without difficulty and dissension in the early days, when the leaders were split on the wisdom of the strike of 1894 and there were misunderstandings between them and the MFGB, which they had just joined.[109]

Table 6.1 *Membership of trade unions in selected years, 1900–13 (in thousands)*

	1900	1905	1907	1909	1913	Percentages of labour force 1900	1905	1907	1909	1913
Northumberland	24.0	23.4	32.3	35.6	203.4 (Northumberland and Durham)	60	52	66	66	90 (Northumberland and Durham)
Durham	80.3	90.4	105.6	117.3		71	72	77	80	
Yorkshire	54.5	55.2	79.1	83.8	215.9 (Yorkshire and Lancashire & Cheshire)	54	49	64	59	80 (Yorkshire and Lancashire & Cheshire)
Lancashire & Cheshire	37.0	34.2	75.4	62.8		42	36	79	60	
Derbyshire	24.3	26.2	36.1	37.9	(166.6)[a] (Derbyshire and Nottinghamshire)	52	51	65	67	(84)[a] (Derbyshire and Nottinghamshire)
Nottinghamshire	20.0	20.3	28.4	33.1		74	64	85	87	
South Wales	134.7	122.4	158.2	164.5	179.1	91	74	83	80	77
Scotland	49.0	52.5	67.2	73.8	110.0	47	46	54	55	75

[a]All Midlands coalfields.
Source: *Abstract of Labour Statistics*.

By 1900, trade union organisations were at last firmly established in all the main coalfields and in many of the smaller ones, and they had at least a presence everywhere. Their membership in the main districts at the peaks and troughs of the trade cycle from then on is given in table 6.1, and it is at once clear that the days of wildly fluctuating membership were over. Moreover, when one contrasts this table with the one showing total numbers employed in the previous chapter,[110] the high percentage who were union members is obvious at once, more especially when one remembers that surface workers were actively discouraged from joining in some areas.

The at least partially successful agitation for the eight-hour day has already been described in the previous chapter. With its conclusion, together also with the death or retirement of old leaders, or at least the waning of old bitterness, the way was open for the permanent unification of the North-Eastern unions with the MFGB, and this took place even before the Eight Hours Day Act was passed, in fact as soon as it became clear that there would be legislation. One consequence of this was that the Federation felt itself strong enough not only to launch another agitation, but also to carry it to national strike action. This was the question of a minimum wage, and it was itself at least partly a consequence of the restriction on the length of the working day. It had long been the tradition for managers to make up the earnings of pieceworkers who, through no fault of their own, but simply because they were in what came to be known as an 'abnormal place', were unable to achieve the normal weekly level of pay. This custom is mentioned in the first surviving Tyneside bond, one for 1703, which speaks of the 'consideration' to be paid 'at the judgment of the viewer' for work of unusual difficulty.[111] Probably the custom was to be found in all coalfields throughout our period, but it was especially prevalent in South Wales towards the end of the nineteenth century, largely because the majority of seams there were of a less consistent nature than in most other coalfields. Perhaps, too, this prevalence had something to do with the relative shortage of labour there, and with the need to attract and keep particularly good men. The collieries with cutting price lists which were rather on the high side tended to find this easy, whilst those with relatively low lists found it difficult, and were therefore inclined to be generous in the matter of consideration money. According to Jevons, 'in the last quarter of the nineteenth century there were many mines in South Wales in which consideration money was undoubtedly too freely paid'.[112] The development of better techniques of cost accounting, especially in the big companies which were becoming increasingly dominant in the district at that time, showed up the exact cost per ton of allowances of this kind, and with costs in general going up for a variety of reasons, including legislation which the owners opposed, managers were pressed to economise as much as possible. Some owners, in fact, set definite limits to the amount which could be granted each day, and a legal decision in January 1908 made this easier. Judge Bryn Roberts, in a case at Pontypridd involving the Ocean Coal Company, stated that consideration money was not a matter of right, as the courts had previously held if the practice had the sanction of custom, but 'a matter of gratuity'. The legal reduction of hours from January 1909 compounded the difficulty by reducing earnings in all those cases where the hewer was unable or unwilling, by increasing his pace of work, to keep up his output, and by making it impossible for the miner in a bad place, but with no allowance, to work overtime and so maintain his earnings.[113] The issue came to prominence in the Cambrian Combine strike of 1910–11, which began as a dispute over setting the price list in a new seam, but was broadened by the strike committee, which, along with other sympathisers, set itself up, in effect, as an alternative to the official union leadership. Ultimately, the strike, lacking official union support, was beaten. But that it lasted so long was a reflection of a mood which was increasingly affecting the younger miners – a mood of implacable hostility to the owners based on rising belief in the long-run incompatibility of the interests of capital and labour. The alternative leadership gave expression to this mood in the syndicalist programme which they published in 1912, *The Miners' Next Step*; and though the Cambrian battle was lost, the union was now forced by the pressure of its members, stirred up by the ginger group of younger leaders, like Vernon Hartshorn, Noah Ablett, and A.J. Cook, to carry on the war, and, indeed, to extend it into a campaign for a minimum wage for all underground workers.

The initiative from South Wales was taken up by the MFGB at national level, though because of the wide differences in customs and wage levels, its agitation was always for the principle of the minimum wage to be accepted nationally, but for the details to be worked out locally. The employers refused to consider this, and as a result the first national coal strike was called in February 1912. At this, government intervened, and failing to get any quick negotiated settlement, it rushed through a Minimum Wages Act of its own devising. The effects of this Act on wage rates will be considered later. At present all we need to note is that, although it failed to give the union either the specific district minima it had asked for, or even the general minimum figure of 5s, and although the ballot on ending the strike showed a majority of 244,011 to 201,013 in favour of continuing, the Act did end the strike. The union leaders, fearing disintegration of the Federation, and perhaps weighing the fact that the district which had begun the agitation, South Wales, was the only major one which was in favour of resuming work, called it off.

We have not hitherto said much about employers' organisations, except to imply their existence through their participation in joint boards, etc. This existence, outside Durham and Northumberland, was a development of the second half of the nineteenth century to an even greater extent than was the development of the formal trade union. It was not, however, a very powerful development. Sweezy has described the growth of monopolistic organisations in the North-Eastern coalfield, and has shown how unstable they were in the 1820s and 30s, breaking down in 1845 finally, as he believed.[114] A.J. Taylor showed that there were ten further years of attempts to revive the cartel, some of which had a measure of success.[115] But it is evident that the combination had ceased to have long-term viability from the 1820s or 30s onwards, even before the coming of the railway finally destroyed its prospects. But during this prolonged period of its death throes, it is clear that it tended to hang together better on one thing than on any other – cooperation *vis-à-vis* the workers. The union won the 1831 strike because it took the owners by surprise, and a number of individual owners gave way. But a united front in the following year routed the union. Again in 1844, cohesion amongst the employers was effective. This tradition continued thereafter in the form of a joint owners' association which, in Northumberland, became the first to accept formal collective bargaining with a trade union. Durham followed soon afterwards. Some firms remained outside the owners' associations even in the North-East, but they were mostly small and insignificant.

In the other coalfields no such tradition of cooperation between owners existed, and though owners' associations were set up at various times – a few perhaps as early as the 1820s[116] – they were not comprehensive and had little success, except in the limited function of negotiating the sliding scales, where they operated. Taylor attributes this lack of success to 'this country's geographical and geological peculiarities',[117] but this can be put more succinctly in one word – competition. The main problem in this connection was the very great differential in productivity between high-cost and low-cost producers, which led to great differences in the interests of individual owners and so to chronic instability in all owners' associations. To this must be added the labour-intensive character of the industry, which meant that owners could close the pits when demand was slack with relatively little loss to themselves, so that they were not under the same pressure to control output, markets, and prices as more capital-intensive industries, like iron and steel or chemicals. J.T. Woodhouse, a prominent Midlands mining engineer, put the point well to the Select Committee of 1873: 'the resolutions that the masters come to are no more than a rope of sand. If they make a bargain with each other on Monday, they go out and break it on Tuesday, so that there is no cohesion among them.'[118] Though owners' associations did come into more or less permanent existence, with the majority of owners being members in most coalfields, from the early 1870s onwards, they had very little power, for some individual owners were always ready to act on their own if it suited them. Where owners' associations did stick together it was because it suited all members' interests – as in the early weeks of the great 1893 lockout, but not the later ones.

7

WAGES

After the rather brief excursion into the main institutional aspects of the settlement of wages in the last chapter, it is time to turn to the question of what wage rates were at various times. This is a matter which is not only an essential preliminary to an examination of earnings, which were, no doubt, the main ultimate concern of most miners, but one which has importance for its own sake, for it is essential information for assessing in monetary terms the input of labour to the industry.

It is difficult enough to establish an index of wage rates for any industry in the nineteenth century, but it is particularly so for coalmining, as will have become apparent from the outline of the methods of wage payment above. In most occupations there was (and is) a 'going rate for the job', the 'job' being fairly closely defined in universal terms, or at any rate generally understood and having a reasonably uniform character. In coalmining, the physical actions of the job itself may have been closely defined in general terms – swinging a pick below ground, pushing a tub through a narrow tunnel, and so on – but the conditions in which it was done could vary so much, and were so vitally important to the success of the miner's efforts, that each particular job was, as we have seen, subject to a separate bargain. This bargain established the 'price list' for each seam at each pit – the cutting price for coal, the putting price of so much per score, and so on. Christopher Storm-Clark, after interviews in the 1960s with many old miners, was so impressed by the diversity of experience that he wrote that for many 'who started their career in mining before the First World War, the concept of a wage *rate* for this period is almost entirely meaningless'.[1] Nevertheless, it is easy to exaggerate the significance of this point, for the piecework rates were set with the normal current rate of earnings in the district very much in mind. It is possible, therefore, as Rowe said, to regard 'the normal earnings per shift . . . as a rate of wages just the same as with time-workers',[2] always remembering that such a 'rate' was, in practice, a range of rates rather than a single figure – a range which was narrow in the North-East, when the 'county average' system was in operation, but could be quite wide in other places and at other times.[3]

Prior to the establishment of district-wide collective bargaining there is no organised body of information about wage rates, and in attempting to establish an index of wages one has to rely on a wide variety of scattered material, which has been examined with all that we know about the institutional arrangements for wage-fixing always firmly in mind. There is a good deal of this variegated material, at any rate from the 1840s onwards – observations made to Royal Commissions, Select Committees, etc.; surviving wage lists and pay books of particular collieries; a few early trade union records; returns made by various Chambers of Commerce and private firms; statements in pamphlet literature, newspapers, other journals, and contemporary books of various kinds. Not all this evidence is trustworthy, and it certainly does not refer in a uniform way to the particular information which we are at present seeking, namely the wage rate per day (or, in the case of pieceworkers, the average earnings per shift). Storm-Clark dismisses this material altogether, saying that 'because only a proportion of the underground labour force received payments direct from the colliery pay-office, both published and unpublished data on rates of wages are, in practice, useless as indicators of what the miner expected to take home in his pocket at the end of the working week'.[4] This is an exaggeration, even as far as earnings are concerned, but it does not apply to wage rates, which are our present concern. Surprisingly little of the material is contradictory, though, as Anthony Slaven points out in introducing his table of average day rates and weekly earnings in Scotland from 1856 to 1913, 'at best they represent crude averages of numerous rates current in each year'.[5] Moreover, one has to beware of numerous pitfalls, such as the fact that 25 cwt (or some other figure greater than 20) was reckoned to the ton so far as hewers' cutting prices were concerned in some coalfields at the beginning of our period, and in the North-East the chaldron, a measure of capacity, was used rather than any measure of weight. The completely standard 20-cwt ton was not achieved everywhere until the 1872 Coal Mines Regulation Act.[6] But for all its difficulties, this material can be used to construct indices of wage rates, and this has been done in table 7.1. As will be apparent from this discussion, it can only be regarded as a collection of rough approximations, and it is subject to all the usual cautions about indices of nineteenth-century wages.[7] But despite this, it is our impression that the figures in bold type, which are based on contemporary data of one sort or another, are reasonably reliable, though it is fair to say that there were sometimes quite considerable differences in the sources for the same year. These occurred most especially when owners and trade unionists supplied figures for the same date, though this was not invariably the case, and for the later part of the period in the North-East, these sources were in substantial agreement.

Table 7.1 *Shift earnings of hewers on piecework, 1800–86 (1886 = 100)*

	Northum-berland	Durham	Yorkshire South	Yorkshire West	Lancashire & Cheshire	East Midlands	Staffordshire North	Staffordshire South[a]	South Wales	Scotland	UK average
1800/2	**53**	**59**	**55**	**56**	80	**50**	**52**	…	**63**	**71**	**65**
1804/6	…	…	…	…	…	**66**	**57**	…	…	**83**	…
1811	…	…	…	…	…	…	…	…	**102**	…	…
1813/14	**65**	**71**	**58**	**59**	84	**66**	…	…	**68**	**79**	**73**
1818	**52**	**60**	…	…	…	…	…	…	…	…	…
1819	**48**	**54**	**49**	**50**	75	…	…	…	59	77	**64**
1822	**52**	**60**	54	55	80	**56**	…	…	**61**	**89**	**69**
1825	**75**	**81**	**64**	**66**	86	64	…	…	**86**	**119**	**86**
1826	…	…	…	…	…	…	…	…	…	**107**	…
1827	**52**	**60**	…	…	…	…	…	…	**86**	**95**	…
1828	**52**	**60**	…	…	…	…	…	…	…	**95**	…
1829	**48**	**54**	55	56	80	**54**	**62**	96	**76**	**95**	**72**
1830	**52**	**60**	55	56	80	55	66	**96**	80	**95**	**74**
1831	**73**	**80**	**55**	**56**	80	55	70	96	**86**	**95**	**80**
1832	**67**	**75**	55	56	80	55	70	96	**81**	**95**	**78**
1833	65	67	55	56	85	55	70	96	**80**	**95**	**77**
1834	65	67	60	61	90	60	83	96	**81**	**95**	**79**
1835	**65**	**67**	65	67	**90**	66	**86**	**96**	**86**	**95**	**81**
1836	72	74	75	78	95	70	86	96	86	**101**	**86**
1837	75	77	75	78	95	70	86	96	**86**	**107**	**87**
1838	75	77	75	78	90	70	83	96	86	**95**	**85**
1839	75	77	75	78	90	70	83	96	**89**	**95**	**86**
1840	72	75	75	78	83	70	**83**	96	**86**	**89**	**82**
1841	**72**	**75**	**68**	**70**	**75**	**66**	**72**	84	**81**	**83**	**77**
1842	**71**	**73**	68	70	**73**	**56**	**62**	84	**76**	**71**	**72**
1843	**66**	**66**	66	68	73	56	**55**	**84**	**71**	**65**	**69**
1844	**71**	**71**	**66**	**68**	**80**	60	63	84	**71**	**76**	**73**
1845	75	75	70	73	80	**75**	73	96	**86**	**79**	**80**
1846	82	**86**	**77**	**80**	**80**	**75**	80	**120**	**102**	**89**	**90**
1847	80	82	77	80	77	73	77	**120**	**102**	**89**	**89**
1848	**67**	70	75	77	67	66	73	108	**89**	**65**	**75**

1849	**71**	73	73	73	**67**	**66**	73	**108**	**89**	**65**	**75**
1850	**72**	75	72	74	**67**	**63**	73	108	**89**	**68**	**76**
1851	**72**	75	72	74	67	**63**	75	**112**	**85**	**68**	**76**
1852	72	75	72	74	**70**	66	77	112	**85**	**77**	**78**
1853	**77**	79	76	78	75	70	83	112	**107**	**77**	**84**
1854	101	99	**85**	**88**	**95**	**85**	**98**	112	**107**	**83**	**96**
1855	105	105	95	98	110	96	**98**	**112**	**107**	**95**	**103**
1856	101	99	100	103	**115**	100	**98**	**132**	**107**	**101**	**107**
1857	**92**	92	95	98	100	95	**88**	108	**107**	**87**	**97**
1858	90	91	**91**	**94**	83	90	**83**	108	**95**	**83**	**90**
1859	90	91	**92**	**95**	**83**	90	**83**	108	**95**	**83**	**90**
1860	92	93	**94**	**97**	90	95	**83**	112	**100**	**89**	**94**
1861	94	95	**98**	**102**	100	100	**83**	112	**100**	**89**	**97**
1862	**98**	97	**102**	**105**	**100**	100	**83**	112	**100**	**95**	**99**
1863	**101**	99	**97**	**100**	100	100	85	112	**102**	**95**	**99**
1864	102	100	**105**	**102**	105	105	89	**126**	**113**	**101**	**106**
1865	**106**	103	**114**	**111**	115	110	98	**120**	**117**	**107**	**111**
1866	**111**	110	**123**	**120**	125	120	103	**120**	**117**	**125**	**119**
1867	**95**	97	**108**	**105**	**115**	108	100	**120**	**117**	**101**	**108**
1868	**95**	**95**	**100**	**100**	**100**	100	93	**112**	**102**	**89**	**97**
1869	95	95	**100**	**100**	**100**	100	93	**108**	**100**	**89**	**97**
1870	**101**	**99**	**100**	**100**	**105**	**100**	**93**	**122**	**105**	**100**	**101**
1871	**117**	**102**	**109**	**111**	115	**109**	**100**	**134**	**107**	**120**	**110**
1872	**141**	**129**	**134**	**135**	140	**134**	**129**	**144**	**124**	**163**	**138**
1873	**171**	**160**	**163**	**164**	**180**	**163**	**145**	**157**	**163**	**178**	**166**
1874	**163**	**147**	**145**	**145**	**150**	**145**	**135**	**137**	**134**	**148**	**146**
1875	**138**	**128**	**130**	**130**	130	**130**	120	**111**	**119**	**120**	**125**
1876	129	114	120	120	**120**	120	**108**	106	**107**	**115**	**116**
1877	117	103	113	112	**110**	113	100	99	107	115	**109**
1878	**99**	**101**	**105**	**105**	105	**105**	95	**87**	**103**	**97**	**101**
1879	**95**	**92**	**100**	**100**	100	**100**	90	**84**	**97**	**100**	**97**
1880	**96**	**95**	**97**	**97**	**98**	**97**	92	**91**	**97**	**100**	**97**
1881	**97**	**99**	**97**	**97**	**107**	**97**	95	**102**	**100**	**101**	**101**
1882	**103**	**103**	**103**	**103**	**110**	**103**	102	**105**	**105**	**110**	**105**
1883	**106**	**105**	**100**	**100**	**110**	**100**	103	**102**	**107**	**115**	**106**
1884	**102**	**104**	**98**	**98**	**100**	**98**	100	**100**	**110**	**110**	**103**
1885	**104**	**102**	**100**	**100**	**100**	**100**	100	**100**	**105**	**105**	**102**
1886	**100**	**100**	**100**	**100**	**100**	**100**	**100**	**100**	**100**	**100**	**100**

Where such divergences occurred, a compromise estimate has usually been made, unless there appears to be good reason for preferring one figure to the other.

The remaining figures in table 7.1, those in roman type, are of a different nature. They are no more than our guesses, made on the basis of a mixture of guides, as to the likely course of wages. It is, of course, open to anyone to take a different view. The main factors taken into consideration in making these guesses have been changes in the price of coal, changes in the price of iron, changes in other local wage rates, and the movement of wages in coalfields for which there is information. It proved possible to push the table back to 1829 continuously on this basis, but prior to that the data are too scattered and uncertain to permit even the sort of speculation that is incorporated in the table.

The final column, it should be noted, is simply a weighted average of the other columns, together with estimates for the missing figures prior to 1830. The weights are provided by the estimated proportions of the workforce in each group of coalfields as derived from tables 5.3 and 5.4 above. It has been added for the convenience of those who want a single expression of the movement of miners' wages in the country as a whole.

There is insufficient data on the smaller coalfields prior to 1886 to warrant an attempt to deal with them in the same way for the whole

Notes to Table 7.1

(i) The 1886 Wage Census has been used as the base, except for Northumberland and Durham, where Rowe (1923) has been followed, and except also for Yorkshire, where figures from local newspapers and from colliery records have been used to effect the division between South and West.

(ii) The 'raw' figures for 1886 are as follows: Northumberland – 5s 2¼d; Durham – 4s 8d; South Yorkshire – 5s 6d; West Yorkshire – 5s 4d; Lancashire & Cheshire – 5s 0d; East Midlands – 5s 4d; North Staffordshire – 4s 10d; South Staffordshire – 4s 2d; South Wales – 4s 11; and Scotland – 4s 2½d.

(iii) Nothing has been added for the value of free or low-rent housing, which was almost universal for married men in the North-East, but relatively unusual elsewhere. Similarly, nothing has been added for the value of free or cheap coal, which was fairly generally allowed.

Deductions for candles, powder, etc., are, in principle, excluded, but it is not always clear whether gross or net pay is alluded to in the sources. This is likely to be most important in South Wales.

[a]Ten Yard seam only.

Sources: virtually every Royal Commission, Select Committee, etc., which reported in the nineteenth century contained some material on wage rates, and all of it has been taken into account here, along with the returns in *SP* 1887 LXXXIX and *SP* 1890–1 LXXVIII. The paybooks of a number of collieries have also been used. In addition, information was derived from the following: Anderson (1964); Arnot (1949–53), (1955), and (1967); Brassey (1879); Cunningham (1925); Griffin (1962) and (1971); Griffin (1969); Hardy (1956); Machin (1958); Raybould (1973); Scott (1949); Slaven (1966); Williams (1962).

period. Shropshire, where the piecework hewers were paid by the stint, as in South Staffordshire, has been well covered in the 1830s and 40s by H. Scott. The stint rates in this period changed as follows:

January 1831	1s 10d	January 1839	2s 10d	January 1844	2s 3d
January 1832	2s 1d	January 1840	2s 7d	April 1845	2s 9d
May 1836	3s 1d	January 1841	2s 3d	August 1845	2s 6d
January 1837	2s 7d	January 1842	2s 0d	October 1845	2s 9d

The normal number of stints done in a week was eight.[8] Since Shropshire was a district where output was stationary or declining from an early stage, and since it was close to the Black Country, it is probably reasonable to assume that the movement of wages behaved in a somewhat similar fashion in the two districts. For Cumberland, apart from one or two isolated figures at the very beginning of the period,[9] and a little in the *Report* of the Children's Employment Commissioner, there is nothing until the 1870s. What there is suggests that wages were closely related to those of Northumberland, but at a slightly lower level. The Forest of Dean wage rates probably moved in much the same way as those of neighbouring South Wales, but at a level about 6d per day lower,[10] though the gap may well have been larger at the peaks of booms. The Cannock Chase coalfield was only opened up around 1870, and though the levels of wages were rather higher than in the Black Country initially, they came down more rapidly in the middle 70s. Finally, money wages for hewers in Somerset and Bristol appear to have been at the same very low level in 1886 as they had been forty-five years earlier when Dr Stewart was reporting for the Children's Employment Commission, namely 3s 3d per day.[11] In the boom years of the middle 1850s they were reportedly as low as 3s a day, and for the depressed years of 1858–61 a range of 2s 6d to 3s 4d was given.[12] But they must surely have risen in the early 70s, though the likelihood of very much change at other times seems to be fairly low.

One element in the wages of miners which was not without importance was a variety of allowances in kind. These have not been included in the statistics shown in the tables in this chapter, but they certainly should not be forgotten, especially in making comparisons between different coalfields. In most districts, adult miners who were householders (and others in certain circumstances) received free or cheap coal for their own use. In the early part of the nineteenth century, indeed, they often received more than they needed, and some of this coal was sold to add to their income.[13] But the monetary value of this was quite small in any case and probably only amounted to an addition of 1d or 2d a shift to their wages. Of substantially greater value in Durham and Northumberland, and occasionally elsewhere, was the provision of a free cottage (or an allowance in lieu) to all married men. This was usually reckoned to be worth 10d or 1s per shift at the beginning of the twentieth century, and whilst it was probably worth a penny or two less fifty years earlier, it was almost certainly of greater importance in proportion to wages at that time. The practice was unusual outside the North-East and some parts of Scotland, and in most coalfields houses were only provided free for colliery officials. However, houses for a substantial proportion of miners were provided by the coalowners in most districts away from existing urban areas, although a rent was normally charged for them which fully covered their cost.

An additional allowance in some districts, but an offset in others, was the provision of tools, candles, powder, etc. Perhaps this should not be properly treated as an allowance, an addition to income, for these things were necessary to the miner's job. However, where deductions were made for them, which was most notably the case in South Wales, and also, as was almost universal, for pick-sharpening, they were certainly a factor in reckoning the hewer's net income.

Table 7.1 is confined to the shift earnings of piecework hewers, since there is much less information available about other grades prior to 1886, except for that which was provided by the Children's Employment Commissioners, and for one or two other odd years in some districts. Perhaps the best way of conveying what is known is to show the relationship of the wages of these other grades to those of hewers in 1886, and then discuss such evidence as there is for earlier years. Table 7.2 is the first step in this. The normal weekly earnings which it refers to are, of course, rather different from the figures we have been dealing with so far, for, even though absenteeism is ignored, there are differences in the normal number of shifts worked by different types of worker. Deputies, general labourers, and surface workers were generally expected to work six days a week, even where other underground workers had a shorter working week. And it is as well to remember, too, that hours differed considerably for the different grades.

Before discussing the districts separately, there is one general point. All the evidence that we have suggests, what indeed we should expect, that the weekly earnings of pieceworkers fluctuated more over the course of the trade cycle than did those of men paid by the day.[14] Since 1886 was a year of deep depression, we must reckon, therefore, that the pro-

Notes to Table 7.2

[a]The proportion for time-rate hewers was 82% in South Staffordshire, etc., and 84% in Gloucestershire & Somerset.

[b]The proportions relate to time-rate hewers in Shropshire, no figures being given for hewers on piecework.

[c]The hewers' assistants were in two groups according to their rate of pay.

[d]The proportion for those working less than 60 hours a week was 77%.

[e]The bank and screen boys were in two groups according to their rate of pay, the proportions being 58% and 35%.

Source: *SP* 1887 LXXXIX.

Table 7.2 *Weekly earnings of different types of colliery labour as percentages of hewers' earnings, 1886*

	Cumber-land	Durham & North-umberland	York-shire	Lanca-shire & Cheshire	North Wales	East Midlands	North Stafford-shire	South Staffordshire,[a] Leicestershire & Warwickshire	Shrop-shire[b]	Glouces-tershire & Somerset[a]	South Wales	Mid Scotland	West Scotland	East Scotland
Underground adults														
Deputies	106	111	104	114	…	100	115	95	121	92	120	98	107	100
Putters (piece)	…	76	81	76	…	…	{68}	62	…	95	99	89	…	92
Putters (time)	78	…	64	84	…	64		60	88	…	86	87	90	78
Fillers (piece)	…	…	…	{92}	…	91	94	82	…	…	…	…	…	…
Fillers (time)	…	…	…		…	77	85	78	90	…	75	…	…	…
Timbermen, shifters, setters, etc.	88	80	90	100	93	90	91	77	94	…	99	91	…	…
Rolleywaymen, roadmen, etc.	91	81	78	100	…	81	88	74	95	…	92	99	100	94
Stonemen, driftsmen, etc.	115	90	99	110	…	…	94	88	…	95	103	99	94	110
Engine tenters	100	94	90	94	86	78	90	80	…	…	86	103	98	98
General labourers	77	74	73	86	75	76	72	68	…	67	78	68	78	79
Boys														
Putters (piece)	53	48	48	{47}	{45}	…	57	…	…	66	58[c]	57	{56}	70[c]
Putters (time)	46	…	38			…	49	49	55	40	35[c]	57		52[c]
Drivers	…	29	39	36		37	36	41	46	…	44	55	55	48
Trappers	23	23	20	…		…	…	25	…	…	38	…	…	…
Surface adults														
Banksmen	82	114	75	83	78	{68}	74	63	95	{77}	85	77	90	85
Screensmen	76	66	68	72	…		72	63	82		81	71	84	75
Engine tenters	94	94	92	111	106	97	98	101[d]	104	95	103	94	107	107
Smiths	102	85	95	113	118	99	97	87	…	86	93	103	109	110
General labourers	75	70	70	76	75	67	68	63	98	66	77	69	78	75
Boys														
Bank & screen-boys	43	32	29	41	43	36	42	38	38	…	[e]	53	46	…
Women	35	…	…	34	…	…	…	21	…	…	36	36	…	40

portions shown in table 7.2 are generally higher than they would have been had the Wage Census been taken in a year of more normal trade. Bearing this in mind, it seems possible to say that in most coalfields at this period, the skilled underground jobs were paid at slightly lower rates than the average hewer got working on piece rates. In a depressed year, like 1886, this meant that their earnings were very similar, but in most years the hewer would get rather more, having a margin of something like 5 to 15%. In foregoing this margin, and in working slightly longer hours, the skilled oncostman obtained greater regularity in his earnings, and often had a rather easier job in terms of physical effort, though not of responsibility, since most such jobs were concerned with the safety of the pit.[15] Certain responsible and/or skilled jobs at the surface, typified in our table by the engine tenters and the smiths, were also paid at roughly similar weekly rates to those of the hewers, but in these cases both the hours worked per day and the number of days worked per week tended to be higher, often a lot higher for the enginemen. Other adult faceworkers, even though not particularly skilled or responsible, generally got over four-fifths of the hewers' average earnings, whilst less skilled work elsewhere underground was usually paid at something like two-thirds to three-quarters of the hewers' rate, though it was more than that in some cases. The two most obvious of such cases were in Lancashire and South Wales, both districts in which hewers were generally less far ahead of most other miners in their earnings. All these proportions, of course, were lower in boom periods, though the divergence was probably no more than a few percentage points, except perhaps in 1872–3, when there was a sufficient gap to tempt even deputies to want to return to hewing.[16] Earnings of most skilled surface workers were seldom much above three-quarters of those of the hewers, and in boom years they must have gone down closer to a half. Earnings of the older boys underground were usually about half those of hewers, but younger ones got more like a third as much, and some few even less.

The evidence for earlier years is very patchy, though there is a good deal about the wages of children in 1841. What there is suggests that there had been no fundamental change over the half-century previous to 1886 – and probably not before that – except for that produced by the gradual elimination of the younger boys; and, reinforcing this, an increase in the average age of those engaged in underground haulage brought about by changing technology. The earnings of putters working on their own were much the same proportion of those of hewers in 1841 and 1886 in Lancashire, Yorkshire, the West Midlands, and South Wales. They were slightly higher in Scotland, and quite markedly higher in the North-East of England. It was there, of course, that the average age of putters seems to have gone up the most. The proportionate rise in the putters' earnings would seem to have taken place by the early 1860s, and this reinforces the view that it was linked with the development of the district's shift system, whereby the putting came to be done mainly by 'boys' of eighteen to twenty-one years of age.

The relationship of the earnings of the skilled underground oncostmen to those of the hewers was very much the same as in 1886 in all cases where we have information, which covers pretty well every coalfield for at least once in the previous half-century. For general labourers underground and for surface workers there is hardly any evidence, but what there is – mainly for Durham and Northumberland – suggests that, whilst there was little change at the pithead, the relative position of the underground labourer almost certainly improved after 1870, though perhaps not before then.[17]

From 1886 onwards there is a good deal more information about wages available, but it is often of a rather intractable nature. Because of this, we have relegated to an appendix the description of how the indices presented in tables 7.3 to 7.6 have been formulated. At this point perhaps all that needs to be said about them is that, though partly based on detailed and reliable data about the movements of the percentages on or off 'the standard' basis rates, they are also partly based on more intermittent and sometimes more obscure data on those basis rates. This has involved making assumptions about which workers were assigned to which category by the compilers of the late nineteenth-century statistics, and there is a possibility that these assumptions may be mistaken. Assumptions have also had to be made about exactly when the changes in basis rates which have been uncovered took place, and in this case there is a virtual certainty that they are not precisely correct. However, the influence of errors of this nature on the indices is unlikely to have been of much significance; and whilst it is not possible to attach a formal and precise margin of error to the figures, we do not think that it would exceed 10%. It should be added at this point that, as in table 7.1, the UK average column in table 7.3 is put in for the convenience of those wanting a single expression of the movement of miners' wages in the UK. It is derived in the same way as in the former table.

One further point about tables 7.3 to 7.6 before discussing what they and the earlier tables tell us: the four categories of worker cover by far the great majority of underground workers, and can be regarded as representative of those few remaining; and it is possible to work out the relationships of the wages of the different groups to each other for each year. This has not been done here; but in order to provide a sequel to table 7.2, the proportions which the wages of other groups bore to those of piecework hewers are shown here for June 1914 in table 7.7. This also enables us to give some idea of the change in the wages of surface workers, which were

Table 7.3 *Money wages per shift of piecework coal-getters, 1886–1914 (1886 = 100)*

	Cumber-land	North-umberland	Durham	South York-shire	West York-shire	Lancashire & Cheshire	North Wales	East Midlands	Warwick-shire	North Stafford-shire	Black Country	Cannock Chase	Somer-set	Forest of Dean	South Wales	Scotland	UK averag
1886	100	100	100	100	100	100	100	100	100	100	100	100	100	100	100	100	100
1887	100	92	100	100	100	100	100	100	100	100	100	100	100	100	96	100	99
1888	108	88	100	102	103	102	103	103	102	103	103	102	102	102	96	102	100
1889	120	107	104	115	115	115	115	115	115	117	115	115	112	112	109	117	112
1890	139	120	125	130	130	130	130	130	130	132	130	130	123	122	135	145	132
1891	140	126	132	141	140	140	140	142	142	143	140	140	131	130	147	151	141
1892	135	121	128	141	140	141	140	143	143	144	140	140	131	130	130	145	137
1893	127	117	122	141	140	141	140	144	144	145	140	140	131	139	110	132	130
1894	131	117	125	138	137	138	137	141	141	142	137	137	128	125	121	144 (139)[a]	132
1895	123	109	117	132	130	131	130	135	135	136	130	130	118	117	114	122	124
1896	125	104	117	132	131	131	130	136	136	136	130	130	115	115	109	111	121
1897	127	104	119	133	131	131	130	137	137	137	130	130	115	115	110	116	123
1898	129	111	124	133	131	132	130	138	138	138	130	130	116	116	115	132	127
1899	134	122	134	140	137	138	136	146	145	145	136	136	123	123	124	156	138
1900	162	143	154	149	146	148	148	156	156	155	145	145	154	131	155	187	157
1901	164	150	164	165	160	161	165	172	172	170	159	159	162	146	177	170	167
1902	160	132	146	160	156	158	164	169	169	167	155	155	158	139	157	145	154
1903	152	128	144	155	151	153	161	164	165	163	150	150	151	136	153	144	151
1904	147	123	141	149	145	146	157	158	159	156	143	143	144	135	149	141	146
1905	147	119	139	147	141	143	155	154	156	153	140	140	141	132	141	141	142
1906	149	123	143	148	141	143	158	154	158	154	140	140	141	132	140	141	143
1907	160	139	156	158	150	152	171	164	169	164	149	149	153	141	158	170	158
1908	169	154	171	170	160	162	182	174	181	175	158	158	161	149	171	174	169
1909	164	139	155	164	153	155	173	169	175	168	151	151	156	146	163	156	159
1910	165	139	159	164	152	154	172	168	175	168	150	150	155	147	162	155	158
1911	166	136	161	165	152	154	172	168	177	168	150	150	162	147	165	155	159
1912	168	140	163	168	153	155	173	169	180	170	151	151	162	156	170	162	163
1913	183	157	185	182	165	168	187	182	196	185	163	163	174	159	179	189	178
1914	185	164	184	186	167	171	189	184	201	188	165	165	178	151	183	187	180

[a]The bracketed figure includes the four strike months at the lower rate; the other figure is for eight months only.
Sources: see appendix to this chapter.

Table 7.4 *Money wages per shift of putters, etc., 1886–1914 (1886 = 100)*

	Cumber-land	North-umberland	Durham	South York-shire	West York-shire	Lancashire & Cheshire	East Midlands	Warwick-shire	North Stafford-shire	Black Country	Cannock Chase	Somer-set	Forest of Dean	South Wales	Scotland
1886	100	100	100	100	100	100	100	100	100	100	100	100	100	100	100
1887	100	92	100	100	100	100	100	100	100	100	100	100	100	96	100
1888	108	88	100	103	102	102	103	103	104	102	102	102	103	96	102
1889	120	107	104	115	115	115	115	115	118	115	114	112	113	109	117
1890	140	120	124	130	129	130	131	130	133	129	128	123	123	135	145
1891	136	126	131	141	139	140	143	140	145	139	138	131	130	148	151
1892	131	122	127	142	139	140	144	140	146	139	138	131	130	130	145
1893	128	118	121	143	139	140	146	140	147	139	137	131	139	111	132
1894	131	118	124	140	135	137	144	137	145	136	134	128	125	122	144 (139)[a]
1895	124	110	116	134	129	130	138	130	139	129	127	118	117	115	122
1896	124	105	116	134	128	130	140	130	140	129	126	115	115	110	111
1897	125	105	117	135	128	130	141	130	141	129	126	115	115	111	116
1898	127	113	121	136	129	130	143	130	142	129	126	116	116	116	132
1899	131	124	131	143	134	136	151	136	150	135	131	123	123	126	156
1900	161	146	151	153	143	145	162	146	161	144	140	138	135	157	187
1901	162	153	160	168	156	159	179	161	177	157	152	145	145	179	170
1902	157	135	143	164	152	155	176	158	174	153	148	142	137	159	145
1903	148	131	140	160	147	150	172	154	169	148	143	136	133	156	143
1904	142	126	137	153	140	143	166	148	163	141	137	129	132	152	140
1905	141	123	136	152	137	140	162	145	160	138	133	126	129	144	141
1906	142	127	139	154	137	140	162	146	161	138	132	126	128	143	141
1907	153	143	152	165	145	149	172	156	172	146	140	137	136	162	170
1908	162	157	165	177	154	158	183	167	184	155	149	145	143	175	174
1909	157	144	150	172	147	151	179	161	177	148	142	140	140	167	155
1910	156	143	155	173	145	150	177	162	177	143	140	139	140	166	154
1911	151	143	158	175	145	150	177	163	178	143	140	139	140	170	155
1912	159	149	161	178	146	151	178	166	180	144	140	139	147	174	162
1913	173	161	185	195	157	163	193	181	196	155	151	149	148	184	189
1914	174	176	179	200	158	165	195	181	200	160	153	153	141	188	186

[a]See footnote a to table 7.3.
Sources: see appendix to this chapter.

Table 7.5 *Money wages per shift of timbermen, stonemen, repairers, etc., 1886–1914 (1886 = 100)*

	Cumber-land	North-umberland	Durham	North York-shire	West York-shire	Lancashire & Cheshire	North Wales	East Midlands	Warwick-shire	North Stafford-shire	Black Country	Cannock Chase	Somer-set	Forest of Dean	South Wales	Scotland
1886	100	100	100	100	100	100	100	100	100	100	100	100	100	100	100	100
1887	100	92	100	100	100	100	100	100	100	100	100	100	100	100	98	100
1888	108	88	102	102	102	102	102	102	102	102	102	102	102	103	96	102
1889	120	107	104	115	114	114	115	115	114	115	115	114	112	113	109	117
1890	140	120	125	130	128	127	130	130	129	130	130	128	123	123	134	145
1891	142	125	132	140	138	137	140	140	139	140	140	137	131	130	145	151
1892	137	120	128	140	137	137	140	140	139	140	140	137	131	130	129	145
1893	127	116	122	140	137	136	140	140	138	140	140	136	131	139	110	130
1894	131	116	126	137	133	132	137	137	135	137	137	135	127	125	120	144 (139)[a]
1895	123	108	118	130	126	125	130	130	128	130	130	126	117	117	113	122
1896	124	103	118	130	126	125	130	130	128	130	130	125	115	115	108	111
1897	125	102	119	130	126	124	130	130	128	130	130	125	115	115	108	116
1898	126	110	125	130	126	124	130	130	128	130	130	124	116	116	114	132
1899	131	120	134	136	131	129	136	136	133	136	136	130	123	123	123	155
1900	160	141	155	145	139	137	145	145	142	145	145	138	161	131	152	186
1901	169	148	165	159	152	149	159	159	155	159	159	156	170	145	173	169
1902	158	129	147	155	147	145	155	155	151	155	155	146	165	137	154	145
1903	154	125	145	150	142	140	150	150	146	150	150	140	158	134	150	143
1904	149	120	142	143	135	133	143	143	139	143	143	134	151	133	146	140
1905	147	117	141	140	132	129	140	140	135	140	140	130	147	130	138	140
1906	148	120	144	140	131	129	140	140	135	140	140	129	147	129	137	140
1907	160	135	158	149	139	136	149	149	143	149	149	137	160	137	154	169
1908	169	148	172	158	148	145	158	158	152	158	158	151	169	145	167	173
1909	163	135	157	151	141	138	151	151	144	151	151	144	164	142	159	155
1910	162	135	161	150	139	136	150	150	144	150	150	136	162	142	158	154
1911	163	133	163	150	138	135	150	150	144	150	150	136	171	142	159	154
1912	165	137	165	151	139	135	151	151	144	151	151	136	172	150	163	161
1913	179	153	188	163	149	145	163	163	155	163	163	146	184	151	173	188
1914	171	160	187	165	150	147	165	165	158	165	165	151	189	145	178	185

[a]See footnote a to table 7.3

Sources: see appendix to this chapter.

Table 7.6 *Money wages per shift of underground labourers, 1886–1914 (1886 = 100)*

	Cumber-land	North-umberland	Durham	South York-shire	West York-shire	Lancashire & Cheshire	North Wales	East Midlands	Warwick-shire	North Stafford-shire	Black Country	Cannock Chase	Somer-set	Forest of Dean	South Wales	Scotland
1886	100	100	100	100	100	100	100	100	100	100	100	100	100	100	100	100
1887	100	92	100	100	100	100	100	100	100	100	100	100	100	100	96	100
1888	108	88	100	102	102	102	102	102	102	102	102	102	102	103	96	102
1889	120	107	104	115	115	115	115	115	115	115	115	115	112	113	109	117
1890	140	120	123	130	130	130	130	130	130	130	130	130	123	123	134	146
1891	136	125	129	140	140	140	140	140	140	140	140	140	131	130	147	151
1892	130	119	125	140	140	140	140	140	142	140	140	140	131	130	151	145
1893	127	114	118	140	140	140	140	140	143	140	140	140	131	139	129	132
1894	134	114	120	137	137	137	137	137	140	137	137	137	128	125	120	145 (140)[a]
1895	123	106	111	130	130	130	130	130	136	130	130	130	118	117	113	123
1896	123	100	110	130	130	130	130	130	135	130	130	130	115	115	108	112
1897	124	100	111	130	130	130	130	130	136	130	130	130	115	115	109	117
1898	125	107	115	130	130	130	130	130	137	130	130	130	116	116	114	133
1899	130	116	123	136	136	136	136	136	144	136	136	136	123	123	123	158
1900	159	135	140	145	145	145	148	145	154	145	145	145	153	132	153	189
1901	159	141	148	159	159	159	164	159	169	159	159	159	162	147	174	172
1902	154	124	131	155	155	155	163	155	166	155	155	155	157	140	154	147
1903	145	119	128	150	150	150	160	150	162	150	150	150	151	138	150	146
1904	140	114	124	143	143	143	155	143	156	143	143	143	144	138	146	143
1905	138	110	122	140	140	140	154	140	156	140	140	140	140	135	138	143
1906	139	113	124	140	140	140	156	140	157	140	140	140	140	135	137	143
1907	150	127	135	149	149	149	168	149	168	149	149	149	153	144	155	173
1908	159	138	146	158	158	158	179	158	180	158	158	158	161	153	167	177
1909	153	126	132	151	151	151	171	151	175	151	151	151	156	151	159	159
1910	152	127	134	150	150	150	170	150	176	150	150	150	154	151	158	158
1911	153	125	135	150	150	150	170	150	179	150	150	150	162	150	160	158
1912	156	130	135	151	151	151	172	151	183	151	154	151	168	161	167	169
1913	171	146	153	163	163	163	188	163	200	163	170	163	185	163	177	200
1914	174	154	150	165	165	165	193	165	206	165	175	165	194	159	183	200

[a]See footnote a to table 7.3.
Sources: see appendix to this chapter.

Table 7.7 *Average shift earnings of different types of colliery labour as percentages of hewers' earnings, June 1914*

	Cumberland	Northumberland	Durham	West Yorkshire	South Yorkshire	Lancashire & Cheshire	North Wales	East Midlands	Warwickshire	North Staffordshire	Cannock Chase	Black Country	Shropshire	Forest of Dean	Somerset	South Wales	Scotland
Underground adults																	
Deputies, etc.	89	97	90	91	87	87	90	82	80	94	91	99	107	88	90	83	87
Haulage men	73	76	83	62	70	73	72	68	68	70	60	75	68	69	77	71	91
Timbermen, stonemen, etc.	86	79	90	81	81	78	77	76	68	75	77	...	...	83	98	82	97
General labourers	71	64	62	70	65	68	68	58	62	61	66	82	81	66	76	62	73
Boys																	
Putters and drivers	42	38	35	40	36	42	43	36	37	49	35	39	47	48	51	40	57
Surface																	
Winding enginemen	72	85	90	80	81	84	85	84	71	73	79	87	99	87	104	78	79
Screensmen	53	54	61	51	48	53	50	51	42	39	50	54	57	64	83	56	48
Engine tenters	66	73	73	61	59	59	65	59	52	50	58	62	77	68	78	60	68
General labourers	53	55	61	53	48	48	50	47	48	48	51	54	54	63	59	62	52

Source: CIC, appendix 49.

not governed by the same percentages on or off standard basis rates as applied below ground. Their wages did, in fact, move at very much the same times and in the same direction as those of the underground workers, but the fluctuations were weaker, and there was a marked tendency for the average level of wages for surface workers to decline relative to those of the men below ground. This began to be apparent in the 1890s, but became more pronounced after 1900.

Inspection of tables 7.1 and 7.3 to 7.6 and of figure 7.1, which presents some of the same material in graphic form, brings out very clearly the fluctuating nature of wages in the coal industry, most especially from the 1840s onwards. The smooth lines of the cotton workers' and the building workers' wage indices in figure 7.1 are in marked contrast to those of the hewers, and there would be a similar contrast if the index for farmworkers, engineers, printers, or, indeed, any group of workers for which we have information, were added. Wage rates in coalmining were exceptionally volatile, and the reason is not far to seek. The price of coal, as we shall see in a subsequent chapter, was extremely volatile, and wages formed well over half of the cost of producing coal. Most of the other costs were relatively inflexible in the short run, so that unless profits were to take the whole of the impact of price fluctuations, wages had to move more or less *pari passu* with prices. But in so far as the industry was a competitive one, there was no possibility of profits absorbing the whole of the price movements. On the one hand, competition in the labour market inevitably drove up wages as prices rose, given that the supply of skilled miners could not adjust immediately; and on the other hand, profits were not sufficiently high to accommodate the whole of the variations in revenue resulting from the wide changes in price. It is significant that wage fluctuations became more violent after the 1830s, when the coming of the railway broke down the local monopolies or oligopolies enjoyed in most coalfields by existing coalowners who were well placed for water transport. It is significant, too, that probably the least volatile coalfield up to the 1830s, so far as wages were concerned, was that of Durham and Northumberland, where monopolistic control was strongest. So long as profits were relatively high over the average of boom and slump, coalowners did not seek to overturn the force of custom which prevented wage cuts in many downswings.[18] The same custom, coupled with miners' ignorance of profit levels, or even of prices until long after they might have changed if markets were distant, kept wages from rising far in the upswings, so long as the expansion of output was restrained and did not require much new labour to be attracted into the pits. The yearly bond also introduced time lags into wage changes. But when competition cut into profits, and better communications began to dispel ignorance, then the owners sought to change the customs.

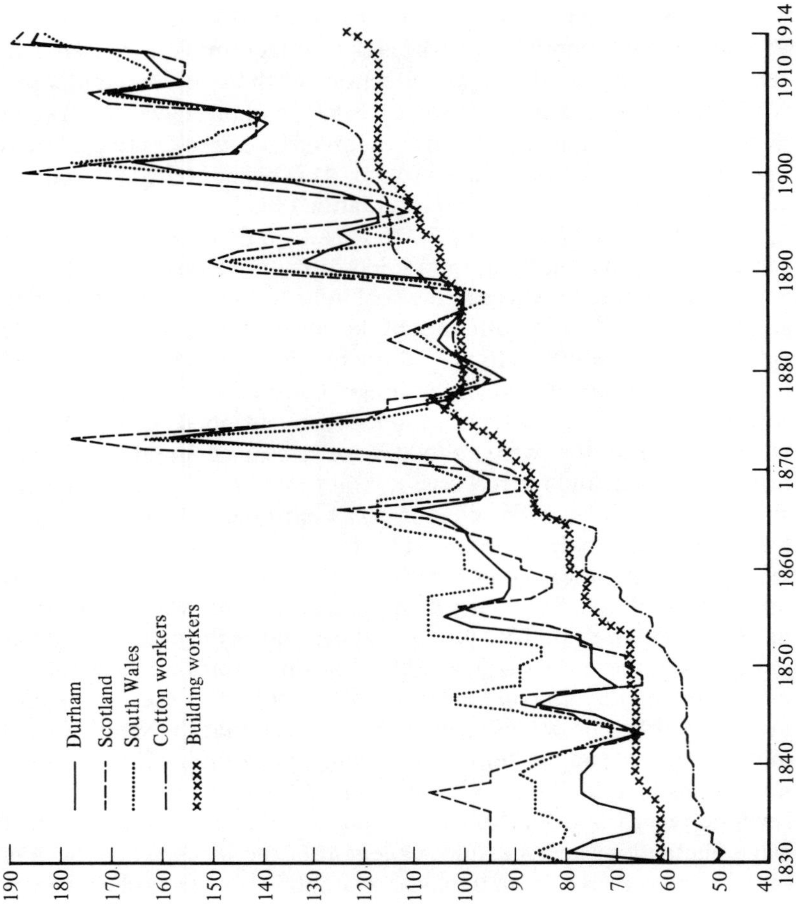

Durham
Scotland
South Wales
Cotton workers
Building workers
190
180
170
160
150
140
130
120
110
100
90
80
70
60
50
40
1830
1840
1850
1860
1870
1880
1890
1900
1910
1914

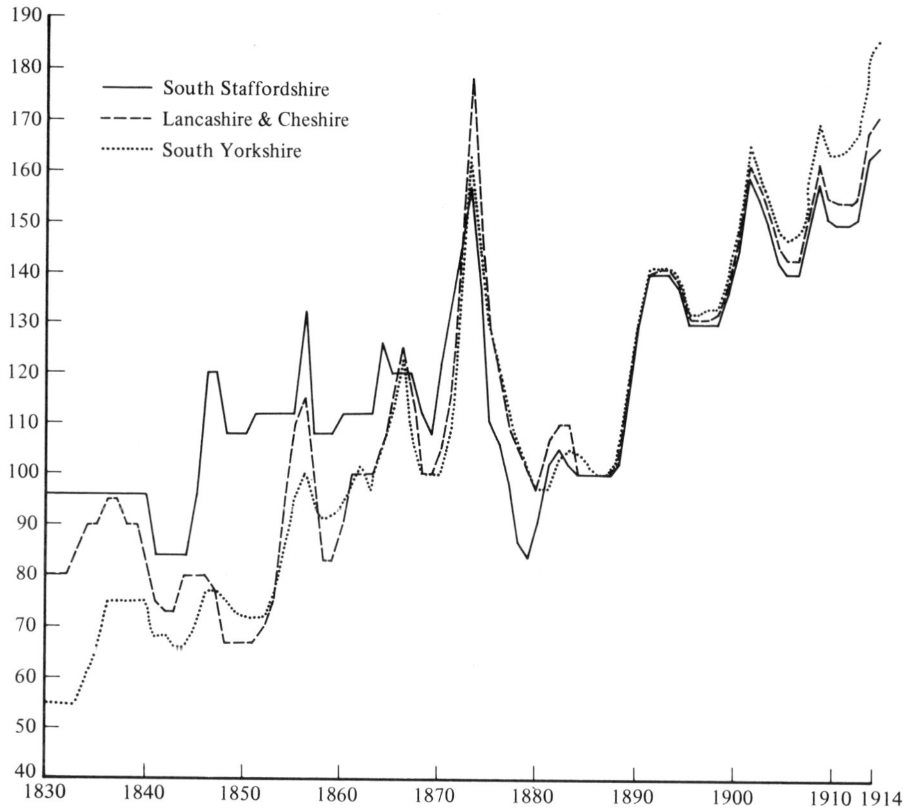

Figure 7.1 Shift wages of piecework hewers, 1830–1914 (1886 = 100)

The Napoleonic Wars produced a change in the previous situation of general stability in miners' wage rates, partly because of the shortage of labour which accompanied the rise in the demand for coal, and partly because of the general inflation of the period. There appears to have been a sharp rise in wages, most particularly in the period 1800–4 and again in 1810, in which the North-East shared, though not quite to the same extent as in those districts where the coal was suitable for smelting and iron ore was found in the coal measures.[19] Wage rates did decline in the next two or three years from the high level of 1810, but it was not until the end of the decade that they returned to something like that of 1800, which had itself been perhaps 20% above that of prewar.

It is possible to regard the behaviour of wage rates during the French Wars as an isolated episode, linked to peculiar wartime conditions; and, indeed, there is some merit in this view so far as Durham and Northumberland are concerned, for wage rates there returned to relative stability for another fifteen years after the Wars ended. It is perhaps significant that the only major strike of the war period in that coalfield, in 1810, was more akin to that of 1765 than to later strikes, for it was not for an advance of wages or against a reduction, but rather in opposition to the owners' move to change a custom of the industry. In this case the owners were demanding a change in the time of the annual hiring from October, when sales were most brisk, to January, when winter weather was liable to reduce sailings. Clearly this would have improved the employers' bargaining position, and was not unconnected with wages. But it is likely that the compromise date, April, which was eventually adopted, was almost as favourable to the owners as January; yet there was no immediate wage reduction, and when reduction did come, it was 'not considerable', if John Buddle is to be believed.[20] Even the strike of 1831 – after which wages did rise quite sharply, from an average of perhaps 2s 9d to 3s a shift for hewers in the late 1820s to 3s 9d – was launched primarily to stop a practice which had grown up recently and which the men thought both a violation of custom and a means of evading the spirit of the obligation in the bond to provide a minimum number of days' work in a week. The wage rise was, in effect, a means of buying off the strike (along with the cut in the boys' hours and the abolition of truck). However, the lockout of the following year was definitely an attack on the new wage level, as well as an attempt to impose discipline over the practice of restriction of output by the men. The employers had not only broken the force of custom by violating the spirit of the bond, but also by buying off their workers with a large wage increase, and by successfully colluding to impose a subsequent, though lesser, wage cut. In a way, therefore, it is possible to see the events of 1831–2 in the North-East as inaugurating the period of widely fluctuating wage rates in that district, a period which became fully established in the deep slump of the early 1840s, and prevailed thereafter until 1914.

The relative stability of wage rates which had been temporarily restored in the North-East following the Napoleonic Wars was not followed elsewhere, certainly not fully. Scottish miners in the eighteenth century had had higher average wages than those in England,[21] and this still seems to have been the case into the 1830s, though during the wars there was probably a narrowing of the differential, for wages rose less markedly in Scotland. The postwar fall was also less pronounced, however, and the rise to the peak of 1825 was apparently sharper than anywhere else. It would seem, therefore, that the pattern of wide fluctuation had begun there by the 1820s. This was probably also the case in Yorkshire, for hewing rates as low as 2s 4d to 2s 8d a day were reported there in 1819, compared with 5s 3d in 1825, both supposedly including hurriers' pay, which the hewer had to find.[22] Perhaps there is some significance in the fact that these rates come from an iron-making district, for the other main example of the pattern of wide fluctuations having been established as early as this comes from South Wales, where the link between the coal and iron industries was, at that time, at its strongest. There, the wartime peak rate in 1810 was double or more the wage rate of 1801, or 1816, or 1822; and whilst the peak rate of 1825 was lower than in 1810, it was still 50% above that of the surrounding troughs.

Elsewhere, where the outlets for coal were not so overwhelmingly in the iron industry, we may conjecture that the wage stability which had characterised the eighteenth century was at least partly restored after 1815, until improvements in transport and communications began to widen both the market opportunities and the scope for competition, and dispel ignorance about the movements of prices, if not about profits. These improvements were mainly associated with the spread of railways in the 1830s and 40s.[23] Once these had changed the circumstances of the coal industry and brought about a high degree of competition in practically all coalfields, a degree of competition which was enhanced in the second half of the nineteenth century in the exporting coalfields by the expansion of coal production overseas, then it is hard to visualise any other outcome so far as wages are concerned than that which actually occurred. Competition in the markets for coal ensured that wages followed prices downward in slumps, and competition for labour was at least part of the reason for wages following prices upward in booms.[24] The MFGB did, indeed, endeavour to limit the downward volatility of wage rates in and after the 1890s, with some measure of success, especially in the mid-90s.[25] But there is doubt as to whether this policy had much effect on the fluctuations of earnings (as opposed to wage rates), and it was not a policy which had any appeal for the majority of union leaders in the major exporting coalfields. There it was accepted, as it had been virtually everywhere before 1889, that wages must move with prices.

The experience of wages following the violent fluctuations of prices for forty years or more, often to the accompaniment of strikes and lockouts, which were commonly marked by violence and almost always by much suffering and misery, reached an extreme in the period 1871–7. In that period wage rates rose in Northumberland, for example, by almost 70% in the first two years, and fell by over 30% in the next four years. In South Wales the rise from 1871 to 1873 was 55%, and the fall in the following two years alone was nearly 30%. In almost every coalfield there were long and bitter strikes (or lockouts) in 1874–7 in vain attempts to retain the wage gains of the early years of the decade. These disputes were almost always entered into against the advice of union leaders, and when, as was inevitable in the circumstances, they failed, those leaders, anxious to find some way of avoiding the losses which such disputes imposed on both the union funds and on their members personally,[26] enthusiastically embraced the principle of the automatic sliding scale – and their members agreed with them. Later, there came to be extensive grass-roots hostility to sliding scales, and it is no doubt true that they had weaknesses from the men's point of view. But there is no doubt either about their initial popularity,[27] and they continued to satisfy the majority in South Wales until late in the 1890s, and one was even established as late as 1914. They were not without advantages, especially when a minimum was incorporated. The economist William Smart claimed for the sliding scale that 'it formulates the system which is already followed in the coal trade, though with much friction, that wages follow price. It thus substitutes a clear principle for seemingly arbitrary or half-informed action.'[28] That the men later came to see their interest in such arbitrary action does not mean that they did not see it in the automatic principle in the 1870s or 80s, nor does it mean that their interest did not in fact lie in the scales at that time. As Evans has argued about South Wales at that time, 'until a strong organisation was created they had to preserve the sliding scale at any price, since it was their only safeguard against unlimited reductions. In such circumstances, and lacking the funds needed to support a strike, their bargaining position was extremely weak unless market conditions were in their favour.'[29]

The existence of sliding scales in some coalfields at some times naturally raises the question as to what were their effects. Their critics point to the incentive given to employers by sliding scales with no minimum wage provision to produce as much coal as possible, since however far prices fell, profits would be almost maintained by the proportionate fall in wages. Similarly, there was an incentive to enter into long-term contracts at below market prices. Later critics have also condemned sliding scales for making collective bargaining unnecessary and leading to weak trade unions.[30] However, sliding scale agreements had to be renegotiated, usually at something like two-year intervals, and colliery price lists still had to be bargained over. Moreover, sliding scales do not appear to have affected the strength of the unions in Northumberland and Durham, and if they contributed to the weakness of South Wales unionism prior to 1898, there were other factors, as we have seen,[31] which were more important, above all the race/language difficulties.

Perhaps the best way of examining the effects of sliding scales is to see what happened to wage rates in different coalfields, and in the South Wales coalfield in the periods before and after the scales were adopted. If one looks at figure 7.1 there seems to be very little difference in the cyclical movement of hewers' wages in South Wales and in the other coalfields shown. The amplitude of the fluctuations was slightly greater in the case of Scotland, and the trough of the 1890s was definitely less marked in South Yorkshire and South Staffordshire – in fact, one can say in the Federated Districts, where the MFGB was operating its policy of fighting for a 'living wage' as an inflexible cost of the industry – than in any of the other coalfields. But in most cases the fluctuations are not very different in all districts. Despite the criticism that, as one miner is alleged to have put it (and doubtless thousands more did too), 'the bloody thing has no bottom', the reduction of wages in South Wales in the slumps of the late 70s, the mid-80s, and the mid-90s was not markedly greater than anywhere else; though it must be said that it was just a little greater in 1888 and 1896, whereas it had not been in 1879, at a time when the initial sliding scale, which included a minimum wage provision, had been in force. As a compensation, wages rose more in South Wales in 1890 and in 1900–1 than anywhere else except Scotland. There is, therefore, some evidence to suggest that wages fluctuated a little more amply where a sliding scale with no minimum was operating than in most other coalfields, even those, like Durham, with substantial export markets. But the difference does not really amount to much. There is, perhaps, one further qualification to be made, and that is that in the two trade cycles prior to the great boom of the early 70s, South Wales miners got their biggest wage rise one or two years before the other coalfields and their biggest wage cut came one year after the rest. This pattern was not repeated subsequently. Indeed, South Wales lagged in 1870 to 1871. But in view of the fact that there was no sliding scale until *after* the big wage cut of 1875, it seems difficult to attribute the change in pattern to the scale, and it is more plausible to argue that the earlier pattern arose from the more acute shortage of labour, and subsequent surplus, at that time in South Wales than anywhere else.

Turning from fluctuations to the trends in wage rates, it is useful to turn our indices into moving averages, and this has been done in figure 7.2 for the same five coalfields as before. This reveals considerable differences in the period up to 1870, and especially up to the 1840s, but thereafter patterns are fairly similar, at least up to the 1900s. Scottish hewers' wages were on a declining trend from the peak of 1825, when money wages reached a level not subsequently reached for over forty years, to the middle 40s. In the other coalfields, however, whilst wages fell in the decade following 1810, the long-term trend was more nearly stationary than downwards over the first four or five decades of the nineteenth century. We have already quoted John Buddle in 1830, describing the fall in wages since the Napoleonic Wars as 'not considerable'.[32] The rising trend for Durham wages, visible in figure 7.2, can be similarly described up to the late 40s. In Yorkshire, and probably the Midlands generally, except in the Ten Yard seam of South Staffordshire,[33] hewers' normal wages were round about 3s per shift (and occasionally lower) in all except the most active years right through to the middle 40s. Information is lacking about Lancashire wages in the early part of the century, but it is probable that they moved in a similar fashion to those in the other inland coalfields, though at a level 3d or 6d a day higher, probably because of the greater competition for labour in the district. In South Wales, too, the trend was probably flat through to the 1840s, though when expressed in nine-year moving averages it apears to undulate more, owing to the sharper fluctuations at intervals somewhat longer than nine years.

From the late 1840s to the early 70s the trend of money wages rose more or less steadily in all the coalfields shown in figure 7.2, though less steeply in the Black Country and in South Wales (and one could add to these Lancashire, which is not shown, and probably most of the smaller coalfields). The trend of agricultural and building industry wages have been added to the figure to represent other occupations which supplied the coal mines with probably more incomers than any others, and it is clear that miners were improving their relative position throughout this period *vis-à-vis* the builders (and, one can add, the cotton workers) but not the farm labourers. Between the mid-70s and the mid-80s, miners' wages everywhere show a steeply declining trend, which is matched in a milder way by farm wages, but is not reflected at all in those of builders. Finally, from the late 1880s right through to 1914 the trend of hewers' wages was once again quite strongly upwards, though with interesting minor variations between coalfields. Durham and South Wales – the exporting districts, in fact – had a less steeply rising trend in the decade 1885–95 than did Scotland or the Federated Districts, but they caught up subsequently (though this did not happen in the case of Northumberland). And amongst the coalfields of the Federated Districts, South Staffordshire wages trended upwards less steeply from the late 90s than

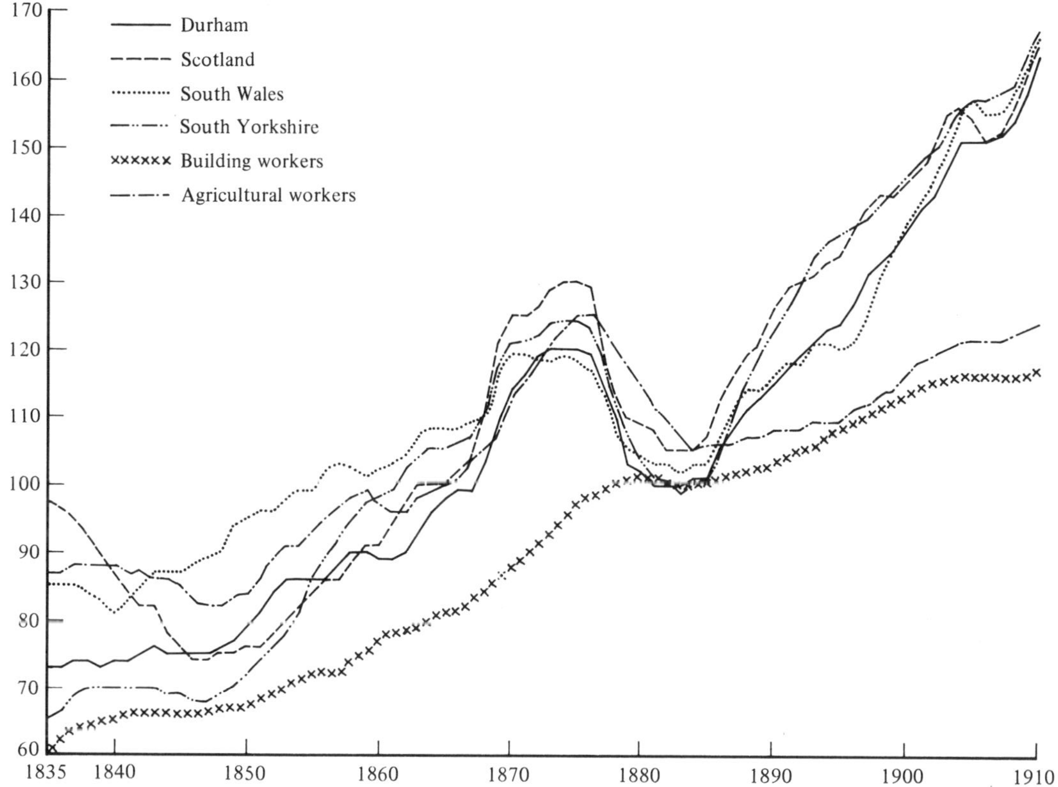

Figure 7.2 Shift wages of piecework hewers, 1835–1910 (1886 = 100) – five-year moving averages (centred on year shown)

did those of South Yorkshire. The former was typical of the movement in Lancashire, West Yorkshire, and the Forest of Dean, whilst the latter typified the remaining districts.[34] During this final period the wages of miners showed a marked divergence upwards from the trend of both building and farm wages.

We have spoken mostly hitherto of hewers' wages alone, largely because there are inadequate data about other groups of miners to trace the course of their wages in detail prior to 1886. For the last part of our period, however, it is possible to look at these other groups; but it is not worth doing so at any length, both because the basic picture is very much the same as for hewers, and because such minor differences as there were are implicit in our earlier discussion of tables 7.2 and 7.7. Suffice it to say that (i) the wage rates of putters and others who might be described as semi-skilled faceworkers and haulage hands varied very little in their trends from those of hewers in the same district; (ii) the wage rates of skilled daywage men also showed similar trends to those of hewers in all except the Federated Districts, where the upward slope of the trend was rather less steep; and (iii) general underground labourers' rates trended upwards at about the same rate as those of their own district's hewers in all coalfields except Durham, Yorkshire, Lancashire, the East Midlands, and North Staffordshire. In these coalfields the trend resembled that of the Black Country hewers.

Having established, so far as it is possible to do so, the 'facts' about miners' wages, it is time to try to explain more fully why they were what they were. It is likely that this could not be done in any complete sense, even if there were much more data available about the nineteenth-century economy than is in fact the case; and it is certainly not the intention here to embark on an explanation of the general level of wages and of the miners' place within that. The general level of economic magnitudes will be accepted as given, and we shall merely attempt to explain the behaviour of miners' wages in the short and in the long run.

Taking the short run first, it is evident from the close correlation between prices and wages, and the fact that the former always led the latter, that one key to the year-to-year (or, at times, month-to-month) fluctuations lies in changes in the demand for coal, and hence for miners' labour. These changes took place in the context of a much less volatile supply of coal – that is, in effect, a less volatile supply of miners *and* the other factors of production. In practice, this meant the capital embodied in shafts, machinery, headings, roadways, tubs, ponies, etc., etc. Not only does this key lie in changes in the demand for coal, but also in its nature, that is in its generally price-inelastic character in the short run. As the level of activity in the economy (and, for that matter, in the economies to which coal was exported) rose, so the demand for coal rose, even though prices might also rise. The immediate response was to work the existing factors of production more intensively – the number of days on which the pits were open was raised until it came as close as accidents and maintenance needs allowed to the theoretical maximum (or its conventional substitute, where six days' work every week had ceased to be accepted as the norm): that is, the rightward-shifting demand curves moved along existing supply curves which were relatively, even for a time perfectly, elastic. At this stage, in the conditions of competition which generally obtained from the 1830s and 40s onwards (if not earlier), neither the price of coal nor wages would change very much,[35] though it is possible that collusion by owners could have forced up prices in the face of buoyant demand, and it is also possible that a strong trade union could have forced up wages, and hence prices, against little resistance from employers who were well aware of the buoyancy of demand, for their profits were gong up, as overheads were spread over increased output. There is, however, nothing to suggest that either of these possibilities was explored, at any rate after the ending of the Limitation of the Vend; and there is nothing to suggest that any union leader ever considered the second possibility – perhaps because there was no national union unity until 1908 so that the fundamental conditions were not present.

As the upswing in activity continued and demand for coal rose, the existing capacity began to become less and less adequate to meet it, and coal prices would then begin to rise. The immediate effect was to bring about such expansion of capacity as could be achieved in the relatively short run, by opening new faces, sinking shallow pits at the outcrop, and reopening marginal collieries. But, with only one exception between 1840 and 1914,[36] this proved inadequate to meet the rising demand of the booms, and new deeper sinkings, which took a year or two (or more) to come into production, were required. Whilst this increase in capacity was in the making, owners of existing mines were in what may perhaps be described as a quasi-oligopolistic position, and prices rose increasingly steeply as the boom reached its peak. Such new openings as could take place in the middle stages of the upswing, together with the actual operation of sinking new pits, might well begin to create a shortage of labour, more particularly in those districts where the long-term trend of output was most steeply upward. But, in the light of the prevalence of severe short-time working in depressions, and of the continuing year-by-year growth of the community from which miners were more or less automatically recruited, it is hard to believe that such shortages did generally become evident in most coalfields at this stage of most upswings. Yet the rise in wages normally began very soon after the rise in prices – in 1864, for example, or in 1870 – two or more years after the trough of the depression and before the peak of the boom.

It is not immediately obvious what was the cause of this, if it was not shortage of labour; that is, in economic terms, the result of the movement of the demand curve to the right meeting a more steeply sloping part of the supply curve for labour. It is possible that one cause was the men's demand for higher wages, whether expressed through a formal trade union or not, meeting employers' unwillingness to resist such a demand in the prevailing circumstances of buoyant trade. The accounts of the 1831 strike in the North-East are consistent with this hypothesis; and there are many instances in the literature of demands for wage increases in the upswing, even at a comparatively early stage of it, being met with little or no resistance.[37] Another possible cause is the force of tradition, though in view of the persistence of wage stability into the 1830s in some districts, this could only apply generally to the later part of our period. On the whole, it seems most likely that the main cause was the collective feeling on the part of the men that active trade meant higher profits, even before prices had risen substantially, and that they should have a share of them. This led to a collective willingness to threaten the owners' prosperity by laying the pits idle. The employers themselves probably accepted the idea that the miners should have some part in the prosperity,[38] and, in any case, as Page Arnot put it about 1872–3, 'the market was too good for the coal-owners to face the prospect of prolonged stoppages'.A good individual example of this comes from the letter-book of the Niddrie & Benhar Coal Company, operating in the Lothians, on 15 January 1891, when prices were at their peak, the managing director wrote to the manager of the Benhar colliery: 'You quite understand that no strike must be allowed to occur – if 3d or 4d a shift will satisfy the demand, just judge for yourself when it is necessary to concede it.' The contrast with his attitude in a letter sent two-and-a-half years later, in the slump, is stark. On 27 June 1893 he told the manager: 'The Directors think the men are unreasonable and they cannot agree that this company should pay a higher rate of wage than is current in the district. The colliery has not been paying for many months back, and it will not make much difference keeping it idle for a time.'[39] In economic terms, one can say that the threat of the miners to reduce the supply of labour to nil at a wage level below that which they were demanding was particularly effective, because the employers' demand curve for labour was relatively inelastic over the relevant range at such times of quasi-monopoly for owners of existing capacity. Moreover, this essentially static economics conclusion is reinforced when it is remembered that the demand curve for labour was moving outwards to the right during the boom.

As the boom proceeded to its peak, prices normally rose sharply and wages followed soon afterwards. In this phase, shortage of labour probably did become a significant factor in raising wages in most coalfields up to and including the early 1870s, and in some districts later. Unfortunately, it is impossible to distinguish the effects of labour scarcity from the effects of collective pressure for wage increases or of custom. A contemporary economist, R.H. Hooker, argued firmly in the early 1890s that shortage of labour was not significant by then. The coalowner, he said, 'appears to be able to secure sufficient labour by the rise in wages which he is compelled to give, not because he wants the extra hands, but because it is demanded by the old hands'.[40] But there is no way of testing this opinion. Techniques which relate wage experience to levels of union membership are inappropriate in the nineteenth-century British coal industry, both because of the absence of adequate data and, more important, because of the presence of a collective attitude amongst miners which made formal union membership a matter of much less significance than in other occupations. As Challinor has noted, 'the non-unionists were frequently more militant than the union members'; and, as he also points out, 'in an important sense, the union never dies. By the middle of the nineteenth century, organization among miners had reached a sufficient level of maturity to ensure that, whatever happened nationally, at individual collieries men retained some semblance of solidarity.'[41] If anything, this is put in too deprecating a manner; but it calls attention to an important characteristic of the industry from the 1840s onwards, and to some extent earlier. Collective solidarity *within* a coalfield was exceedingly well-developed in the North-East and in Scotland even before the nineteenth century began, and it quickly became almost as strong elsewhere, although the Irish were a disruptive influence on it. And if it was still possible to get some miners to blackleg in *other* coalfields as late as the 1870s, this was only of any significance in periods of depression, and had no influence on bargaining power in upswings.

In the downswing the order of events was reversed. Prices fell, not so much initially because demand fell (except, perhaps, in the second half of 1873), as because new capacity began to come into operation, and the new labour attracted into the industry, though less productive, was increasing the output. With prices falling, employers' resistance to pressure for wage increases stiffened, and, frequently, such pressure was eased. Trouble only really came when the fall in prices had gone so far as to force some employers, at least, to demand that wages be cut. In the days before permanent union organisations came into existence, such cuts, like the rises which had preceded them, were usually quite moderate, and they did not in themselves lead to much more than grumbling – unless, that is, they were associated with attacks on customary practices, as in the North-East in 1765, or with food shortages, as in South Wales in 1816.[42] But as formal unions came into being and the oscillations of wages became wider, the miners began to show less disposition to accept the loss of the gains of the boom without resistance, and some unions owed their very existence to such resistance.

This continued to be the feeling of rank-and-file miners throughout the rest of our period, despite the fact that union leaders became increasingly cautious about such resistance in downswings, and were almost uniformly against the strikes and lockouts which did occur at such periods. Most of these disputes, which tended to be both unusually lengthy and almost entirely unsuccessful, at any rate until the equivocal case of the Federated Districts in 1893, were entered into against the advice of union leaders, at least from the 1860s onwards. Leaders like William Pickard in Lancashire, John Normansell in South Yorkshire, and John Dixon in West Yorkshire had founded their unions in partially successful resistance to wage cuts in 1858, when the owners had demanded what resistance showed to be excessive reductions. But by the late 1860s and 70s they were arguing forcefully against resistance which they thought was foolish.[43] Normansell's circular in May 1874 expresses their attitude well: 'We are now in a falling market, when it would be madness to strike, when every other district is down, and we must make the best of it.'[44] Later critics, like the Webbs, who have accused such leaders of adopting employers' economics, seem to have ignored what was the reality of the situation, especially in the later 70s, when many employers were making losses and few were achieving even modest profits. Those leaders who ignored the reality at the time, like Halliday of the Amalgamated Association of Miners, smashed their unions. It is hard to believe that the 'forward policies', which latter-day authors, like Keith Burgess, have clearly regarded as likely to have served the men's interests better than those of moderation,[45] really would have done so in the circumstances of 1874–9 or the mid-1880s. The eight weeks' strike in Northumberland in 1877 against a cut of 12% off standard (7½d a shift for hewers), which failed, must have cost the average hewer something like £10 in wages foregone. If he had been successful it would have taken over a year to make up for the loss, disregarding additional losses incurred by borrowing at high rates of interest or by selling possessions at a discount. With the failure, predicted by the leaders whom he ignored, this represented a dead loss of around a fifth of a year's earnings. The same consideration applies *a fortiori* to the seventeen weeks' strike of 1887.

Despite the leaders, however, miners continued to resist wage reductions in downswings, though with no major success until the 1890s – and that not a clear-cut one – and usually at the cost of much suffering and bitterness. It is, of course, true that these long drawn-out, defensive battles provided the stuff out of which myth was made to encourage solidarity and unionism in the future, and that in this way they had an imponderable effect on the later history of the industry. But this was in the minds of the strikers of 1877 probably to an even less extent than the maintenance of regimental traditions was in the minds of the majority of the men who charged at Balaklava.

With the 1890s the situation did change a little, in that the leaders of a stronger union – stronger because it covered a much wider area – themselves adopted the policy of resisting wage cuts in a depression. As has been suggested above, they had a measure of success, though much less with regard to earnings than to wage rates; and this was repeated in a rather more muted fashion after the cyclical peaks of 1900–1 and 1907–8. Unthinking rank-and-file resistance to wage cuts was transformed into the principle of the 'living wage', and the defence of the principle was conducted more coherently than in previous downswing disputes. The MFGB certainly exerted some power in such an unfavourable situation, even though its 'victory' was, as we have seen, far from indisputable.[46] But up to that time it is hard not to believe that formal unions had had very little influence on the course of wage rates, that rises would not have occurred more or less when they did in any case, and that union resistance in downswings meant little but loss and suffering. At the most, as Charles Markham, an 'enlightened' employer in the East Midlands, put it, 'combination might accelerate or retard the rise or fall in the price of labour, but it must ultimately depend on supply and demand'.[47] The wider, more powerful union put together by Benjamin Pickard, William Ashton, and the rest did have more defensive power than the old-time combinations; but it would still be a mistake to exaggerate it, firstly until all the coalfields were included in it, and secondly so long as the employers were up against competition in the markets for coal. The experience of the 1920s brings home the importance of this last point.

So much for the reasons behind the short-run movement of miners' wages: what about the long term? Here one can begin by pointing to the strong upward trend of the demand for coal up to 1873, and to the fact that the lower rate of growth of demand which followed was offset, to some extent, after 1888 by falling product per man, so that both the labour force and wage rates returned to rates of growth more nearly like those of the thirty years before 1873. The supply of labour, as has already been noted, did prove inadequate at existing wage rates in some coalfields at various times, and higher wages – for whatever the reason they were given – were required to attract the newcomers to the industry whom we noted in chapter 5. However, after 1873, this ceased to be of much importance outside South Wales and areas like South Yorkshire in the 1900s or Nottinghamshire for a few years on either side of 1900. For the rest, as we saw, the potential increase in the labour force from the mining community was more than adequate to supply the men required – at any rate until the extraordinary need for more labour in the aftermath of the Eight Hours

Day Act. Even South Wales, the Doncaster area, Warwickshire, and so on, got most of their newcomers from migrants from coalfields with surplus labour after 1873. Hooker, writing about twenty years after that time, was quite clear that 'the supply of labour in coal-mining seems now to have hardly any effect in settling the remuneration, although no doubt it would have if the supply were limited. But there appear to be now sufficient applicants for the work to enable us to consider the supply as almost constantly in excess of the number required.'[48] Of course, this did not apply so universally in the peak years, like 1900 and 1907, and at other times not all Welsh, South Yorkshire, Warwickshire, or even South Durham coalowners saw it that way – and the steeper rise in the trend of wages for these districts than for Lancashire, West Yorkshire, or South Staffordshire does reflect their need to attract more men for elsewhere.

If shortage of labour and the need to attract newcomers can be seen as the main positive reason for the rising trend of wages up to 1873, and as an influence, albeit a weaker one, in the period from roughly 1899 onwards, one is still left with the question why *excess* labour in cyclical downswings did not wholly negative the rising wages of the upswings – why, in fact, there was a ratchet effect, which, when smoothed out by moving averages, produces the rising trend. This is not at all an easy question to answer. On the face of it, there was nothing to prevent employers cutting the wage rate in one depression to what it had been in the previous depression, in both of which there was (in effect, by definition) excess capacity as well as excess labour, working short time. If one looks solely at money wage rates this, indeed, is what generally did happen for long parts of our period, though with variations from time to time and from coalfield to coalfield in the speed with which the cuts came. In South Staffordshire this happened in the three depressions from 1847 to 1869, and the money wage rate was even lower in 1879 and 1886. In Somerset the trend of wages was apparently level right up to the end of the nineteenth century. In Durham hewers' wages returned in each succeeding slump to about 3s a day up to and including 1858–9, and the slump level of 1868–9 was matched in 1879–80 and 1886–8. This last statement was true of every coalfield; and whilst in all coalfields not yet mentioned wages in the depression of 1858–9, as well as 1868–9, improved on the preceding depression, these were the only times prior to the 1890s when this happened. The rising trend of money wages up to the early 1870s, then, depends heavily on what happened to wage rates in the two depressions of the so-called 'Mid-Victorian Boom', and on the extreme height of the upswing culminating in 1873.

It is arguable that it is misleading to look solely at money wages, and that real wage rates are what should be considered. Figure 7.3 is an attempt to do this, though it must be regarded as an impressionistic affair,

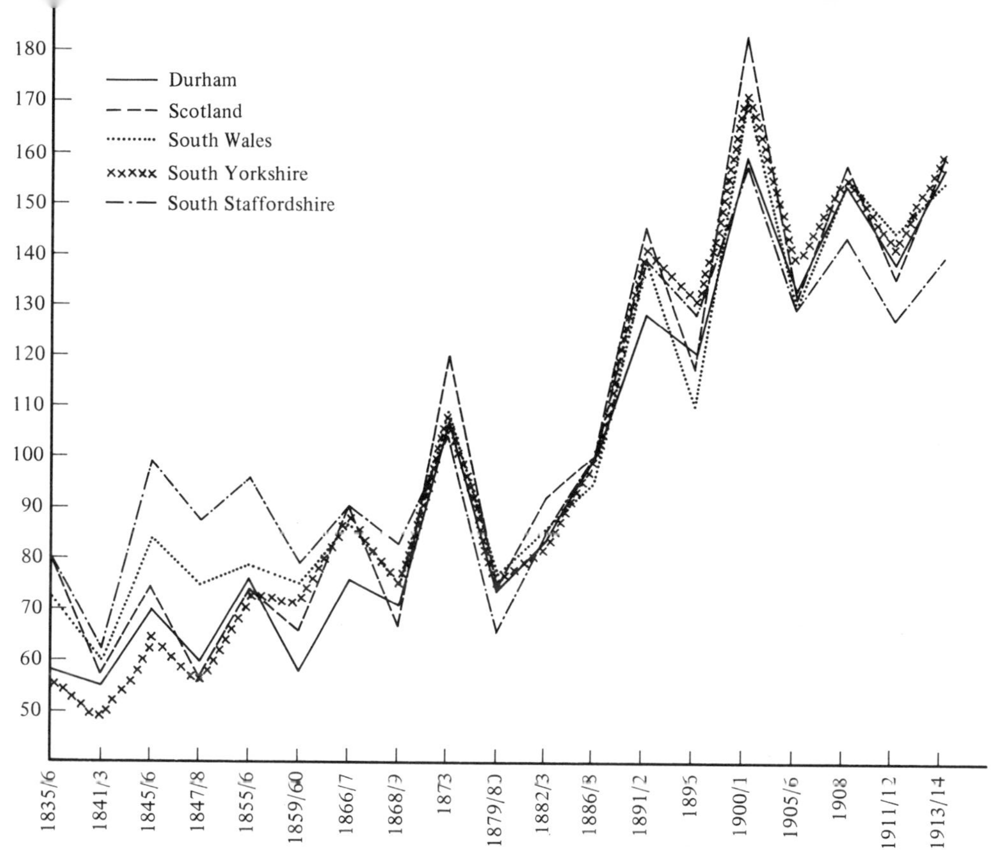

Figure 7.3 Real wages of piecework hewers at cyclical peaks and troughs, 1835/6–1913/14 (1886 = 100)

because of the absence of a retail price index prior to 1891, and the consequent necessity to use wholesale prices,[49] as well as because of the uncertain reliability of some of the earlier wage data. Partly because of this, and partly to bring out the extremes in the fluctuations, a slightly different method has been used from that used for the money wage figures. Only the cyclical peak and trough rates have been plotted, and these are unlikely to have been affected too much by short-term divergences between retail and wholesale prices. It is clear from this figure that even when real wages are examined, there still remains the period in the 1850s and 60s (and following 1879 – instead of 1886, as in the money wage series) – when, in the great majority of coalfields,[50] wages were not cut back in each succeeding depression to the level of the previous one. Let us concentrate first on the mid-Victorian period.

According to the simple neo-classical model of the labour market, the widespread failure of employers to cut wages in the downswing so far as in the previous downswing must have been the result of either a change in the level and/or elasticity of the supply curve of labour, or a change in the demand curve for labour, or some imperfection in the market mechanism affecting either or both curves. The first of these seems inherently unlikely in view of the recurrence of widespread short-time working in each depression; though the fact that a substantial number of workers left the industry between 1874 and 1878 indicates that there was, perhaps, more elasticity over the relevant range of the supply curve at that time than at any other. However, it did not prevent wages returning to roughly the 1868–9 level, so some other factor was at work to offset it.

If some autonomous change in the supply curve of labour is to be rejected, what of the possibility of an artificial interference with the supply curve, such as would fit also into so-called 'institutional' models of the labour market – collective imposition of a barrier to wage reduction beyond a certain point (higher than the previous slump level) by a formal or informal organisation of existing workers, who were in a position to restrict entry by newcomers? Put this way, such an explanation stands out as implausible for an era when railways were making competition between coalfields readily possible, yet when no national trade union existed, other than in name, for more than a very few years, and none ever succeeded in formulating, let alone in pursuing, a national policy.[51] It may well have some application for 1893 and afterwards, but it can scarcely have had much validity in the middle of the century.[52] And even if, because of group solidarity (say), it did carry some weight, it would still be necessary to look at the demand for labour, for the wage level at which workers set their 'barrier' could surely only be held provided it was not so high as to drive substantial numbers of employers out of business; for one cannot believe that any union could hold its position in the face of a massive increase in unemployment, which was sure to come until competition between coalfields – and, for that matter, from overseas producers – was eliminated. No union did, in fact, try to do this, though the MFGB indulged in what might be called 'brinkmanship' in the 1890s.[53]

What, then, of the demand for labour? According to the neo-classical model, it is set by the marginal value (or, if there is imperfection of competition, the marginal revenue) product, which is assumed to be derived independently of the particular labour market in question by the demand for its product, as expressed in prices, coupled with labour's marginal physical product.[54] Given the highly competitive conditions in coal markets in times of downswing, and especially in export markets, it seems reasonable to accept that the price of coal was independent of the miners' wage level at such times, and that thus far the neo-classical model will be appropriate. There are those, however, who would question the applicability of the whole concept of a downward-sloping marginal product curve to the coal industry, on the grounds that the variability of geological conditions in most, if not all, mines is such that the production of the marginal man or hour is just as likely to be above the average production as below it, provided that the marginal man is of much the same skill as all the others. This proviso might not be met when labour is scarce and more men are being taken on, but it is likely to apply in the downswing, so far at least as each individual colliery is concerned, given the industry's custom of spreading unemployment through short-time working for established workers.[55] The force of this argument is largely lost, however, when it is recalled that the quality of individual pits differed quite considerably, and that there were unquestionably marginal *mines* in which the productivity of all factors of production, including labour, was below average. Their marginality was reinforced by the tendency of the better, more productive workers to go to the collieries which could afford to give the better cutting price lists.[56] Since one of the reasons for a mine's marginality was the inferior quality of its coal, such a mine was under particular pressure in a downswing to reduce its prices, and hence to need to reduce its costs. Mines whose marginality sprang from difficult working conditions would be under less pressure to initiate a price reduction, but would have equal need to reduce costs once prices had fallen. In the light of all this, the concept of marginal product does not seem inapplicable to the coal industry; but it must remain an essentially theoretical matter, for there is no way in practice of discovering what the marginal product is.

Let us leave theory for the moment and look at what happened in a downswing in more concrete terms. The owners opened their pits and produced as much coal as possible so long as they could sell it at the price given by the market – having met their long-term contracts, of course, which might be a very large part of total output in the case of pits pro-

Table 7.8 *Movements in indices of hewers' wages and of prices between cyclical turning points (first differences)*

	North-East		Yorkshire		Lancashire		South Wales		Scotland	
	Wages	Prices	Wages	Prices	Wages	Prices	Wages	Prices	Wages	Prices
1836/9–1842/6	−10	−26	−10	−17	−17	−23	−18	−13	−42	−40
1842/6–1846/8	+18	+15	+12	+2	+7	+17	+31	+13	+24	+36
1846/8–1849/51	−12	−24	−6	−15	−13	−30	−17	−7	−21	−37
1849/51–1854/6	+33	+50	+29	+29	+48	+30	+22	+19	+33	+47
1854/6–1858	−15	−26	−9	−15	−32	−21	−12	−7	−18	−20
1858–1866/70	+20	+33	+28	+11	+42	+29	+22	+13	+32	+35
1866/7–1868/70	−15	−15	−21	−8	−25	−13	−17	−7	−14	−39
186686/70–1873	+71	+132	+64	+132	+80	+124	+63	+125	+77	+152
1873–1879/80	−73	−142	−67	−137	−82	−133	−66	−140	−81	−148
1879/80–1882/3	+12	+10	+6	+16	+12	+5	+13	+14	+18	+13
1882/3–1886/8	−11	−14	−3	−32	−10	−16	−14	−22	−15	−31
1886/8–1890/1	+35	+59	+41	+62	+41	+58	+51	+101	+51	+73
1890/1–1896/7	−18	−44	−10	−35	−10	−42	−38	−83	−40	−44
1896/7–1900/1	+47	+117	+30	+89	+30	+92	+68	+118	+76	+144
1900/1–1905/6	−28	−88	−19	−79	−18	−66	−37	−65	−46	−127
1905/6–1907/8	+34	+52	+19	+36	+19	+34	+31	+55	+33	+75
1907/8–1909/11	−14	−29	−8	−18	−8	−13	−9	−16	−19	−54
1909/11–1913/14	+26	+69	+16	+36	+17	+35	+21	+23	+34	+75

Sources: see text.

ducing higher quality coal. If, after a limited amount of stockpiling, they could not sell all they produced, they – meaning, effectively, some marginal producers – cut the price, and hoped thereby to clear the market. This made them put downward pressure on wages, which might or might not meet resistance; but in the absence of a strong union it was unlikely to be very serious resistance so long as wages were well above the previous slump level.[57] The whole operation, in fact, might or might not succeed. But undoubtedly, in many cases it did, for prices and wages were reduced in every depression by this sort of process. In these cases, presumably, the marginal producers did sell a larger amount at lower prices, and non-marginal producers found that they, too, could reduce wages, using the fall in prices as the justification, just as it was both justification and *necessity* for the marginal producers. The problem to explain is not these, but the cases where it did not succeed, or was not tried, to the same degree as in the previous depression. And if the reason does not lie in changes on the supply side of the labour market, it must be that demand in these depressions did not fall so far, in relation to capacity to produce, as in others. In other words, we may, if we so wish, deduce that marginal revenue product did not fall so far, either because marginal physical product rose, or because prices fell less, or some combination of the two.

Of course, as we noted earlier, this argument is entirely theoretical, for there is no way of identifying marginal physical product. However, this

Table 7.9 *Movements in indices of hewers' wages and of prices between cyclical troughs (first differences)*

	North-East		Yorkshire		Lancashire		South Wales		Scotland	
	Wages	Prices	Wages	Prices	Wages	Prices	Wages	Prices	Wages	Prices
1842/6–1849/51	+6	−9	+6	−13	−6	−13	+14	+6	+3	−1
1849/51–1858	+18	+24	+20	+14	+16	+9	+10	+12	+15	+27
1858–1868/70	+5	+18	+7	+3	+17	+16	+5	+6	+18	−4
1868/70–1878/80	−2	−10	−3	−5	−2	−9	−3	−15	−4	−4
1878–80	+1	−4	+3	−32	+2	−11	−1	−8	+3	−18
1886/8–1896/7	+17	+15	+31	+27	+31	+16	+13	+18	+11	+29
1896/7–1905/6	+19	+29	+11	+10	+12	+26	+21	+53	+30	+17
1905/6–1909–11	+20	+23	+11	+18	+11	+21	+22	+39	+14	+21

Sources: see text.

may not matter so much, for even if it did rise more over some cycles than others, it would simply lead to faster-falling prices if demand did not hold up more in the downswings of those cycles. In this connection it is worth looking at the relationship between price movements and wage movements, using the data presented above and the pitmouth price indices developed in chapter 9, below. This is done in table 7.8, measuring from trough to peak of cycles (and *vice versa*) for the five coalfields for which the data are best, taking whichever year is appropriate in each individual coalfield series.

Whilst it is possible to look at downswings and upswings in isolation, it is pretty clear that what happened in one was a major influence on the other which followed it. In considering the price change in any depression, therefore, and its relationship to the wage change, it is not enough just to compare depressions. For example, the very large price falls of the depression of the late 1870s are not just, or even primarily, a reflection of extremely weak demand. They are mainly associated with the enormous rises of the preceding boom. Because of this, the data of table 7.8 have been regrouped in table 7.9 on a trough-to-trough cycle basis; this being preferred to peak-to-peak, despite the loss of the first and last lines of the data, because our focus is on the situation at the bottom of successive depressions.

Where there is a rise in prices from one depression to the next it implies that demand for coal (as expressed in monetary units) has risen relative to supply, and where there is a fall in prices the opposite is implied. If wages rise by more than prices (or fall by less), then the implication is that demand for labour has been affected by something other than the strength of the demand for coal. How important such other factors were, when taken together, depends on the difference between the changes in wages and in prices.

Concentrating first on the period up to 1870, it would appear that wages did better than prices in all five coalfields in the 1840s. Demand for coal, in relation to supply, fell in the depression of the late 40s and early 50s compared with that of the beginning of the decade in four out of the five coalfields, yet wage levels were higher in three, and a good deal less low in Lancashire. In the fifth district, South Wales, wages rose quite a lot more than prices. Unless there was some major change in profits or in the costs of production other than wages, the implication here is of rising productivity rather more than offsetting the effects of a weak demand for coal in relation to the increased supply. There are no precise data on costs of production for this period, and the information on profits is based on the income tax assessments, which were probably defective to an unknown extent in the early days of the reimposed tax. But it is likely that technical progress in the 1840s did have some effect in reducing the proportion of non-wage costs as well as in raising productivity; and there is a strong possibility, too, that the rate of profit fell with the virtual ending of the oligopolistic arrangements in Northumberland and Durham, and with the general widening of the field of competition brought about by the massive extension of the railways at that time.

In the two cycles which followed, wages and prices moved fairly closely together, except in the North-East in the 1860s, and provided one takes the two cycles together in the case of Scotland.[58] In the two inland coalfields, wages did rise by an amount significantly, but not greatly, more than prices, but in Scotland and South Wales the margin was slight. In the latter two, therefore, as well as in Durham and Northumberland, it appears that the strength of the demand for coal was by far the main factor in the continuing rising trend of wages; whilst in the inland districts it was probably also the most important influence. However, this conclusion needs some modification in the light of the fact that, except in the North-East, the rise in coal prices from 1849/51 to 1868/70 was below that of the Sauerbeck wholesale price index, substantially below in the cases of Yorkshire and South Wales.[59] This clearly points to reductions in the cost of supply of coal. Despite this, we may tentatively conclude that up to 1870, since the 1840s at any rate, whilst other factors were clearly influential, especially in the 1840s and in the inland coalfields, the main reason for the rising trend of wages was probably the growing strength of demand, as expressed in monetary terms, in relation to supply.

After 1870 there were two cycles when the level of prices fell from trough to trough, with wages falling too in the first one, though usually by a substantially smaller amount. The net effect of the whole period was to leave money wages in all coalfields at practically the same level, whilst the price of coal had fallen by between 14 to 37 points. To put this in proper perspective, however, note that the Sauerbeck index fell by 40 points.

What seems clear, therefore, is that demand was weak relative to supply – a point which is emphasised by the very low rise in prices in the upswing to 1882/3. This clearly had an effect on wage levels, which changed very little in either trough compared with its predecessor. But it looks very much as if, far from being the only influence on wages, it was not even the main one. Money wages held up better than prices, perhaps partly because of lower profit levels after the mid-70s, but probably more because of rising productivity and an even bigger fall in other costs than in wage costs. Of these latter two factors, rising productivity was more significant in the second cycle, and the relative fall in non-labour costs in the first.[60] When one looks at real wages, the stability of money wages in the face of the large fall in overall prices is the outstanding influence, as many have pointed out before in relation to wage levels in general.[61]

From the trough of the later 1880s through to the end of our period all coalfields show successive trough-to-trough rises in both prices and wages, with the price rises being the greater in the majority of cases, and with only two cases when wages rose by substantially more than prices, in both cases preceded or followed by a substantial movement of the opposite kind. Leaving aside these last two instances, the only times when price rises were considerably higher than wage rises were in the last cycle (except in the North-East), and in the preceding one in South Wales and the North-East. It seems reasonable to conclude, therefore, that the strength of demand for coal was the main influence on the trend of money wages for much of this period, but with other factors coming in increasingly towards the end. Probably the most important of these was the decline in productivity, which accelerated in the later 1900s, and which had the effect of reducing the elasticity of supply of coal (i.e. of reducing the *length* of the elastic portion of the short-run supply curve, rather than of changing its slope) as well as having a directly depressing effect on wages. Another possible influence, related again to the elasticity of supply of coal, may have been a slight rise in the profit rate: certainly profit per ton did not fall in the downswing of the 1900s to anything like the level of 1896.[62] However, even towards the end of the period, the rise in coal prices was very much higher than in the overall wholesale price index, so that it will not do to lay too much emphasis on other influences than the strength of demand for coal except as they affected the supply possibilities.[63]

Hitherto in this chapter we have concentrated entirely on the rate of wages per shift worked, because it was on that that the employer's attention was mainly fixed, and it was that which was the object of wage negotiations. Moreover, it was probably the first, if not the ultimate, consideration to most workers at most times, though only because desired weekly earnings for the individual could normally be achieved by varying the number of shifts he worked. When the pits were not open long enough for that to be done, the question of earnings came more to the front of men's minds.

It might be possible to construct annual indices of average weekly earnings on the lines of the wage indices above, but it would involve making precise assumptions about the regularity of working which our data scarcely warrant except in a very few cases, at any rate before the 1890s.[64] Moreover, the range of individuals' earnings around the average must inevitably have been much wider than in the case of shift wages, for it was influenced not only by skill, physical ability, and luck on five or six days, instead of one, but also by the regularity with which the pits were open and the frequency with which different workers missed a shift. Because of these considerations, we shall confine ourselves to table 7.10, which covers the period from 1894 or 1895 onwards, and to a few remarks about the relationship between wages and earnings, and not attempt anything more ambitious.

The data on the average number of days the pits were open, which were collected from the mid-1890s, and which we presented in table 5.17, enable us, in conjunction with our wage data, to make estimates of average full-time weekly earnings for the last two decades of our period, on the assumption that the average man was never absent from work when the pits were open.[65] To arrive at a more realistic figure we have to have information about absenteeism. And since this is not available on a regular basis for most coalfields, further estimation is required about this matter. In making the calculations for table 7.10, absenteeism has been estimated for districts other than Durham and Northumberland on the assumption that it fluctuated between (and prior to) the fixed points given by table 5.21 for 1899, 1905, and 1913/14 (suitably adjusted to bring them into comparability) in the same proportions as in the average of Durham and Northumberland. The resulting figures must not be regarded as very precise estimates of weekly earnings, and it must be realised that the 'average' hewer is in some ways a mythical, or at least a symbolical, figure. Nevertheless, the representation does have some value. It is presented only for hewers, both because they could scarcely work when the pits were not fully open and because information about absenteeism is more ample in their case. Their earnings certainly fluctuated more than those of other groups, and their trend over this period was almost certainly more steeply upwards. However, it probably did not differ in this respect from the trends of money wage rates already discussed. In converting the money estimates into an index of real earnings, the Labour Department indices of retail prices for food and clothing have been averaged.

The main factors affecting weekly earnings – or, for that matter, earn-

Table 7.10 *Estimated money and real weekly earnings of hewers, 1894–1913 (money earnings in shillings, real earnings 1900 = 100)*

	Northumberland		Durham		South Yorkshire		Lancashire		East Midlands		South Wales		Scotland	
	Money earnings	Real earnings	Money earnings	Real earnings	Money earnings	Real earnings	Money earnings	Real earnings	Money earnings	Real earnings	Money earnings	Real earnings	Money earnings	Real earnings
1894	...	...	...	...	33.6	93	30.2	94	...	...	...	...	...	...
1895	23.1	71	24.2	78	28.7	81	26.1	83	27.0	76	27.3	73	25.2	69
1896	23.5	72	24.8	79	30.6	86	27.6	88	27.0	76	26.6	71	23.2	63
1897	25.0	75	25.7	81	32.8	91	28.4	89	29.0	80	27.6	73	25.0	67
1898	26.8	79	27.0	84	35.3	96	29.5	91	31.2	85	...	...	28.2	74
1899	29.7	90	29.1	93	36.6	102	31.8	101	34.8	97	32.7	87	32.7	88
1900	34.4	100	32.8	100	37.3	100	33.0	100	37.5	100	39.2	100	38.7	100
1901	35.2	102	33.7	102	37.1	99	33.1	100	38.0	101	42.4	108	33.6	86
1902	31.6	91	30.9	94	36.5	97	32.5	98	38.4	102	40.3	102	29.7	76
1903	30.6	88	30.5	92	35.3	93	29.7	89	34.5	91	39.0	98	29.3	75
1904	29.5	84	29.8	89	33.9	89	28.1	83	32.2	84	38.2	95	28.4	72
1905	28.5	81	29.9	89	33.9	88	27.5	81	31.2	81	34.5	86	28.5	72
1906	30.1	85	30.8	91	36.3	94	28.5	84	33.0	85	35.8	88	29.6	74
1907	33.9	93	33.7	97	40.6	103	32.7	94	39.3	99	41.1	99	34.4	84
1908	36.8	100	36.6	104	41.6	104	33.9	96	38.5	96	43.6	104	34.5	83
1909	33.6	90	33.5	95	39.7	99	32.0	90	37.3	92	41.8	99	30.3	72
1910	34.0	90	34.3	95	41.1	100	32.6	90	38.5	93	38.4	89	28.1	66
1911	33.7	88	35.1	96	41.9	101	34.4	94	37.9	91	42.6	98	29.3	68
1912	35.6	90	36.2	96	46.2	108	37.4	99	42.2	98	46.5	103	31.2	70
1913	39.6	100	41.3	109	49.3	115	40.2	106	47.4	110	48.3	107	35.8	80

Sources: see text.

ings over a longer period – were, as has just been implied, the regularity with which work was offered and the frequency with which shifts were missed. There was a well-established, but by no means precise, inverse relationship between these two, epitomised by thebehaviour of the Forest of Dean miners described by the managing director of Trafalgar Colliery to the Eight Hours Day Committee: 'They are very good in their attendance, for this reason, that they get so much play time in the summer in consequence of there being no demand for our house coals, that when they do get work they are only too glad to attend.'[66] It was notorious, too, as was brought out in chapter 5, that during booms, when wage rates were high and the pits were open full time, absenteeism rose. It is clear, therefore, that the average number of shifts actually worked per year varied rather less than did the number of days on which the pits were open. Indeed, putting together the latter data, as given in table 5.17, and the information on absenteeism in the North-East supplied to the Eight Hours Day Committee and to the Sankey Commission,[67] it seems that the average number of shifts worked in Durham between 1895 and 1913 ranged from 240 to 260 per year (ignoring the years with major strikes), with about two-thirds of the years in the even narrower range from 246 to 252. The range may well have been greater in other coalfields, though there is no particular ground for saying so, and it almost certainly was greater in earlier periods. However, whilst years of deep depression like 1879 and 1886 may have seen more short-time working than any subsequently, it is doubtful whether any higher figure than the 260 shifts of 1913 in Durham was ever achieved; and it is unlikely, too, that there were many years which diverged as far as did 1879 and 1886 from a mean figure which may have been lower than in 1895–1913, but was still probably above 230 days a year in an unseasonal district.[68]

Years of deep depression, when low wage rates coincided with a great deal of short-time working, must have brought hewers' earnings down to levels not far above those of agricultural workers and others with relatively low-paid, but steadier, employment.[69] These years, it must be emphasised, were exceptional, but they serve to bring out the point that the large fluctuations noted above in relation to wage rates were not generally modified so far as earnings were concerned – by more assiduous attendance – at the extreme lower turning points of the trade cycle, however much they may have been – by increased absenteeism – at the peaks. The coal industry was one in which the industrious worker experienced a very wide range in his earnings in the course of every few years, and in which even the less industrious experienced a range with only a few degrees less spread.

Apart from the wage rate and the regularity of work, there was a number of other influences on the miner's earnings which were of some importance, though decreasingly as time went on. Amongst these may be mentioned deductions, fines, and the system of payment, including the emotive subject of truck. All these had two things in common: they were a known feature of the wage bargain, and they were all subject to abuse by unscrupulous employers. Largely because of the latter, but also because ideas of what was right and proper changed, they were all, to a greater or lesser degree, fruitful sources of discontent and of bad industrial relations. Deductions – that is, withholding of part of the miner's pay on account of tools, candles, powder, etc. – were the least important and the least contentious. They varied between different coalfields, from charges for everything in South Wales to nothing but pick sharpening in Scotland;[70] but everywhere they were a known factor in the wage bargain, and they do not seem to have caused any serious problems or resentment, or to have varied greatly over time. The same cannot be said of the fines and/or confiscation of coal cut, which were exacted by employers for inadequately filled corves or tubs, or for the inclusion or too much dirt or slack. According to Machin, 'dissatisfaction over these matters did more than anything else to poison and embitter relationships within the industry'.[71] Despite this, it is fair to say that the fines were a fully recognised part of the wage bargain; that sensible employers realised, as one of them put it to the Royal Commission on Labour, that 'it is in the employer's interest to have a scale of deductions adjusted so as not to bear upon the men who are able and anxious to do their work honestly';[72] and that sensible miners recognised the need for some sanction against sending out 'rubbish',[73] otherwise, as Machin himself put it, 'had there been no penalties many colliers would certainly have been less careful than they were'.[74] The trouble was that owners of marginal mines, under enormous pressure to reduce costs in depressions, often could not take a sensible long-term view. Instead, they clutched at the straw of cheating by unjustifiably stepping up the rate of condemnation of tubs for inadequacy. No doubt there were other employers who cheated out of greed and shortsightedness; and probably there were still more banksmen who abused their position when, as sometimes happened, they were paid a commission on confiscated tubs. It was these abuses, and suspicions that they were even more extensive than was probably the case, as much as doubts about accurate weighing, which lay behind the mid-century agitation for miners to have the right to appoint checkweighmen. Fines were the most widely used, by the middle of the nineteenth century, of the various means by which employers tried to ensure that they only paid for the large coal, which was what the wage bargain was usually in terms of – though sometimes a much lower payment was made for slack. Another means which lingered on to the early 1870s, often, even, where it had been superseded by fines, was the custom of requiring more than 20 cwt to be sent out for each ton paid for. This, along with the use of forks for filling the tubs,[75] was a crude way of keeping down the output of slack and ensuring that payment was for roughly a ton of large coal.[76] It was not, as it often appears to modern eyes, a form of cheating on an agreed bargain, for the bargain was in terms of the 'big' ton. But it was not as efficient a method of ensuring quality as fines were, and this was particularly true when screening-and-weighing machines, which measured the large coal more exactly, came into use around 1850. The 'big' ton was then gradually eliminated from those coalfields which had used it. Not only did it no longer serve its original purpose, but in a changing climate of opinion it came to seem unfair, and its removal was a particularly easy way for miners to seek a wage increase. By the early 1870s, when the 20 cwt ton became standard practice everywhere, it had been continued at only a few backward pits.[77]

The screening-and-weighing machines which helped to eliminate the 'big' ton were called 'Billy Fairplays', and themselves became objects of suspicion and dislike by the late 70s and the 80s, as no doubt they had been on their first introduction.[78] However, properly used, they eliminated the arbitrary judgment of the banksman as the criterion for confiscating tubs, and there seems to have been no sustained opposition to them until the depressions of the late 70s and the 80s, when, perhaps, their screen meshes were widened, openly or otherwise, by owners who had become extremely cost-conscious. One prominent owner, George Elliot, MP, who was a member of the 1873 Select Committee, eulogised the Billy Fairplay in evidence to it in terms which were probably overdrawn, saying that it was the 'means of getting rid of all the squabbles and uproars which were constantly arising anterior to its introduction'.[79] But, if overdrawn and over-optimistic about the decline of petty 'uproars', he was surely correct in saying that the machine was fairer than the banksman's rule-of-thumb, even when there was a checkweighman looking over his shoulder.

It is necessary to try to get this question of fines and confiscations into perspective. On the one hand, union officials tended to play them down at the various enquiries at which they gave evidence,[80] and examination of surviving nineteenth-century paybooks suggests that the great majority of hewers were not fined very often. On the other hand, it is fair to say that the firms whose records have survived probably do not include the 'bad' employers. Moreover, it can be argued that the coalowners as a group fought long and hard in resistance to the principle of allowing checkweighmen to be employed, and that this resistance suggests that they had something to lose. In support of this may be cited the belief of John Normansell, first secretary of the South Yorkshire union, that the establishment of checkweighmen meant 7s 6d to 9s 6d extra per week to the hewer.[81] It is hard to believe the amount stated, since it implies something like a 40% increase in earnings in the early 60s, but that he put it so high is suggestive that there were plenty of abuses by owners. It is impossible to come to any firm conclusion. One can be sure that up to the 1860s or 70s, and beyond in some places, pieceworkers were subject to some amount of what can only properly be described as cheating; that, as A.J. Youngson Brown put it, some 'employers pursued the . . . policy of self-seeking at any moral price with regard to weighing, and . . . exercised their ingenuity to thwart the will of Parliament'.[82] One can be equally sure that some miners got away with short weight in the earlier years of our period and with excessive dirt at all times. The balance was certainly to the owners' advantage, as is perhaps implied by parliamentary interference; but no one can say by how much.

An aspect of the system of paying wages which has not yet been mentioned, but which had a bearing on earnings, was the length of the pay period, and whether or not payment was made in cash. Both were probably affected, in the beginning, by the eighteenth-century shortage of coin, though other influences were also at work.[83] Ashton and Sykes describe the common system of payment in the eighteenth century, whereby a 'subsistence' was paid every week or fortnight, but a full reckoning up of net pay took place at much less frequent intervals, and often left the miner in debt to his employer.[84] Long pays of this kind undoubtedly continued well into the nineteenth century, but it is doubtful whether they were regarded as much of a grievance by the miners, except when employers insisted on arrears of debt being paid off. As Ashton and Sykes point out, as well as enabling the employer to defer a part of his wages bill when cash was short, 'subsistence' did do something to protect the miner against the effects of seasonal and other short-term fluctuations in the state of employment. By the middle of the nineteenth century fortnightly pays were most common, though longer periods remained, especially in South Wales and South Staffordshire. The weekly pay was uncommon until the early 1870s, but was general in South Yorkshire from the middle 1860s.[85]

Much more disliked, at any rate by the second quarter of the nineteenth century, was the system of payment partly in kind or in vouchers on a colliery shop – the notorious truck system. Ashton and Sykes follow George Unwin[86] in attributing the spread of truck to the currency difficulties which accompanied the French Wars, but in some areas, at least, it was encouraged by the opening of mines in previously sparsely peopled areas and the consequent lack of established retailers. In such places, a company shop was almost as much of a necessity in attracting a labour force as were company-built houses, though it should be noted that it is compulsion to deal with the colliery shop, rather than the mere existence of such a shop as a matter of convenience, which constituted truck.[87] Another reason for the establishment of the truck system ties in with the existence of long pay periods and the issuing of advances against the reckoning at the end of the period. Such advances, coupled with compulsion to use them at the colliery shop, were sometimes intended as a device to improve the efficiency of the labour force by cutting down pay-day drunkenness through limiting the amount of cash received on pay-days.[88] Yet a further incentive to adopt the truck system may have been the one which the Webbs saw as the main reason – the desire to pay wage rates lower than those accepted conventionally as normal (or those set by collective bargaining).[89] This hardly seems to have much general application to an industry with such fluctuating wage rates and with such possibilities for local variations in basis rates to take account of differences in natural conditions. One district was an exception to this, however, and that is South Staffordshire, where the truck system was more prevalent than in any other coalfield. In that district there were standard rates per stint throughout the district, with variations only between the thick and thin seams, and between east and west of Dudley. In the light of this, and statements such as that of Samuel Lloyd, a Wednesbury owner, that truck helped to compensate him for the thinness of the seams he worked, one must agree with Hilton that 'the Webbs' explanation would appear plausible so far as the Black Country collieries are concerned'.[90]

Whatever its origins, the truck system does not seem to have operated in a particularly oppressive or exploitative manner prior to, perhaps, the 1820s. Or, to be more precise, the system of operating a colliery shop, even with compulsion to use it, which was commonly exercised only in connection with advances against pay, does not appear to have been a matter of serious grievance. E.W. Evans says of South Wales that 'on the whole, the available evidence suggests that truck was neither unduly oppressive nor extremely unpopular during the early years of the nineteenth century', and he found no exploitation in shop prices until the 1820s.[91] Probably the same was true of South Staffordshire, where labour was generally rather short until the 1820s. It is also probably significant that complaints about the truck system began, and were always strongest subsequently, in periods of falling prices.[92] It may be that this was simply the result of the lag of prices in company shops because they were run less efficiently than other shops: such a lag certainly occurred in some cases when prices were rising.[93] But it was probably also a form of wage-cutting when the pressure to reduce costs was unusually great – a form which, in the Black Country at least, may have been easier for the individual butty or owner to take than a direct attack on the standard wage rates, both because it was covert to some degree, and because it could be done at more frequent intervals.

By 1831 truck was a grievance even in Durham and Northumberland, where it had never been well-established or widespread, and its abolition was part of the settlement of the strike of that year. Significantly, tommy shops in the North-East were not colliery-owned, but run by deputies and other colliery officials. The owners lost little or nothing by agreeing to suppress them. Elsewhere, where owners themselves ran such shops, or where butties did, there was less inclination or power to do anything about them, and despite the legislation of 1831, they lingered on for many years. It was the butties' shops which were the greatest object of dislike, mostly because owners generally only used compulsion to use the colliery shop in connection with advances against future earnings, whereas butties used compulsion unconditionally on all their employees. Butties' shops, however, were not nearly so common as public houses kept by butties, and the most frequent, as well as the most extreme, condemnation was reserved for those butties who were also publicans, and who were accused of encouraging, if not actually compelling, the miners' spending of their pay in their pubs.[94]

As George Hilton concludes, 'there can be little doubt that Alfred Marshall was correct in describing the truck system as a device for employers "getting back by underhand ways part of the wages which they nominally paid away" ',[95] though there were certainly honourable exceptions where colliery shops were operated as a convenience to workers and not on exploitative lines. It is necessary, however, as with the other abuses referred to above, to get the matter in perspective. Hilton's considered estimate was that 'the loss in annual income as a result of the truck system was almost surely less than ten per cent for virtually all labourers who were subjected to it'.[96] If 10% was a maximum figure for Black Country nailers and the employees of butties, it must have been quite a lot less in the other, less badly affected districts. Moreover, market forces operated to mitigate the system, at least partially. The high prices charged to captive customers in colliery shops in both Scotland and South Wales were often, if not always, offset by higher wage rates. Two examples may be quoted to illustrate this, though there is plenty of other evidence to confirm it.[97] The first comes from Scotland in a period of depression, and is contained in a letter of July 1843 from the manager of Govan collieries to his agent, Archibald Russell of Whiterigg. He wrote as follows:

> Mr. Dixon has come to the following resolution as to the colliers' and miners' wages. Seeing that the wages is not completely settled in the mining district some working at 3s. and some at 3s. 6d. and not wishing to take a lead in bringing the wages below his neighbours he has fixed the wages in the meantime at 3s. per day cash and fortnightly pay'ts, which he considers equal to 3s. 4½d. per day with months pays and lines to a shop, or goods from a store. And should the wages become fixed in the country at 3s., a reduction to 2s. 9d. a day will take place upon the same principle immediately thereafter.[98]

The second example comes from South Wales in the subsequent boom, in a letter of April 1846 from Anthony Hill of the Plymouth Works to John Hutchins of Dowlais: 'Now that the men can get work when and where they please, the *truck* works *must* pay higher wages . . . if they get 10 per cent by truck into one hand they must pay it out from the other. But we, who do not truck, would be silly to adopt their scale of wages.'[99] In addition to this direct way in which the effects of truck on earnings were offset, there was an indirect way, which had more relevance in South Staffordshire with its standard wage rates. This was the tendency for the best workmen to avoid collieries where truck was practised. It was noticeable in the way that the best miners were attracted to the Thick Coal mines, and truck was rarely, if ever, found except in thin-seam pits; but even amongst the latter, the more able and provident men avoided the tommy shop owners and the publican butties.[100]

The gradual disappearance of truck from coal mines probably owed little to the legislation against it. The 1831 Act was largely a dead letter, and though there were enquiries in the early 1840s, the early 50s, and in 1870–1, no major legislation emerged until the Truck Amendment Act of 1887, by which time it was largely unnecessary so far as the coal industry was concerned. The zenith of the system was probably in the 1840s. In the 50s it seems to have faded quickly away from South Wales, the West of Scotland, and South Staffordshire, especially the last two. In a few places in all these districts it lingered on into the 1880s, but it was not much found away from the remoter areas, such as South Ayrshire, and the butty mines of South Staffordshire and those few which survived in South Wales, after about 1860.[101] The National Association of Miners paid a good deal of attention to it, but it probably only affected a small proportion of their members, and the attention may well have sprung primarily from the fact that Alexander Macdonald was an Ayrshire man.[102] The connection between the butty system and truck in England and Wales was clearly very close after 1850 or so; whilst the survival of truck in parts of Scotland probably owed something to the powers which Scottish law gave to independent shopkeepers (but not to tommy shop owners) over indebted customers.[103]

We earlier promised to return to the question of the effectiveness of workers' combinations in changing wages from what they would have been in the absence of collective action.[104] It is now time to take up this matter, although the question has already been partly answered by implication in earlier aspects of the discussion, and particularly in our attempt to explain the rising trend of wages between cyclical troughs in the last twenty years or so of our period. Broadly speaking, the answer up to 1890 must be that unions, or less formal combinations for that matter, appeared to be successful in raising wages on the upswing of trade cycles;

but whether they really should receive much credit for it is extremely doubtful, for rising demand and the temporary shortage of labour would usually have raised wages in any case. Probably the most that can safely be claimed for them is that they started the upward movement a bit sooner in the upswing than might otherwise have been the case – a strike of a few days, or even the threat of a strike, being enough to persuade employers that it was sensible to let the workers have a share of rising profits.[105] A parallel claim, that unions delayed wage cuts in depressions, receives a small amount of support, but certainly could not be held to be universally true.[106] Even where it was true, when lengthy strikes were necessary to bring it about, the loss to the miners in the short and medium term was considerable.

After 1890, trade unions had a little more influence on wage rates, most particularly in the Federated Districts in the cyclical downswings. The policy of the MFGB has already been described, and the comparison made between the course of wage rates in its districts and in the other major coalfields. In discussing figure 7.1 the comparative success of the MFGB in moderating the fall in wage rates, especially in the mid-1890s, both compared with earlier depressions and with the current experience of other coalfields, was brought out. When we look at real earnings, however, the success does not seem quite so great. The indices of real earnings given in table 7.8 can be extended back to 1890 on a rather more approximate basis.[107] If this is done, the following results are obtained for 1890 (1900 = 100): Northumberland, 86; Durham, 84; South Yorkshire, 87; Lancashire, 88; East Midlands, 88; South Wales, 77; Scotland, 76. This implies that the reduction of real earnings from 1890 to 1895 or 1896 (whichever was the lower) was not very much different in South Yorkshire or Lancashire from Durham or South Wales. It was greater in the East Midlands, and greater still in Northumberland and in Scotland. The secular trend of expansion in Durham and in South Wales seems to have kept them up with similarly growing South Yorkshire. However, the relatively good performance of the slow-growing Lancashire coalfield may well be put down as a success for MFGB policy. But against this, the fall in employment opportunities in the East Midlands, despite the trend of relatively rapid growth, meant that earnings fell at much the same pace as in Scotland. It is perhaps not surprising that it is the historian of the Derbyshire miners who has raised the biggest doubt about the merits of the MFGB's policy.[108] The figures here do not match up to those which he quotes for a single colliery, but they do lend a measure of support to his case. It is fair to add, however, that earnings in the two coalfields for which we have information for the year 1894 about days worked, Yorkshire and Lancashire, seem to have been at a relatively high level, probably significantly above the levels of South Wales or Durham. Thus the

Federation may well have succeeded in imposing a useful delay in the fall in earnings in that particular downswing.

In the next two downswings there was much less talk of the new policy of the MFGB and, as was pointed out above,[109] it was implicitly abandoned in 1909. However, a similar, if shorter, lag in the decline in earnings occurred in the Federated Districts after the 1900 peak (if not that of 1907/8), the main fall coming after 1902 rather than in that year. But the amount of the fall from peak to trough was dependent less on union policy than on the long-term trend of output, and so of demand for labour, in the coalfield concerned.

It may be objected that we have been comparing the Federated Districts and the other coalfields as if only the former were influenced by the existence of a trade union, and that the comparison should really be made between periods when unions existed and periods when they did not in each coalfield. This is extremely difficult owing to scarcity of data, but we have already shown, in figure 7.2, how the trend of money wages was roughly the same between the 1840s and the early 70s as it was from the later 1880s to 1914; and in figure 7.3 we have shown how a gently rising trend of real wages between the 1840s and the 70s was abruptly shifted upwards by the fall in general prices in the 1880s and 90s, only to finish our period with a decade of stagnation. There does not seem to be any reason to believe that the trend of real earnings would be very much different, though the abruptness of the shift in the 1880s would almost certainly be somewhat modified, for short-time working was unusually prevalent then. By the same token, improved regularity of work in the 1900s made real earnings in that decade, taken as a whole, more of an improvement on the 90s than was the case with real wage rates. It is hard to see, however, what part the unions played in this, for the 1880s was a period of weakness for them. All told, there is no trade cycle period, taken as a whole, when it is possible to say that real earnings were substantially better because of union influence than they would otherwise probably have been, with the exception of some parts, but not all, of the Federated Districts in the 1890s, and with the possible exception, too, of the period 1908–14, when, however, the effects of the Eight Hours Day Act make a firm judgment infeasible. At most times, the most that unions, or collective action generally, achieved was some acceleration of rising earnings in upswings of the trade cycle and some delay in falling earnings in downswings. It must be said that of course this achievement was worthwhile in itself, as the attitude of owners to unions testifies. Moreover, it would be foolish to deny either that unions did bring non-pecuniary benefits to the miners (including, eventually, fairer treatment as between individual workers), or that there was 'psychic income' in the feeling of not being entirely at the mercy of employers or of impersonal economic forces, even if it was, to a fair extent, an illusion. Solidarity was not without its satisfactions.

APPENDIX
Wages per shift, 1879/86 to 1914

As was mentioned in the main text, wages in the coal industry were determined in all districts by the late 1880s (if not earlier) by two rates: the basis rate, typified by the pieceworker's price list, and the percentages on or off the basis rates of a particular year, known as 'the standard'. The particular year chosen as the standard in different coalfields depended on the way in which collective bargaining developed. The districts where it was established first, Northumberland and Durham, had the earliest standard – 1879. So, too, did Cumberland and South Wales, which adopted it in the revised sliding scale agreement of 1880 and stuck to it thereafter until 1915. Most other coalfields later adopted 1888 as their standard year, for that was the one chosen for the so-called Federated Districts, those covered by the Miners' Federation of Great Britain. The Forest of Dean and Somerset followed suit deliberately, and Scotland also probably took 1888 because of its early connection with the MFGB. From 1888 (or shortly afterwards), then, there was established in all coalfields a system whereby district-wide changes in wage rates were regulated by changes in the percentages. Given no alteration in basis rates, it is easily possible to calculate wage indices for all subsequent years. But, of course, the average basis rates of a district could easily change even if every single established price list remained unaltered: new working faces were constantly being opened and old ones closed. Moreover, existing basis rates were sometimes changed, most especially with the adoption of technical changes such as electricity, coal-cutting machinery, or new haulage systems. The percentage adjustments on or off standard cannot, therefore, be used without first ascertaining what changes took place in the basis rates. Since this cannot be done for each and every pit and seam, it has to be done by contrasting the known normal wages per shift at the end of our period, as revealed by the enquiries of the Sankey Commission, with the hypothetical wages that would have been paid if the only adjustment to the known wages of an earlier date had been the percentages on or off standard. This is done in table 7.A1, adopting the technique used in J.W.F. Rowe's exceedingly useful pioneering work, *Wages in the Coal Industry*, published in 1923.[110] The first line for each of the coalfields covered shows the wages per shift in 1879 (where that was the standard year) or in 1886. The latter are derived from the 1886 Wage Census for the

Table 7.A1 *Wages per shift, 1879/86 and 1914 (in shillings and pence)*

	Piecework coal-getters	Putters, etc.	Repairers, stonemen, timbermen, etc.	General underground labourers
Northumberland				
1879	4 11	3 6	4 0	3 4
1914 hypothetical	7 6	5 4	6 1	5 1
1914 actual	8 5½	6 5½	6 8½	5 4½
Durham				
1879	4 3½	3 8	4 0	3 3
1914 hypothetical	7 2	5 9½	6 3½	5 1½
1914 actual	8 3	6 10	7 9½	5 1
Cumberland				
1886	4 5	3 5	4 1	3 3
1914 hypothetical	6 5	4 11½	5 11½	4 8½
1914 actual	8 2	5 11½	7 0	5 8
South Yorkshire				
1886	5 6	3 7	5 0	4 0
1914 hypothetical	9 1	5 11	8 3	6 7
1914 actual	10 3	7 2	8 3½	6 8
West Yorkshire				
1886	5 4	3 6	4 10	3 10
1914 hypothetical	8 9½	5 9½	7 11½	5 9½
1914 actual	8 11	5 6½	7 3	5 6½
Lancashire & Cheshire				
1886	5 0	3 9	4 6	3 9
1914 hypothetical	8 3	6 2½	7 5	6 2½
1914 actual	8 6½	6 2½	6 7½	5 10
East Midlands				
1886	5 4	3 5	4 6	3 5
1914 hypothetical	8 9½	5 7½	7 5	5 7½
1914 actual	9 10	6 8	7 6	5 8½
Warwickshire				
1886	5 0	3 1	4 4	3 0
1914 hypothetical	8 3	5 1	7 2	4 11½
1914 actual	10 0½	5 7	6 10	6 2
North Staffordshire				
1886	4 10	3 2	4 2	3 4
1914 hypothetical	7 11½	5 2½	6 10½	5 6
1914 actual	9 1	6 4	6 9½	5 7
South Staffordshire (i.e. Black Country)[a]				
1886	4 5	3 4	4 7	3 4
1914 hypothetical	7 3½	5 6	7 7	5 6
1914 actual	7 1½	5 4	…	5 10
Cannock Chase				
1886	5 1½	3 4	4 4	3 5
1914 hypothetical	8 5½	5 6	7 2	5 7½
1914 actual	8 5½	5 1	6 6½	5 7
Somerset				
1886	3 3	2 11	3 0	2 3
1914 hypothetical	4 11½	4 5½	4 7	3 5
1914 actual	5 9½	4 5½	5 8	4 4½

Table 7.A1 (*cont.*)

	Piecework coal-getters	Putters, etc.	Repairers, stonemen, timbermen, etc.	General underground labourers
Forest of Dean				
1886	4 6	3 4	3 11	2 10
1914 hypothetical	6 1	4 6	5 3	3 10
1914 actual	6 9½	4 8½	5 8	4 6
North Wales				
1886	4 3	...	3 9	2 10
1914 hypothetical	7 0	...	6 2	4 8
1914 actual	8 0½	5 9½	6 2	5 5½
South Wales				
1879	4 9	3 3½	4 0	2 11
1914 hypothetical	7 7	5 3	6 5	4 8
1914 actual	9 4	6 8	7 8	5 9
Scotland				
1888	4 5	3 6	4 4	3 0
1914 hypothetical	7 11½	6 3½	7 9½	5 5
1914 actual	8 3	6 6	8 0	6 0

[a]The first column is for time-rate getters. The 1914 figure is that attributed by Finlay Gibson to piecework getters, but this is clearly the result of a misunderstanding of the system of payment in the district. The numbers of repairers, etc., covered by the 1914 returns are too low to make the average wage figure meaningful.
Sources: see text of appendix.

most part, with the normal weekly earnings given there usually converted into shift wages on the assumption that the number of days worked per week was the same as given in great detail in a return published in 1890, but collected earlier.[111] In addition to this, a variety of other information has been used, especially that given in Rowe's book. Like Rowe, we have found it impossible to deal with the Leicestershire and South Derbyshire coalfield, which in 1886 was lumped in with South Staffordshire and Warwickshire, and in 1914 had two separate wage negotiating bodies. The information on Shropshire is too scanty to be worth using. However, we have managed to separate the two districts of Yorkshire, which were taken together in the 1886 Wage Census, and to improve on Rowe's separation of South Staffordshire and Warwickshire. The second line for each coalfield shows the hypothetical wages in 1914, which are those which would have obtained if there had been no other change than the percentages on or off standard. The last line gives the actual average wages per shift in June 1914 according to returns provided by Finlay Gibson for the Sankey Commission,[112] modified where the classification clearly does not correspond with that which we have used. The only important examples of this are the repairers, etc., and the labourers in

Table 7.A2 *Apparent changes in basis wage rates, 1879/86 and 1914 (in pence; positive unless indicated)*

	Piecework coal-getters	Putters, hurriers, etc.	Repairers, stonemen, timbermen, etc.	Underground labourers
Northumberland	7½	9	5	2½
Durham	11½	10	11½	0
Cumberland	9	8	8½	7½
South Yorkshire	8½	9	0	0
West Yorkshire	1	−1½	−5	0
Lancashire & Cheshire	2	0	−6	−2½
North Wales	7½	...	0	5½
East Midlands	7½	7½	0	0
Warwickshire	13	3½	−2½	9
North Staffordshire	8	8	0	0
Black Country	0	−1	...	2½
Cannock Chase	0	−3	−5½	0
Somerset	6½	0	8½	7½
Forest of Dean	6½	2	3½	6
South Wales	13	10½	9½	8
Scotland	2	1½	1½	4

Source: table 7.A1.

Scotland, for whom Finlay Gibson gives figures of 9s 0d and 6s 6d per shift. Rowe (1923, p. 74) indicates that he thinks the former figure may apply only to oncost workers on piece rates, and the Govan Colliery Pay Books suggest that this may be the case. If this is so, the figure we want cannot be more than 8s 0d per shift, and we have accordingly substituted this here. Similarly the labourers' figure may be too high because it includes some semi-skilled daywage men, and a figure of 6s 0d is substituted here.

It is clear from table 7.A1 that even allowing for our inability to be sure that we have eliminated all errors due to misclassification, and for possible errors both by the collectors of the data and in the process of turning weekly into shift earnings, there were some grades and districts where changes in basis rates must have taken place. Assuming our data to be correct, the changes in basis rates over the period were those given in table 7.A2. (Changes of less than 1d have been ignored.)

Before commenting on this table it must be stressed that some of the classes in some of the small coalfields are based on meagre numbers, and the figures for these cannot be regarded as at all reliable. In fact, only the piecework getters can be regarded as reasonably trustworthy in every case. In all the big coalfields most classes are reliable enough, but there are probably misclassifications in Lancashire and the Midlands, and perhaps also in Yorkshire. Having sounded this general warning, it does nevertheless seem possible to make some general statements. In particular, it would appear that the hewers had achieved some increase in their basis wages in every coalfield except the two parts of South Staffordshire. Another exception, not apparent from the table, was the class of daywage hewers in the East Midlands, but this probably resulted from the gradual change to a higher proportion of pieceworkers in a district where they had been a distinct minority earlier. The hewers' most immediate assistants, the putters (hurriers, hauliers, drawers, etc.), who were also often on piece rates, also experienced an increase in basis rates in most coalfields, differing in proportion a bit, but usually of a similar order of magnitude. The advance of the hewers' and putters' basis rates varied a good deal between different districts, but it is not doing much violence to reality to put them into three groups: South Wales, Warwickshire, and Durham, with increases for hewers of around 1s; Lancashire, West Yorkshire, Cannock Chase, and the Black Country, with little or no increase; and the rest, with increases for hewers varying from 4½d to 9d. In general, the districts with the biggest advances were either ones where expansion of output was particularly rapid or smaller ones which were especially affected by the decline of sub-contracting (North Staffordshire) or the spread of trade unionism (North Wales) or the effects of the Minimum Wages Act (Somerset). The districts with little or no change in basis rates were ones where these last three factors did not operate and where expansion of output, especially in the 1900s, was relatively small. (In South Staffordshire, of course, output declined.)

The relative position of the piecework hewers and of the putters improved *vis-à-vis* all other classes in the majority of coalfields.[113] There are a few minor exceptions to this statement, but none where there is not rather more than the usual measure of doubt about the reliability of the statistics. The unskilled underground labourers mostly enjoyed some rise in their basis rates, but this was by no means generally true in the larger inland coalfields or in Durham; and, outside the Federated Districts, the wages of the more skilled daywage men generally kept in closer touch with those of the faceworkers. In the Federated Districts, however, the skilled daywage men appear to have had no change in their basis rates, or else to have suffered reductions. It is hard to believe that such reductions did actually occur, though it is possible that a combination of a decline in exploratory work and a lengthening of underground roadways, with consequential decline in the more highly paid and increase in the less highly paid members of the group, may have produced a decline in *average* basis rates for what is not all that homogeneous a category. However, there is no way of resolving our doubts on this question, and we must perforce accept the indicated declines in basis rates. But the possibility of error must be borne in mind, and it seems most likely that, if there is such an error, it is in the 1914 classification.[114]

Having established, as far as possible, the changes in basis rates between 1879/86 and 1914, it remains to try to decide just when they occurred. There is no way of doing this in any precise fashion, and the best that can be done is to consider the general and particular influences at work and then make assumptions on the basis of them.

The major influences on basis rates were: (1) the development (or exhaustion) of collieries and, often more important once district-wide collective bargaining was established, particular seams of coal. For example, the Ten Yard seam in the Black Country had had a higher price list than other seams from time immemorial, and as it was exhausted or flooded out in the last quarter of the nineteenth century one would expect average basis rates to decline, *ceteris paribus*. In the opposite direction, the development of the Barnsley Bed/Top Hard seam in South Yorkshire and the East Midlands would tend to raise average basis rates. (2) Changes in sub-contracting, which, with the decline of the full butty system largely complete outside the Black Country by 1886, meant the switch from the little butty system to direct employment. North Staffordshire was the only district much influenced by this. (3) Legislation, especially the Minimum Wages Act of 1912. The Eight Hours Day Act also probably had some effect in those districts where output was expanding most strongly and a shortage of unskilled labour was felt. This was mostly effective so far as the lower paid were concerned. (4) The growth of trade union power, especially through the acceptance of district-wide collective bargaining in new areas, which usually had the consequence of tending to level up basis rates in the collieries with low price lists. The smaller coalfields were the ones most affected by this, most notably, perhaps, North Wales and the Forest of Dean, but it probably had some importance too in Scotland. (5) Technical changes which could be used by trade unions as grounds for arguing for a rise in basis rates. Since such bargaining was done in respect of an individual colliery or seam, a well-organised union was usually in a stronger position *vis-à-vis* the employer than in the case of the district-wide bargaining on the percentages on or off standard.

In the light of these influences and of the particular circumstances of each coalfield, we make the following assumptions about the shift in basis rates between 1879/86 and 1914. We do not discuss changes of less than 3d, because the figures are not sufficiently accurate to justify taking them into any special sort of account. It is assumed that they occurred at an even rate over the whole period.

Piecework getters (putters *pro rata*, except in Cumberland)

Durham and Northumberland

The increases probably came mostly through new openings, though there may well have been an unusual amount of upward revision of rates associated with changes in the shift system following the Eight Hours Day Act. The new openings were fairly evenly spread throughout the period after 1889. It is here assumed that (i) in the case of Northumberland, both hewers' and putters' basis rates rose evenly between 1890 and 1909 so that they were 6d a shift higher by the later date; and that the remaining 1½d or 3d increases came in 1910–14; and (ii) in the case of Durham, hewers' and putters' basis rates rose evenly between 1890 and 1909 in such a way that the former were 9d a shift and the latter 6d a shift higher by the later date; and that the remaining 2½d and 4d increases came in 1910–14.

Cumberland

The hewers' basis rate was reduced by 4d in the settlement of the 1892 strike. The putters' rate remained the same, but subsequently the actual percentages paid were always 12% below nominal. It is assumed that the later increases were spread evenly throughout the rest of the period.

South Yorkshire

The increases probably came mostly through the development of the Doncaster district after 1905, though new openings earlier must have had some effect. It is here assumed that both hewers' and putters' basis rates rose evenly between 1890 and 1904 so that they were 2½d and 3d a shift respectively higher at the later date, and that the remaining 6d a shift was similarly spread over the period 1905–14.

East Midlands

The increases probably came mostly through the development of new collieries working the Top Hard seam. There was a lull in such sinking in 1905–14. A further small boost to basis rates was associated with the 1908 Derbyshire agreement for a 'make-up' rate for abnormal places. It is here assumed that all but 1d a shift of the increases was spread evenly between 1890 and 1905, and that the 1d was added in 1909.

Warwickshire

The increases probably came mostly through the development of new collieries, which took place mainly after 1899. It is here assumed that

(i) hewers' basis rates rose evenly between 1890 and 1899, between 1900 and 1908, and between 1909 and 1914, so that they were 4d, 4½d, and 4½d a shift respectively higher at the three later dates; and (ii) that putters' basis rates rose evenly between 1900 and 1908 and between 1909 and 1914 so that they were 2d and 2½d a shift respectively higher at the two later dates.

North Staffordshire

The increases were probably mostly owed to the decline in sub-contracting, though some may have been the result of new developments. It is assumed here that they were spread evenly throughout the period.

Somerset

The only class for which there was a recognised basis rate in 1886 was day wage hewers. This rate was increased between then and 1914 on two occasions – in 1900 by 6d a shift, and in 1911 by 3d. It is here assumed that piecework hewers' rates were increased similarly and approximately *pro rata*, viz. by 4½d and 2d a shift.

Forest of Dean

The development of new steam coal collieries, mostly after 1899, is probably the main cause of the increase in basis rates. It is here assumed that the increases were evenly spread over the period 1900–14.

North Wales

The establishment of agreed price lists at pits in this coalfield followed rapidly on the establishment of a trade union organisation in 1900, and the district list of 1907 was little changed later. It is here assumed that all the increases took place between 1900 and 1907.

South Wales

The increases probably came mostly through the opening of new collieries and the expansion of steam coal output, but the Minimum Wages Act had an unusually large impact on the earnings of the higher paid workers in this coalfield owing to its peculiar geological characteristics. It is here assumed that 10d of the hewers' and 8d of the putters' increases in basis rates was evenly spread over the period 1890–1911, and that the remaining 3d and 2½d came in 1912–14.

Scotland

It is assumed that the increases were spread evenly throughout the period 1890–1914.

Skilled daywage men

It is assumed that all declines in basis rates were spread evenly throughout the period 1886–1914, and that all increases were spread evenly throughout 1890–1914, with the following exceptions: Northumberland – 1½d of the increase occurred in1910–14, the remainder in 1890–1909; Durham – 2½d of the increase occurred in 1910–14, the remainder in 1890–1909; Somerset – 6d of the increase occurred in 1900, the remainder in 1911; Forest of Dean – the increase occurred in 1900–14; South Wales – 2½d of the increase occurred in 1913/14, the remainder in 1890–1912.

Labourers underground

The influences which operated on the wages of other miners in the various coalfields also affected those of this unskilled group, though usually in a minor degree. But there was an additional major influence in the legislation of1912, for this was usually the group with most to gain from the Minimum Wages Act. Where the increase in basis rates was 3d or under it is here assumed that it all came in the period 1912–14 (or 1910–14 in the case of Northumberland). The assumptions for other coalfields are as follows: Cumberland – 5½d of the increase occurred in 1890–1911, and 2d in 1912–14; Warwickshire – 2d of the increase occurred in 1890–9, 3d in 1900–8, and 4d in 1909–14; Somerset – 3d of the increase occurred in 1900, and the remainder in 1911–14; Forest of Dean – 4d of the increase occurred in 1900–9, and 2d in 1912–14; North Wales – 4½d of the increase occurred in 1900–7, and 1d in 1912–14; South Wales – 5d of the increase occurred in 1890–1911, and 3d in 1912–14; Scotland – 2d of the increase occurred in 1890–1911, and 2d in 1912–14. On the basis of these assumptions and the percentages on or off standard it is possible to work out indices of the wage rates of the various major groups of underground workers for the period 1886–1914, and this has been done and the results presented in tables 7.3 to 7.6 in the main text of this chapter.

8

ROYALTIES

The amount of capital and labour employed in the coal industry, and the value placed on the latter, have been the subject of earlier chapters. This chapter will deal, more briefly, with the third factor of production, conventionally referred to as 'land'. In manufacturing industry, land enters the production function in two ways: in a literal way, as an area on which the factory is situated, and secondly – and usually more importantly – as raw materials to be processed. In agriculture, land in the literal sense is incomparably more important than raw materials, and extractive industry shares this characteristic to some degree. Payment for land is an almost negligible feature of the costs of a manufacturing industry, but it is a major expense of the tenant farmer, and it can be quite substantial in the case of mining industries. To explain such payment, the theory of rent was developed, by Ricardo above all others, and it has remained one of the most stable parts of the corpus of economic doctrine. According to the theory, rent is determined by the costs of the products capable of being produced on a given piece of land and their relation to the prices of those products. Marginal land, capable of producing products at a cost just equal to the prices, will pay no rent; and the rent of other land will be determined by the margin between the costs of production on it and the prices of the products. According to this theory, rent does not enter into costs of production, and hence into equilibrium prices, but, like profits, it is determined by prices.

The theory of agricultural rents has, to a large extent, been carried over into mine rents, and this may be accepted as correct in principle, but only up to a point. There is one conspicuous difference, and another which may also have some significance, between a mine and a farm. A mine differs from a farm in that it is exhausted by use. Its operation also tends to disfigure its neighbourhood, whereas farming has certainly been held by some people to enhance natural amenities. For one or both of these reasons, a marginal mine must be expected to pay some rent, though the exact amount will probably be impossible to determine at all accurately, and it is likely to have varied with both differences in the way in which amenities are regarded by landlords, and with their expectations about the future value of known but untapped minerals and the way in which they themselves discount the future welfare of themselves or their heirs. The exhaustible nature of a mine also has a significant influence through the way in which it has affected the system by which mine rents came to be paid. It is true that as late as the early years of the nineteenth century there were still cases of mines being let on the same basis as farm land, at a simple rental of so much per year.[1] But long before then, the majority of mine leases required payment of royalties, varying with the amount of coal-bearing land used up. A more straightforward rent element was usually retained, in the form of a fixed ('certain' or 'dead') rent, which had to be paid no matter how much coal was extracted. The purpose of this was to prevent the lessee from failing to work a mine. But the fixed rent was usually smaller than the lessee's total payment. The fixed rent allowed him to take out a certain quantity of coal, which was exceeded in most years, and beyond that a royalty was paid according to the amount extracted. Usually, too, there was provision for working out 'shorts' – that is, for making up the quantity by which output in any year fell short of that allowed by the fixed rent in succeeding years. The royalty form which rental payment took meant that, in effect, mine rents became a payment for the principal raw material of the industry – coal-bearing land – with the size of the payment varying more or less proportionately to the amount of the raw material used. However, the theory of rent still remained applicable, with modification, because the payment was not related to the cost of production of the raw material, but to its quality and its position. If the marginal mines paid a certain minimum rate, determined by the factors suggested above, the amount by which payments in other mines exceeded that was the result of the quality of the coal, the ease (i.e. the cost) with which it could be got, and the mine's position in relation to markets. But in addition to these factors, individual royalty payments were also the result of the circumstances and expectations of the industry at the particular time at which the lease on which they were based had been granted.

The effect of the quality of coal on the amount of the royalty payment which it would fetch is illustrated most graphically by the difference between the royalty in Scotland on common coal and cannel coal in the period before the invention and widespread use of the gas mantle, when cannel was used by gasworks to produce the yellow flame, desirable to produce the maximum possible light from fish-tail burners. Around 1890 the royalty on common coal averaged 6d per ton, whilst it was double that amount on cannel.[2] In a way, the almost universal difference in royalty between large and small coal is another illustration of this same point. It is more difficult to isolate the effect of variations in difficulty in working the coal on the royalty paid, perhaps because such variations were rarely, if ever, properly anticipated when a lease was drawn up before a colliery was embarked upon. The best evidence for them is probably the extent to which mineral owners gave abatements to lessees on previously agreed terms, when the working of a colliery was proving to be unprofitable. However, there were other reasons for granting such abatements, most notably the failure to anticipate correctly at the time a lease was taken a future downward trend in the price of coal,[3] and there is really no means of saying how important an influence on the size of royalties was the degree of ease or difficulty of working a particular colliery. Our impression is that it was a good deal less significant than either quality or the third factor mentioned, position in relation to markets. That greater distance from markets should operate to reduce rent is both theoretically and intuitively obvious, and it can be shown to have happened in fact. But it is not possible to show this very often, because of the difficulty of separating it from other factors. On the rare occasions when similarity of other conditions does make demonstration possible, however, geographic situation seems to have had a very substantial effect. In one case cited, collieries 40% closer to a shipping port than other similar ones opened very little later paid about the same proportion more in royalties.[4]

Royalty payments were assessed in a number of different ways, though all amounted to a charge for depletion. In most of the inland coalfields it became common early in the nineteenth century to lease mineral land by the acre and thickness of seams, though in Yorkshire the simple acreage was the normal unit, and other systems were also found. Perhaps the acreage system sprang from the relative uniformity of the seams and from the practice of longwall working, which made intermittent measurement easy and reliable. In other coalfields, where pillar and stall prevailed longer, or where there was more faulting or seams tended to thin out, a royalty on the actual coal extracted (or, more commonly, sold) was the normal system which came to be used – a payment of so much per ton, or in the North-East up to the 1830s per chaldron. However, an older system, which related the royalty to the value of sales, prevailed in Scotland, Cumberland, and Lancashire at least until the spread of railways, and long afterwards in some cases. Such a system survived in many Scottish leases as late as 1873, when it was said to be coming back into favour again, though this proved to be a temporary fashion caused by landlords trying to get some advantage from the high prices of the early 70s. The difficulties of assessing average sales value when railways opened up a multiplicity of markets, and when owners were unwilling to open their books to their landlords, militated against this type of sliding scale system of royalties.[5] However, it was said to have been universal in at least one part of Lancashire, the Oldham district, in the late 1850s, and it was certainly known in other parts of the inland coalfields earlier in the century, and may have survived in Cumberland, as well as Lancashire and Scotland, in 1873.[6]

The high prices of 1872–3 resulted in some return to favour of a system which linked the level of royalty to coal prices, on the sliding scale principle, if not in a precisely proportionate way. In this form the system spread to coalfields where it had not previously been much used, such as South Wales, as well as undergoing a revival in Scotland. It did not become widespread, for the number of old leases always greatly exceeded new ones, and the low prices of the later 70s and 80s quickly brought landlords back to a preference for fixed royalties. But as late as 1918, the Marquess of Bute (whose lands, admittedly, included some in Scotland and the North-East, as well as South Wales) got nearly 15% of his royalty income from sliding scale leases.[7]

In the Forest of Dean, which was Crown property, a survival or development of an even older system operated, whereby any working inhabitant could qualify to become a 'free miner', and thereafter be able to stake a mining claim, rather after the fashion of the American West, with the Crown having the right to a fifth of the value produced.

The length of tenure, as well as the system of payment, could vary, though in England and Wales it was usually for a multiple of twenty-one years. The twenty-one-year lease, in fact, seems to have been normal until the depth of sinking began to require substantial outlays of capital, which, in some cases, could only be attracted by the greater security of tenure provided by longer leases. This probably began in the eighteenth century, for there is extant a copy of a lease for sixty-three years for Killingworth Colliery in Northumberland from as early as 1801,[8] though as late as 1830, twenty-one years remained the usual term in the North-East.[9] By the 1850s and 60s, however, forty-two years was more normal, and sixty-three years was not uncommon, though there were still plenty of leases for the old terms of twenty-one years.[10] In South Wales, the longer terms were usual, apparently, at a somewhat earlier date.[11] In Scotland, a seventy-year lease is recorded from the middle of the eighteenth century,[12] but the usual term did not exceed thirty-one years even in the early twentieth century,[13] and it may be conjectured that there was less change in this respect in Scotland during the nineteenth century than in most other districts. Scotland may have shared, however, in the general tendency for leases to get longer, which was noted by a knowledgeable witness, R.F. Pawsey, to the Sankey Commission in 1919.[14]

The length of mineral leases had an effect on the incidence of royalty payments on individual colliery firms, and by the later part of the nineteenth century it is possible that this may have had almost as much to do with variations in the amount of royalties between different mines as did the quality of the coal or the convenience of their situation. Morris and Williams say of South Wales that 'the length of the mining leases . . . often meant that in a developing district the less favourably placed mines, which were started later, paid as high a royalty as better situated collieries, which had been opened earlier, when the prospects of the district were less assured'.[15] They appear, however, to have placed excessive weight on the evidence of one aggrieved coalowner to the Royal Commission on Mining Royalties.[16] The evidence of a Welsh mining engineer on average royalties of collieries in the upper and lower Rhondda Valleys should have carried more weight. Four large collieries at the top of the valley, twenty-six miles on average from a port, paid 5.2d per ton royalty (including a very small wayleave), whilst four other collieries at the bottom of the valley, eleven miles nearer the port, paid an average of 9.1d per ton. He explained that royalties in the early days of exploitation of the district had tended to be set at the same level everywhere, but they had been much reduced in the upper collieries over time.[17]

In the eighteenth century, certainly, there had been some leases in the early years of exploitation of the heads of the valleys at very low rentals, such as the well-known case of Dowlais, where the Marquess of Bute had leased an extensive area to the company for a hundred years at a certain rent of only £100 per year. When this was renewed in the middle of the nineteenth century on the royalty system, the first year's payment was reported as £20,000.[18] It is also true that minerals were leased in the big boom of the early 1870s, and perhaps in other booms too, at rates which were higher than were common in existing leases for similar coal-bearing land.[19] Variations such as these give rise to Morris and Williams's assertion that they 'appear to have arisen as much from *ad hoc* causes as from theoretical principles'. But it is clear that anomalous variations were normally to be found only in the early stages of exploitation of a district, and that others which arose in especially vigorous booms were less inclined to persist for very long in the face of pressure from loss-making lessees – including the ultimate pressure of bankruptcy. Moreover, not all lessees were so lacking in foresight as to take leases with no provision for revision. In fact, theoretical principles would seem to have considerable justification in the normal case in South Wales. There is no reason to suppose the same was not generally true elsewhere, though it was said in 1830 that royalties in Durham tended to be unusually uniform because of the policy of the largest landowner, the Dean and Chapter of Durham.[20] When the Ecclesiastical Commissioners took over, however, their agents appear to have pursued a more discriminating policy.[21]

In addition to rental payments for coal-bearing land, landowners were able to make a number of other charges on colliery entrepreneurs which had the nature of rents. These were wayleaves, both underground and on the surface, instroke and outstroke charges, and shaft rents. The underground wayleave was a charge for moving coal from some other owner's property along the underground passages made in the wayleave owner's. It was accompanied by the analogous shaft rent, for raising 'alien' coal up the shaft on this property, and, on the surface perhaps, by another wayleave. A necessary preliminary to charging an underground wayleave was the moving of coal from one owner's property to another's, for which outstroke and instroke were charged. All these sound to be onerous burdens on the industry, but in practice they were not really so. Outside the North-East of England they never amounted to very much in any case, and even there they were much less than royalties. But in any case, they did not amount to an addition to royalties, but were merely a way of redistributing rental charges amongst different property owners. If a lessee had to pay wayleave to landlord A for taking coal from landlord B's property through A's, then B could get that much less royalty. Similarly, if a surface wayleave was charged for carting coal from a colliery to (say) the river Tyne, then the royalty owners would receive less than if the carting could take place entirely over his own property.[22]

The whole system of payment for coal-bearing land came under question in the later nineteenth century, beginning with the long period of depressed coal prices in the later 1870s and 80s, when royalty owners took an increasing proportion of the value added by the industry, and were resented by entrepreneur and miner alike. A Royal Commission on mining royalties sat in 1889–91, and received opinions hostile to the system of private ownership from most trade unionists and some coalowners, and more ambiguous views from the rest of the latter. Economists pointed out that royalties 'had the effect of producing a certain balance between different districts unequally favoured in the distribution of natural advantages', and that abolition would give windfall gains to intramarginal lessees, but would only reduce prices by the amount of the minimum royalty and would not, in theory, have any effect at all on wages.[23] But despite this, despite the Royal Commission's report in favour of maintaining the *status quo*, and despite the return of prosperity in 1889 and from the late 90s onwards, the system remained under attack in some quarters, especially amongst trade unionists. This attack reached a crescendo in the hostile questioning, by union representatives and their allies on the Sankey Commission, of big royalty owners and their agents. However, all that they succeeded in doing, apart from demonstrating the huge size of some individual incomes from royalties,[24] was showing once again what the Royal Commission of 1889 had shown, that royalties formed a comparatively unimportant fraction of the total costs of the coal industry by the late nineteenth and early twentieth centuries.

It is, of course, impossible to be precise about the average amount of royalty payments, at least until towards the very end of our period, when the imposition of a special tax on mineral royalties furnishes a precise national average – it was, in fact, about 6¾d per ton. In the eighteenth century, according to Ashton and Sykes, the average royalty was higher, and the range covered by individual leases was a good deal wider.[25] But it seems likely both that the average had already fallen by the beginning of the nineteenth century to a figure which was little changed between then and the 1890s, and that the range was not really very much narrower until about the same period. Ashton and Sykes quote a royalty in Shropshire in the middle of the eighteenth century which was as high as 26% of the selling price of coal, and another which may have been even higher, for it was nearly 50% of the tonnage rate which the butty received. At the opposite end of the scale was the very low rental incorporated into Anthony Bacon's (later the Dowlais Company's) first lease from the Marquess of Bute, which has already been referred to. Very low rents, such as the Dowlais Company's was prior to its renewal in the 1840s, were a survival of the early days of exploitation of a new coalfield, as well as of inexperience on the part of landlords in dealing with the letting of their coal-bearing lands. Few, if any, can have survived beyond the middle of the nineteenth century – not many, indeed, beyond the end of the eighteenth century. At the top end of the range, the highest royalties in cash terms were in boom periods for those lessors whose receipts were fixed in terms of a percentage of market price, which was very common in Lancashire and Scotland up to about the middle of the nineteenth century, as we have seen. Those leases of this kind which were still operative in the early 1870s must have pushed the range of royalty payments in cash terms, though not in proportion to price, wider, probably, than it had ever been.[26] The few leases which still fixed royalties as a percentage of profits must have gone even higher, though it is doubtful if any of these existed in which the landlord had not, in effect, provided some of the equity capital.

Prior to the enquiries of the Royal Commission on Mining Royalties, there is little to go on in seeking to discover what was the average level of royalties other than the impressions of informed contemporaries, and there are few enough of these before the furore over the high prices of 1872–3. The well-known North Country viewer, John Buddle, told the House of Commons Select Committee of 1830 that royalties in the North-East ranged from 5d to 1s 6d per London chaldron, which was from nearly 4d to 1s 2d per ton. No average was given, but the implication of another witness's evidence was that it would be about 9d per chaldron (or 7d per ton).[27] Since the pitmouth price of round coal at that time probably averaged around 9s per ton, and round coal was about four-fifths of output and much of the small coal was unsold, we can estimate that the royalty averaged somewhere between 8% and 10% or 11% of sales.[28] Even before 1830, the isolated figures we have from the North-East indicate the same sort of proportion, and though these may not have included any pits in plum sites, it seems reasonably certain that the average royalty in Northumberland and Durham at the beginning of the nineteenth century was not very different from what it was in the early 1870s. By then it was said to average 6d a ton in Northumberland, and 6d to 9d a ton in Durham, which was, again, somewhere between 8% and 11% of the average pitmouth price before the massive increase of 1872–3.[29]

It is possible that Scotland was one coalfield where the average level of royalties did fall in the first part of the nineteenth century. Ashton and Sykes found that many Scottish coalowners took leases involving royalties of a fifth of sales value in the second half of the eighteenth century, but by 1850 or so such a high figure was a thing of the past. Royalties fixed on a proportional basis could still go as high as a sixth, but a more typical figure was an eighth or a ninth; and fixed tonnage royalties, which had been increasing in importance from the 1830s onwards, seem to have averaged about 8d per ton from the 1850s or earlier through to the early 1870s,[30] which was about an eighth of the pitmouth price. Another coalfield where royalties seem to have fallen substantially between the middle of the eighteenth and the middle of the nineteenth centuries was Shropshire, which provided Ashton and Sykes's highest examples. In this case, though, the fall mostly took place in the 1770s or 80s,[31] and it may very well have been the result of the spread of canals in the Black Country opening up competition for the Severnside industry of Coalbrookdale and neighbourhood. The rates of over 20% of prices found up to the early 1770s were, it seems, reduced by the end of that decade, in one case by ten percentage points. A century later, the average of between 6d and 9d a ton reported to the Royal Commission on Mining Royalties[32] cannot have represented a very much lower fraction of normal prices.

For the remaining coalfields, whilst evidence is thin it points in the direction of little change in levels of royalty up to the 1880s or 90s, apart from the variations in proportion produced by the relation between cyclical price movements and fixed royalty payments. In South Wales, where major exploitation of the coalfield dates only from the second half of the eighteenth century, it may be that royalties were higher around 1800 than they had been earlier, the prospects of the district having become more clearly established by then. Perhaps this process even continued a little after 1800, with the renewal of earlier long leases at low rates; but both the length and level of payment of the Dowlais lease of 1748 was exceptional, and it is doubtful if it really was the case that average royalties rose after 1800. Morris and Williams take the owners' evidence to the Royal Commission on Royalties as applying to the mid-century period, and support this with evidence from the 1860s and 70s.[33] There seems to be no reason to disagree with this: indeed, we would conjecture that these levels applied even before 1840. What they amounted to was a range for steam coal of from 6d to 10d a ton for large coal and 3d to 5d a ton for small (ignoring one or two isolated examples on either side of this range), with an average of 8d a ton for large and 4d a ton for small coal. On house-coal, the royalty was usually a bit higher: the royalty owners' figures to the Royal Commission (which gave an average for steam coal of 6d a ton) indicate that it was, in fact, 1d a ton higher on average.[34] If that is so, then assuming about three-quarters of the output was large coal, and in the light of the much greater proportion of steam coal than of house-coal in total output,[35] the overall average for South Wales was about 7½d a ton.

In Yorkshire, the only one of the remaining coalfields about which there is any significant amount of information for the beginning of our period, it seems likely that royalties stayed at around the figure of 6d a ton on average through to the early 1870s, and had returned to that level by the late 80s after a small temporary rise occasioned by leases taken at high rates in 1872–3. Ashton and Sykes's figures of acreage royalties in the thin seams near Bradford and Sheffield in the late eighteenth and early nineteenth centuries, which are in the range from £35 to £53 an acre, imply approximately the figure of 6d a ton, or even less.[36] They do also give one tonnage figure for South Yorkshire in 1796, which was 8d, but since this was for coal sold at 7s 11d, it is a reasonable assumption that this was for the superior house-coal seam, the Silkstone.[37] At any rate, an approximate figure of 10% or less of price seems to be a justifiable deduction; and with the prices of 1870, an average royalty of 6d per ton at that date implies the same sort of proportion then. Figures from the books of one large colliery company in West Yorkshire, showing royalties as 10% to 13% of costs according to seam in 1860 and as a little over 10% of costs overall in 1867–8, bear out our conclusion.[38] The conclusion can, perhaps, be extended to the East Midlands on the basis of Farey's statement that colliery rents in Derbyshire around 1811 were from £50 to £180 an acre.[39] In 1889 they were from £30 to £200 an acre in the East Midlands and Leicestershire.[40] The average in tonnage terms would seem to have been 1d or so below that of Yorkshire, where the coal was of superior quality for coking and the best house-coal fetched at least 3s more per ton.[41]

Whether one can also say that royalties in the West Midlands, North Wales, or Lancashire were also stable up to the 1870s is a moot question. There really is no evidence for the earlier part of the nineteenth century, except for Ashton and Sykes's citation of a royalty of 12.5% of price at

Mold in North Wales in 1825. In the late 1880s, the average royalty of less than 5d per Imperial ton in that district was certainly a lower proportion of price; but there is no means of knowing how typical the earlier figure was. The Lancashire percentage royalties of the first half and more of the century may have gone as high as 20%, though probably an eighth would be a more common figure. By 1880 the Wigan Coal and Iron Company's average of 6½d a ton, or roughly 11%, was probably typical of the bulk of the coalfield, though higher figures probably prevailed in East Lancashire near the main urban markets, and cannel coal certainly paid a good deal more. An overall average of 8½d a ton suggested by an owner to the Royal Commission on Royalties was probably too high, but the figure of 6d a ton suggested by the landlords likewise seems too low.[42]

In South Staffordshire, the Ten Yard seam appears to have attracted a bigger royalty than any other, and as that was worked out or flooded and produced an ever smaller proportion of total output, so the average royalty level of the district was bound to go down. But that did not happen to any major extent until after the early 1870s, though the lower royalties of the Cannock Chase coalfield must have been having some impact on the district average by then. It is not absolutely certain, however, that the high figures per ton sometimes cited for the Thick Coal in the first half of the nineteenth century were really correct. The Earl of Dudley's agent, for example, told the Sankey Commission that the Earl received over 2s per ton on the Ten Yard seam in 1835, and 1s 7d a ton in 1842;[43] but leases in the Thick Coal were normally by the acre, and it may be that the calculation was done taking a rule-of-thumb measurement, such as 120 tons per acre-inch, and failing to allow for colliery consumption, etc., and, more important, for the fact that it was normally impossible to remove more than half the coal from that seam. The 1889 figure of £400 to £500 an acre for the Ten Yard seam implies only 2½d or 3½d a ton if the whole of the seam was removed, whereas the average per ton given for that period was 6d.[44] Since it is hard to believe that royalties fell by three-quarters in the years after 1835, one is tempted to question the earlier figure. It may simply be, of course, that it was untypical. No doubt the Earl's agent would have been anxious in 1919 to minimise the impact of royalties, and show how much lower they were than formerly.

In fine, it is probable that royalty payments per ton did fall between the first half of the nineteenth century and the early 1870s in Lancashire, North Wales, and Staffordshire; but there is really no way of being sure, or of making quantitative estimates.

In the discussion of the level of royalties up to now, the early 1870s has been taken as marking a break in trends. But, as has been mentioned more than once, the break was only a temporary result of the extraordinarily high prices of 1872–3. It took some time thereafter for the correc-

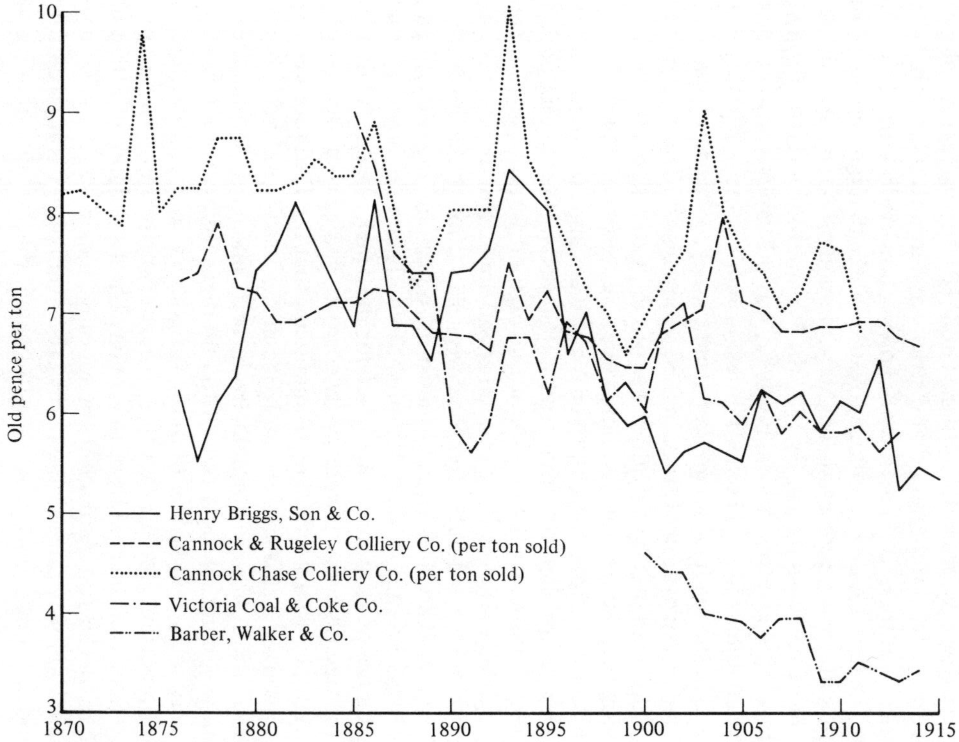

Figure 8.1 Royalties cost per ton for various firms, 1870–1915

Table 8.1 *Average royalties per ton, late 1880s and mid-1920s (in pence)*

	Late 1880s	Mid-1920s		Late 1880s	Mid-1920s
Northumberland	4–4½	5.74	Warwickshire	6	3.82
Durham	5	6.32	South Staffordshire & Worcestershire		6.55
Cumberland	6–7	9.66	Forest of Dean	3	4.35
South Yorkshire	6	4.38	Bristol	6	
West Yorkshire		4.85	Somerset	6	6.93
Lancashire & Cheshire	6–6½	5.25	Denbighshire	4	2.97
Nottinghamshire	4¾–5	4.07	Flintshire		4.00
North Derbyshire		3.74	South Wales & Monmouthshire	6	7.83
South Derbyshire		3.73	Fifeshire & Clackmannanshire	6–7	7.14
Leicestershire		2.86	East & Mid Lothian		6.96
North Staffordshire	5½	3.75	Central Scotland	7–9	6.68
Shropshire	6	3.25	Ayrshire, Argyll, & Dumfries		5.57
Cannock Chase	5½–6	4.06			

Sources: Royal Commission on Mining Royalties (*SP* 1890 XXVI and *SP* 1890–1 XLI) and Royal Commission on the Coal Industry (1925), appendix 8.

tion to come, as is indicated by *The Economist*'s review of 1878, where there is mention of 'appeals for relief made by coalowners everywhere being generally met with the response that the coal will keep *in situ*, and that if the lessee can't afford to pay his royalty rent, the lessor can afford to wait'.[45] The prolonged depression, however, broke down the landlords' resistance in the end, and in the 1880s average royalties came down again to the pre-1872 level. Once started, the trend was continuous in some districts, even accelerating slightly in the 1890s. This can be seen for a number of colliery companies in the inland coalfields in figure 8.1. This diagram, of course, covers only a handful of firms; but there is every reason to believe that these may be taken as typical of the firm which was operating the same pits, or similar pits in the same area, throughout the period covered. The overall average rate of decline of royalties may have been rather faster in those coalfields which were expanding into new areas, for the deeper pits which were required paid royalties at lower rates.[46] The decline in average rates in most of the inland coalfields between the late 1880s and the 1920s is apparent in table 8.1, where it is also clear that there was no similar decline in the other coalfields, except in West and Central Scotland. In Northumberland, Durham, Cumberland, East Scotland, and South Wales the movement of royalty rates was upwards over this period, though it is probable that the Cumberland figure was inflated by the leasing as going concerns of collieries which had been developed by landlords, and the consequent inclusion of a payment for capital in the royalty.[47]

The reason which suggests itself for this divergent behaviour of the major inland coalfields (including West and Central Scotland) and of the others is the disproportionate growth of export demand over the period from the 1880s to 1913, and the superior position of those coalfields near to the coast to satisfy this. The need for the new collieries in the Doncaster and Mansfield areas to find export markets via relatively lengthy rail carriage to Goole, Grimsby, or Hull was a reason for the depression of royalty rates in South Yorkshire and the East Midlands; but the strength of the export demand which made such shipments possible at all meant that royalties on new leases near the coast, in Fife, Northumberland, Durham, or South Wales, could be pushed up. The overall effect on the country as a whole was, as a leading spirit in the Mineral Proprietors' Association claimed in 1919,[48] to bring down the average royalty; but it was only by about 1d per ton.

The question of royalties came by the end of the nineteenth century to assume, as the Memorandum of the Mining Association to the Samuel Commission put it, 'a position in the public eye entirely out of proportion to its real significance'. For most, if not all, of the nineteenth century – and certainly long before its close – the royalty payment was only a small fraction of any colliery's costs. Whatever the merits or demerits of private ownership of mineral lands, which, because of concentration of a majority into a few hands was an offence to anyone of egalitarian inclinations, and was particularly resented by the miners' unions, royalties did not play a large part in the economics of the coal industry. And even if the system had been different – even if royalties had been abolished altogether – it is hard to see that anything about the industry would have changed very much, except the profit levels of intramarginal firms. It is true that this might have had significant secondary effects. Even though marginal profits were unchanged, a higher average level of profits in the industry might have attracted a greater investment of capital, for coal was in some ways a gambler's industry, and no one could be sure how close a new colliery would be to the margin. Such an increase in investment would surely have had consequent effects on output and prices, and the demand for labour, and so on wages. But whilst this might be an initial effect of the abolition of royalties, it would surely prove only temporary for the most part, given the lack of a perfectly elastic demand for coal.

9

PRICES, COSTS, AND PROFITS

We have so far considered labour and its reward, the return to the factor of production 'land', and the quantity of capital employed. It is the purpose of this chapter to examine the return on capital. But since this was in essence a residual, resulting from the difference between prices and costs, it is convenient to look at these latter at this point, before tackling the question of profits. The information on prices and on costs is very different, both in extent and in nature, and the two are best treated separately to begin with.

Prices

It is necessary to begin by asking just what we mean by coal prices. Every single sale of a parcel of coal took place at a price, and of the millions of such transactions which took place in each year, records relating to some hundreds at least have survived for by far the greater part of the period with which we are concerned. Some of these sales took place at the collieries and some at the household consumers' doors or the factory gate (or, at any rate, the office). Many more took place somewhere in between, as it were. In which of all these are we interested? The answer to this question depends on what particular aspect of the coal industry we are concerned with. For the immediate present purpose, which is to analyse the receipts of colliery businesses in relation to their costs, what we ideally require is some annual measure – annual because that is the period conventionally used for most accounting purposes – of the average price per ton received by coalowners in the various districts of the country. To study the development of coal markets, especially from the point of view of the consumer, a different measure of price is desirable – one which relates to the place of use, or at any rate somewhere near it, such as the London Coal Market was for a substantial fraction of household consumption. If we were concerned with the coal merchanting business, the prices paid by final consumers would be the focus of interest; and if an earlier stage of distribution was to be studied, prices f.o.b. ship or in rail wagons at the pit siding would be required. Some data relating to all these kinds of prices have survived; but before going on to discuss them, it may be convenient to say something about the price-fixing process in the industry.

At the beginning of our period and right through to the 1890s, prices in most coalfields were fixed in practice, as well as in theory, entirely on the initiative of the individual entrepreneur. The main exception to this was for coal from Durham and Northumberland shipped by sea, principally to the London market, the price of which was set by the oligopolistic 'Limitation of the Vend' for a large part of the first half of the nineteenth century.[1] In other times and places, the individual owner could, in a sense, set his own price. In so far as competition was imperfect, this left full scope for variations. But the major imperfection was that springing from differences of location, and this, along with the obvious matter of quality variation, was the only important cause of price differences between collieries in the same district. There were, for instance, several prices for coal of the same quality at the same pit, depending entirely on its destination – the result, therefore, of differences in the severity of competition in different markets and of location in relation to those markets. For example, in January 1913, best Haigh Moor housecoal fetched 18s to 19s a ton for local sale in Leeds at all collieries in West Yorkshire. But 'Sharlston Wallsend', reputedly the best West Yorkshire coal, was only priced at 14s a ton when it was for railing to London.[2] In some cases, attempts were made to increase the effect of natural product differentiation by means of advertising, but the scale of such attempts was very limited, and it may be doubted if they ever had much effect.[3] In general, then, we can say that although the individual coalowner was free to set his own price, in practice he was a price-taker, the price being determined by the largely unfettered forces of competition in a nearly perfect market, except for very local sales. But it is worth bearing in mind that this last exception was a very important one in the first three or four decades of the nineteenth century in those parts of the inland coalfields where transport costs remained high. It is also worth pointing out that the very existence of a high-cost coalfield, such as that of Bristol and Somerset, depended on its ability to charge substantially higher pitmouth prices than (say) South Wales collieries, because of the protection afforded by transport costs.

The growing strength of trade unions from the later 1880s onwards – from the late 1860s and early 70s in some districts – may have had more than anything else to do with the setting up of coalowners' associations in the different coalfields (other than that of the North-East, where they had long existed). These were a virtual necessity for collective bargaining purposes; but they also came to try to fix prices as well as wages. They never included all the owners in any area, but in normal times non-members tended to follow the lead of the association, and the change from winter to summer prices, and *vice versa*, was achieved with a high degree of cohesion, especially after about 1900. However, when the policy of the association came into conflict with the individual owner's conception of his interests, he would either openly break the agreement and court expulsion, or, more commonly, he would offer 'special prices' – that is, reductions for the buyer of more than a certain tonnage, or for his regular customers, or for an extra cargo, or, in fact, for any excuse that was handy.[4] Even at the end of our period, therefore, the initiative in setting prices remained very largely in the hands of individual owners, although there was a greater degree of uniformity than there used to be in seasonal price changes, and also in the stages of price rise in times of boom. There may also have been a tendency for prices to be a little stickier in coming down after a boom, but that is far from certain.

As was said earlier, some data relating to all kinds of prices have survived, but to be of much use, relatively long series of comparable prices are required, and these are not too plentiful, at any rate until well into the second half of the nineteenth century. The oldest and longest-continuing series of market prices are those of various varieties of coal from Newcastle and Sunderland in the London market, which are available continuously from 1805, and also for part of the eighteenth century. Even longer series of the contract prices paid by Westminster College and Greenwich Hospital go back into the earlier eighteenth century or beyond. Other lengthy series, covering the years from 1833 onwards, are those of the average declared value of coal exports at the different ports. These are the main general series of prices available prior to the 1870s and 80s, though there is also a good deal of information on prices at the Northumberland and Durham ports for shipment to London. There is little that is systematic, however, about pitmouth prices, even in the North-East, until the middle and later 1870s, when there begin to be both regular quotations of the prices of specific varieties of coal in the trade newspaper, the *Colliery Guardian*, and also official statistics of the average pitmouth prices in the different coalfields. These latter are, of course, exactly what is required for the primary purpose of this chapter, and fortunately some of the earlier material can be used to extrapolate them backwards to the earlier years of the century, though inevitably in a rather speculative fashion. Before doing this, however, it is worth making some attempt to analyse the relations between the prices of different sorts of coal, taking as our starting point the relation between the consumers' price data and average pitmouth receipts.

Two main influences may be expected to have had a bearing on this – changes in the representativeness of the types of coal for which we have price data, and changes in transport costs between colliery and market. The latter, fortunately, do not present any great difficulty so far as

London market prices are concerned, for it is comparatively easy to ascertain and subtract the amount of taxes and transport costs from the North-Eastern ports, at least for the period when the latter changed considerably. And since colliery businesses there generally sold much of their output (and, broadly speaking, the earlier the period the bigger the proportion) on the basis of delivering it onto the ships, this is as far as we need go. The last point also covers the other main price series of the first three-quarters of the nineteenth century, the average export values.

If changes in transport costs do not present a formidable problem, the same cannot be said of the relation between the prices of the types of coal about which we have plentiful data and the average receipts of the coalowners, which include the prices of other types of coal about which much less is known. There are two main questions involved in this: the relationship between the prices of different sorts of coal, and, where this is not stable, the proportions contributed by the different sorts to total sales. No really satisfactory answer can be given to either question, especially the second one; but some attempt must be made to deal with the first, and, where necessary, to make some allowance for the effects of the second.

The impression that coal is a homogeneous commodity will be quickly modified for anyone who looks at the mass of detailed information that is available on prices from the mid-1870s onwards, or even at the several different names under which North-Eastern house-coal was sold in the London market before the end of the Napoleonic Wars. Broadly speaking, coal can be divided into five kinds – coking, gas, house, manufacturing, and steam – but within these groups are many variations. The first type consisted, at the beginning of our period, of any coal from certain highly bituminous seams which possessed the property of caking when fired in an oven. By the end of the period – and, in fact, increasingly from the 1850s, and especially after the high prices of the early 70s – coking coal was almost exclusively the small coal from these bituminous seams, including the very fine slack known as smudge. Gas-coal was divided into a few different qualities, but only a comparatively small number – screened and unscreened, coal and nuts, with slight differences between districts, South Durham having a reputation for the best ordinary gas-coal, and consequently getting a slightly higher price. There was, in addition, cannel coal, found principally in Lancashire, the Lothians, and North Wales, which fetched very much more than any ordinary coal on account of its ability to produce yellow-flaming gas. Even in 1913, long after the invention of the incandescent mantle, Wigan cannel fetched 27s or more per ton at the pitmouth at a time when ordinary Lancashire gas-coal cost 11s to 12s per ton.[5] House-coal was divided into a much greater number of varieties, and was most often graded according to the colliery and/or the seam from which it came. Thus in the details of daily prices in the London market published by the Coal Factors' Society,[6] there is a price for the coal of many individual collieries, and also for some varieties which came from a particular seam common to more than one colliery. The range between the cheapest and the dearest was always quite considerable – of the order of perhaps 25%. By the second half of the century, subdivision had gone even further, and there is, for example, no price for (say) 'Yorkshire house-coal', but a number of prices for Haigh Moor, Silkstone, Warren House, etc., etc., each being further classed as 'best', 'seconds', 'nuts', and so on. Manufacturing coal does not exhibit quite the same variety, though with the spread of screening and its elaboration, and then of washeries, the possibilities of subdivision mounted, and the difficulty of identifying a continuous consistent price series increases. Fortunately, for the main purpose of this chapter this proliferation of grades did not take off until after the official statistics of pitmouth prices began. With steam coal, relative simplicity returns, differences in quality being confined to 'bests' and 'seconds', though the seam was also specified if it had a particularly good reputation. The relationship between the prices of the different major types of coal was not a stationary one. It varied both with the season and with the trade cycles, whilst, more importantly, there were long-term trends during our period. First, the seasonal variations.

Other things being equal, once good all-the-year round transport was available, house-coal prices rose in September–October and fell again in April–May. To a less degree, because much more influenced by long-term contracts, the prices of gas-coal behaved in the same way. The prices of steam coal varied in the opposite way, rising in the summer with the main shipping season, and falling in the winter, when some markets were closed by ice. In the first forty years or so of the nineteenth century, this regular seasonal behaviour had been rather different so far as house-coal is concerned, for the best shipping season before the coming of railways, even for inland collieries, was summer, and this led house-coal prices to behave as steam coal did later. (Steam coal, as distinct from manufacturing coal, was not, of course, a major product until the second half of the nineteenth century.) Naturally, talk of seasonal variations in coal prices applies only to coal sold in the open market. Steam coal sold under contract, such as that taken by the railways, and many export cargoes, was not subject to such variation; nor was most gas-coal, which was also sold under contract, nor coking coal, a large part of which was used in colliery firms' own coke ovens. In the case of manufacturing coal there was a slight tendency to seasonal movements by the 1850s or 60s, caused by the fact that a large part of this class of coal was really a by-product of more highly priced types. Thus, after the spread of railways, there tended to be a relative overproduction of manufacturing fuel in winter and a relative scarcity in summer in some districts where there were no steam or gas-coal seams.[7] However, this tendency was never very strong, and it was, in any case, largely confined to Lancashire and some of the smaller coalfields where round coal output was mostly used for domestic consumption.

Differences between the cyclical movements of prices of the various types of coal were most marked in the relation between manufacturing and coking coal and other types – i.e. between those which were largely by-products and the others.[8] A boom in industry sometimes led to a shortage of manufacturing fuel in its early stages, and a rise in the relative prices of the various types of small coal mainly used in industry. Sooner or later, however, the boom led to expanded demand for house-coal and the larger sorts of steam coal (used mainly in transport) – all the sooner if there was panic buying on the part of householders when prices began to rise. Expansion of output in response to rising prices usually led fairly quickly to relative overproduction of small coal, and the prices of the latter generally fell relative to those of round coal before the peak of the boom was reached. This situation was prolonged throughout the ensuing slump, for the lower demand from industry was not matched by a big fall in household demand. The variation between the relative prices of best coal and slack in the depths of depression and in the run-up to the peak might be as much as by a factor of three – i.e. the price of slack could vary from (say) 15% of that of best coal to 45% of it, though normally it was less than this.

Looking at the longer term, it is not clear that there was any very marked general trend in the relative prices of different sorts of coal before about 1870, except in so far as more elaborate screening produced new intermediate grades between best coal and residual slack. However, the fact that the relative prices of these last two kinds tended to remain much the same, despite the fact that more thoroughly screened slack became finer and less 'nutty', is *prima facie* evidence of a fall in the price of best relative to all other grades taken as a whole. To put it in a different way, the coalowners were succeeding, by more elaborate differentiation and preparation, in finding more (and more remunerative) markets for their third-grade coal. As a result, their average receipts per ton were pushed up closer to the price of best coal. The scope for increasing the revenue from inferior coal varied according to the nature of the seam from which it came and to the nature of local industrial markets – two influences which were not unconnected, since, for example, ironworks were not located on coalfields which produced non-caking coals. Small coal from the latter offered much less scope for improvements by preparation and was, in general, much less saleable than slack which caked easily. This continued to be true right through to the end of our period and beyond, but it was particularly the case before the high prices of the early 1870s had stimulated methods of economising on the more expensive grades. Around 1870, small coal remained of much lower value compared to round coal in the old-developed steam and house coal pits of the Tyne and Wear than in the gas- and coking coal pits of South Durham. In the 1820s and 30s, and probably earlier, small coal around Newcastle was usually sold for not much more than 10% of the price of best round, or about 20% to 30% of the average price of round coal. At many collieries – perhaps most not actually on the river itself – a great deal was 'burnt in the heap', since no market at all could be found for it. By the 1860s such a practice seems to have been less common, though this was probably only because more slack was being left underground. As late as 1861, Robert Hunt's *Mineral Statistics* reckoned that 11% of the Durham and Northumberland output was 'wasted at the pits' – that is, something over a third of the total amount of slack drawn out of the pits, which, since the wastage from coking and gas-coal seams must have been negligible, implies a much higher proportion at some other collieries, probably those concentrating on steam coal. However, by that time the average price of small coal had probably risen a little in relation to round coal even in Northumberland, whilst the growing importance of South Durham must have raised the average for the coalfield as a whole. By 1870 the ratio between the price of small coal and of best coal seems to have been about 30% on average, though the range at individual collieries for which we have records was from less than 20% to over 50%. In the boom of the early 70s, the average, judging by newspaper quotations, rose well over 50%; and though it fell subsequently, it was still around 40% at the end of the decade. In addition to this movement in the relative price of small coal, *tout simple*, grades such as 'beans' and 'nuts' were more common, and fetched up to a third more again.

The same sort of difference between the relative prices of small and round coal as occurred in the steam coal collieries and those producing more bituminous varieties in the North-East also occurred in South Wales, where the coking properties of the small coal were the key to its relative value. However, in South Wales it was the less bituminous steam coals which gradually, from the opening of the central Glamorgan valleys in the 1830s to the 1850s, became a more important part of total output. Despite this offsetting factor, the tendency for small coal to appreciate in relative value was much the same as in the North-East. Its ratio to best coal price in 1870 was about 25% to 30% for steam coal seams and almost twice as much for more bituminous coals. Around the middle of the century it had been about 30% to 40% for bituminous and only 10% to 20%, as in Northumberland, for steam coals.[9] The overall average may be estimated as having risen from about 30% to about 40%.

In other districts, coking properties were less important than nearness to manufacturing markets for coal, and it is significant that the highest ratios of manufacturing small coal to large coal prices that we have evidence of prior to the early 1870s came from industrial Lancashire, where figures exceeding 50% were recorded in 1853 and in 1864–6, and no doubt would be found in other upswing years if records existed.[10] Ratios also seem to have been relatively high in Yorkshire and in the Black Country. But in none of these districts does there seem to have been any significant trend change in the relative prices of best coal and of ordinary slack until the 1870s, and the elaboration of preparatory plant and of varieties of manufacturing coal seems to have been but little developed.

Genuine improvements in quality and more elaborate differentiation of the inferior types of coal proceeded apace after the boom of the early 70s. More and better screening plants were installed, and became standard equipment in all the other coalfields as well as Northumberland and Durham; and washeries were introduced and improved. On the consumers' side of the market there were changes in engines, furnaces, and coke ovens, which enabled relatively cheaper grades of coal to be used, and there was the development, from the 1890s, of a sizeable electricity industry, which took a comparatively poor grade of coal. Demand for inferior varieties was thus rising relatively rapidly, and more exact matching of particular requirements was enabling consumers to avoid 'overclubbing' – the purchase of more expensive coal than was absolutely necessary. A consequence of this was a much clearer and more widespread rising trend than before in the relative prices of manufacturing coal. This can be seen at its most obvious in comparing the changes in the prices of best coal and of plain ordinary small coal; but its fullest extent is best brought out by comparing this with the difference between the price of small coal at an earlier date and that of newer manufacturing grades at a later date. Both comparisons are made where possible in table 9.1, which shows both the changes between 1870 and the period when the official statistics of average pitmouth prices first became available, and the continuation of the trends up to the end of our period.

The districts included in this table have been selected for the availability of data, but none of the major coalfields is omitted. Apart from showing the general increase in the relative price of small coal, table 9.1 also brings out the fact that there was some tendency for house-coals to rise in price relative to steam coals, though both the East Midlands and South Wales were exceptions after 1881. It may be that the explanation lies in the house-coal market rather than in that for steam coal, though the

Table 9.1 *Percentage changes in prices of different types of coal in selected districts, 1870–81 and 1881–1909/13*

	Best house	Best steam	Small coal	Small coal: screened manufacturing
Durham & Northumberland				
1870–81	−8	−8	+20	...
1881–1909/13	+63	+40	+133	...
Yorkshire				
1870–81	+6	−10	+27	...
1881–1909/13	+81	+59	+100	+142
Lancashire				
1870–81	−10	...	+31	...
1881–1909/13	+68	...	+99	+127
East Midlands				
1870–81	+10	−9	+12	...
1881–1909/13	+53	+51	+78	+133
Warwickshire				
1870–81	−12	...	+6	...
1881–1909/13	+80	...	+30	+159
South Staffordshire				
1870–81	+11	...	+23	...
1881–1909/13	+57	...	+25	+81
South Wales				
1870–81	+6	−5	+2	...
1881–1909/13	+84	+31	+114	...
Scotland				
1870–81	−6	−7	+20	...
1881–1909/13	+76	+54	+128	+183

Sources: see text.

expansion of capacity to produce the latter in South Yorkshire, and its increasing move into export markets, where competition was generally fiercer than at home, seems a plausible explanation for that district. However, even there it is the exceptionally large rise in the price of best house-coal which stands out, and it could easily be that the superior and growing repute of Haigh Moor coal is what lies behind the movement of relative prices in this district. Similarly, the advance of Warwickshire house-coal in the London market and the recovery of seaborne sales there from the North-East may explain the movements in those cases.

It is now time to take up the construction of an index of the average pit-mouth price of coal in the various districts, based on the official statistics from 1881 onwards, but extrapolated backwards, partly in the light of the foregoing discussion, but more by reference to the relationship between the different price series that are available. The two main kinds of price data covering the major part of our period are (i) the records of the London Coal Market, which were summarised on an annual basis so far as 'best' coal is concerned in a good many official publications, and (ii) the averaged declared value of coal exported at the major ports in each year from 1833 onwards. The first of these is the only long series covering the first three-quarters of the nineteenth century which relates to market prices pure and simple. The published figures are annual averages of the twelve-monthly means of the prices of certain varieties of coal shipped from Northumberland and Durham which were from time to time regarded as the best. They included freight costs to London, of course, and also the duties on coal brought into London, which were not entirely abolished until 1890. They do not include subsequent distribution charges. The second kind of data, derived from the export statistics, are not of prices strictly speaking, but of average receipts on a mixture of different coals sold for overseas markets. They differ from pitmouth prices in including transport costs into the ship, and also in so far as the mixture of coals exported was not the same as the mixture sold in home markets. There are other, shorter series of price data for periods prior to 1882, some of which will be mentioned and used later; and there are a good many from the 1870s onwards, some of which have already been used in table 9.1. However, no really useful purpose would be served by reproducing this material here, and table 9.2 is confined to the main lengthy series just described.

The coal industry was notorious as one in which prices could be extremely volatile, but it is apparent from these indices, coupled with the long series available in the *Colliery Guardian*, that it also experienced quite extended periods when prices were fairly stable. The characteristic price behaviour of the industry, in fact, at any rate after the development of the railways and the effective ending of monopolistic practices, was a burst of rapid price rise, followed by rather slower decline, and succeeded by a decade or so of comparatively restricted fluctuations, before the next burst of price inflation.

The average receipts per ton of the coalowners, of course, exhibit exactly the same sort of short-run behaviour as the available market (or intermediate) prices, but they have a rather different long-term trend, as would be expected from the differences in the experience for different kinds of coal which were discussed above. For the period when reliable official statistics of average pitmouth prices were collected, from 1882 onwards, there is a persistent tendency for these series to rise at a faster rate than those of all market price series other than those relating to small coal. This reflects both the rise in the relative value of small coal and also the expansion of the output of intermediate varieties between best and small coal, principally for manufacturing purposes. It is most marked in

Table 9.2 *Indices of coal prices in London and at the main exporting ports, 1805–1914 (1886 = 100)*

		Export values at ports				
	Best coal London	North-East	Humber	Liverpool	South Wales	Glasgow & Kirkcaldy
1805	216					
1806	216					
1807	223					
1808	240					
1809	265					
1810	252					
1811	232					
1812	218					
1813	256					
1814	289					
1815	228					
1816	203					
1817	196					
1818	193					
1819	203					
1820	209					
1821	213					
1822	209					
1823	226					
1824	203					
1825	199					
1826	181					
1827	187					
1828	185					
1829	170					
1830	176					
1831	162					
1832	138					
1833	114	74	110	98	90	120
1834	129	72	92	112	81	104
1835	138	67	92	96	79	96
1836	145	71	103	125	89	128
1837	152	74	120	118	100	167
1838	155	72	112	105	98	122
1839	150	83	112	107	100	112
1840	149	78	110	105	106	107
1841	141	81	104	106	106	106
1842	133	79	104	104	100	105
1843	127	85	105	108	101	102
1844	144	91	114	109	94	102
1845	120	85	107	106	98	131
1846	112	83	101	122	103	144
1847	131	82	106	112	105	143

Table 9.2 (*cont.*)

	Best coal London	Export values at ports North-East	Humber	Liverpool	South Wales	Glasgow & Kirkcaldy
1848	113	85	106	114	106	120
1849	110	85	104	102	102	109
1850	106	86	97	98	101	108
1851	99	83	92	93	99	104
1852	102	84	92	96	98	102
1853	133	86	102	106	111	123
1854	150	108	119	124	117	147
1855	138	111	116	118	115	139
1856	118	109	114	122	113	123
1857	117	109	104	117	113	174
1858	115	108	105	117	111	135
1859	114	105	105	112	108	174
1860	126	101	107	124	107	135
1861	122	102	110	129	109	146
1862	109	101	108	121	106	117
1863	113	100	111	115	104	118
1864	126	102	109	120	118	121
1865	127	108	110	120	120	124
1866	126	121	111	120	122	142
1867	130	125	115	125	123	122
1868	117	122	113	120	119	110
1869	117	118	110	112	116	106
1870	115	117	108	112	118	109
1871	120	118	108	114	121	127
1872	158	198	173	162	188	212
1873	207	272	229	229	233	249
1874	164	221	198	179	201	187
1875	151	168	162	150	153	146
1876	134	139	146	127	116	128
1877	122	130	134	113	107	128
1878	112	119	123	111	104	118
1879	112	108	115	111	101	109
1880	99	110	105	108	102	112
1881	109	106	103	104	106	105
1882	107	111	104	104	108	103
1883	115	112	105	114	110	114
1884	104	109	103	113	111	110
1885	104	104	102	113	107	103
1886	100	100	100	100	100	100
1887	101	99	100	108	94	100
1888	101	98	101	110	99	99
1889	112	114	106	126	126	114
1890	126	160	130	137	148	146

Table 9.2 (*cont.*)

		Export values at ports				
	Best coal London	North-East	Humber	Liverpool	South Wales	Glasgow & Kirkcaldy
1891	124	149	128	133	145	141
1892	119	136	128	144	126	132
1893	129	121	125	145	117	121
1894	111	127	124	130	120	127
1895	99	115	113	117	109	103
1896	98	108	109	117	103	100
1897	104	109	107	121	105	102
1898	110	121	113	113	120	120
1899	123	133	114	118	125	137
1900	154	216	163	181	204	195
1901	132	164	146	155	174	158
1902	123	148	132	137	156	143
1903	107	143	129	129	145	138
1904	102	132	120	121	146	128
1905	105	125	116	116	133	121
1906	107	133	114	108	137	126
1907	134	153	128	122	163	162
1908	119	156	133	134	164	140
1909	115	135	121	119	151	129
1910	110	140	121	125	156	133
1911	115	133	119	119	155	128
1912	142	148	135	133	162	160
1913	142	173	151	148	166	174
1914		170			169	

Sources: London figures are derived from data published by the Coal Factors' Society, and taken from papers in the Northumberland Record Office. Export values are based on data in *SP* 1903 LXVIII and *MS*.

the two great exporting coalfields and in Scotland, which was also quite a substantial exporter, from 1900 onwards, despite the fact that export prices rose less than the average of all prices. Presumably it sprang from the greater ability of coalowners in these districts to find more remunerative outlets for their inferior coal, especially from steam coal seams, and this in turn probably owed more to technical developments outside the coal industry than to anything else. These official statistics are given in index form in the second part of table 9.3. The first part is an attempt, necessarily rather speculative, to push the series back before 1882. This cannot be attempted for every coalfield, but something can be done for most of the important ones.

In the case of Durham and Northumberland, the indices of export values and of pitmouth prices move together quite closely for a decade or

Table 9.3 *Indices of average value per ton at the pitmouth, 1805–1914 (1886 = 100)*

Part I: 1805–81

	North-East	Cumber-land	Yorks	Lancashire & Cheshire	Leicester-shire	Warwick-shire	North Wales	South Staffs	Somerset	South Wales	Scotland
1805	51										
1810	84										
1815	103										
1818	92										
1819	123										
1820	130										
1821	135										
1822	131										
1823	127										
1824	134										
1825	131										
1826	122										
1827	123										
1828	129										
1829	113										
1830	122										
1831	98										
1832	123										107
1833	97		120	100						96	107
1834	106		100	114						86	89
1835	111		100	98						84	99
1836	119		112	127						95	113
1837	119		131	120						107	131
1838	113		122	107						105	113
1839	120		122	109						107	101
1840	114		120	107						113	90
1841	123		113	108						113	107
1842	117		113	104						107	99
1843	119		114	107						108	91
1844	105		124	114						100	96

1845	102		116	117						105	133
1846	94		110	124						110	136
1847	109		115	115						112	127
1848	97		115	110						113	104
1849	93		113	101						109	102
1850	92		105	100						108	103
1851	85		100	94						106	99
1852	88		100	99						105	99
1853	111		111	111						118	134
1854	135		129	124						125	143
1855	130		126	119						123	146
1856	113		124	113						121	143
1857	112		113	106						121	125
1858	109		114	103						118	116
1859	107		114	105						115	117
1860	115		116	124						114	111
1861	113		120	129						116	106
1862	102		117	116						113	95
1863	104		121	110	80					111	102
1864	116		119	115	93					126	128
1865	119		120	117	103					128	131
1866	127		121	129	103					130	151
1867	132		125	132	105					131	129
1868	119		123	127	99					127	117
1869	117		120	119	98					124	112
1870	117		117	119	91					126	116
1871	123		117	121	93					129	135
1872	185		188	172	150					201	225
1873	245	273	249	243	208		248	270	192	249	264
1874	197	221	215	190	178	174	193	249	152	214	198
1875	162	135	176	159	166		151	198	123	163	155
1876	141	116	159	135	153		114	180	108	124	136
1877	129	121	146	120	139		101	150	100	114	136
1878	117	116	134	118	119		89	137	96	111	125
1879	108	102	125	118	112	107	91	129	95	108	116
1880	103	105	114	114	98	94	91	114	94	109	119
1881	106	112	112	110	102	96	96	114	93	113	111

Table 9.3 (*cont.*) *Part II: 1882–1914*

	North-East	Cumber-Land	York-shire	Lancashire & Cheshire	Derbyshire & Nottingham-shire	Leicester-shire	Warwick-shire	North Wales	North Stafford-shire	South Stafford-shire	Shrop-shire	Gloucester-shire	Somerset	South Wales	Scotland
1882	107	119	130	112	120	136	122	95	104	145	99	124	99	114	110
1883	113	119	115	115	112	102	103	97	104	115	99	122	100	122	129
1884	106	112	109	112	105	96	97	97	104	115	99	122	100	122	114
1885	103	103	105	106	105	96	97	94	104	107	104	113	95	114	110
1886	100	100	100	100	100	100	100	100	100	100	100	100	100	100	100
1887	100	104	98	99	95	109	100	98	100	95	100	103	100	104	99
1888	99	106	107	104	106	115	102	97	89	107	89	132	108	114	98
1889	117	125	139	124	149	160	142	113	96	128	104	154	115	155	129
1890	158	164	180	157	175	183	163	150	111	148	126	176	144	201	171
1891	158	154	171	156	161	181	158	145	107	148	122	176	143	201	159
1892	141	142	160	142	146	175	156	127	104	148	115	161	133	171	143
1893	126	133	155	145	151	174	166	124	119	148	136	171	131	146	143
1894	128	130	142	132	128	153	137	116	104	137	104	148	124	129	150
1895	117	120	126	123	114	136	120	112	93	123	100	128	114	120	132
1896	114	117	125	120	112	130	112	105	89	117	96	128	114	128	127
1897	120	121	127	115	112	132	114	104	91	120	94	128	114	132	130
1898	133	126	134	123	118	147	120	123	93	122	96	128	114	119	151
1899	156	147	152	145	142	168	137	141	104	137	111	180	157	173	186
1900	231	200	214	207	213	245	217	209	156	197	156	226	189	236	271
1901	193	176	187	186	193	219	198	168	177		126	190	171	228	196
1902	165	156	169	161	166	191	169	154	157		111	190	157	205	165
1903	158	147	157	150	155	170	161	137	147		111	191	148	184	155
1904	145	143	146	142	141	158	147	135	142		107	174	143	177	146
1905	143	141	135	141	134	142	136	134	137		104	161	137	171	144
1906	155	142	136	145	130	140	137	134	133		100	158	135	182	160
1907	195	173	166	163	161	183	167	143	147		111	175	154	226	219
1908	195	168	171	175	164	185	181	152	159		127	187	157	216	191
1909	167	155	150	162	140	168	158	145	149		121	183	143	210	165
1910	177	158	153	163	143	158	139	148	144		115	171	142	216	169
1911	166	155	154	165	148	158	142	151	144		115	171	142	219	169
1912	203	195	166	178	154	179	174	161	150		119	174	152	217	209
1913	235	202	189	197	176	189	189	181	164		126	179	161	232	240
1914	224	184	186	191	176	189	184	174	162		130	179	166	233	223

Sources: see text.

more after 1882, and since it is known[11] that exports earlier included a substantial minority of small coal and a disproportionately small amount of the most expensive, best house-coal, there does not seem to be any objection to using the export values to extrapolate backwards the average pitmouth prices. However, the size of the fluctuations in export values almost certainly fails to reflect properly the greater volatility of house-coal prices. It seems best, therefore, to use a combination of export values and of the London market prices (less duties and freight costs)[12] for the extrapolation.

In the case of South Wales, the average pitmouth prices and the export values move in the same way during the whole of the period for which both are available, though, as in the North-East, the former rose substantially more after 1900 as small coal became increasingly valuable. However, in this district the latter development had begun well before 1900, and it can probably be traced back to the fuel economy in the iron industry which was induced by the high prices of 1872–3, becoming effective during the succeeding depression. If this is so, we do not need to take it into account in our extrapolation backwards from the upswing year 1882. However, there is one characteristic which, unless it is also associated with a once-for-all change in the value of small coal after the great boom of the early 70s, may make it misleading to use the export values in a straightforward fashion as the estimator for earlier pitmouth prices, and that is the fact that average pitmouth prices were at all times after 1882 greater in boom years than those of export values. It becomes important, therefore, to take a view as to whether the value of small coal rose much more rapidly than that of large in 1872–3 and earlier booms. It is reasonably certain that this did not happen to any significant extent so far as the more bituminous seams are concerned.[13] The price of small coal from steam coal seams was both lower and more volatile. In the eleven years 1858–68, the average realised price at Cardiff varied from 1s 6d to 4s 6d per ton, whilst the range for large steam coal was 8s 6d to 12.[14] Such volatility, at a time when steam coal constituted well over half of total output in the 1880s and later, probably goes a long way in accounting for the greater upward movements in average pitmouth prices than in export values which we noted above. But in the 1860s, and *a fortiori* earlier, house-coal was still the most important type sold,[15] and a much larger proportion of small coal from the steam coal seams was stowed underground than subsequently. Consequently, it seems reasonable to assume that, except perhaps to some degree in 1872–3, the export values will probably provide a fair guide to the movement of average pitmouth prices prior to 1882. We have so used them here to extrapolate backwards, without attempting to make any adjustment for 1872–3; but it must be borne in mind that the estimates for those years may be up to (say) 10% too low, and that those for 1864–7 may also be a little bit on the low side.

There are three other coalfields for which average export values at the appropriate ports may be used to extrapolate backwards the official statistics of average pitmouth prices – Scotland, Yorkshire, and Lancashire – though in all cases exports formed a much smaller fraction of sales than in the North-East or South Wales. In all these cases there is the same sort of relationship between the two series as was found in the major exporting coalfields – a markedly faster rise of pitmouth prices after 1900, and a greater propensity to rise of that series in boom years earlier. The latter tendency can almost certainly be put down to the same factor already noticed – the increasing relative value of small coal, dating back to the early 70s. If this is so, it seems reasonable to adopt the same procedure as in the case of South Wales, namely to employ the export values but bear in mind the likelihood of the resulting estimates being somewhat too low in 1872–3 and perhaps some earlier years when prices reached much higher levels than normal. However, the Scottish ports' export values behave extremely oddly in the period 1856–61 and in some occasional years previously. Prior to 1862, therefore, a series of prices at Elphinstone Colliery, provided by the Select Committee of 1873,[16] has been preferred, despite the fact that it is very unlikely that the Lothians colliery was really typical of Scotland as a whole. It does seem, though, that it was more typical than the export values, which were probably affected aberrantly by irregular shipments of cannel coal. In the Lancashire case, too, the export values in the 1850s and 60s seem to fluctuate a good deal less than other evidence on prices would suggest to be reasonable. For the period 1841–66, there are available the prices of the Bridgwater Trustees' coal at their Manchester wharf, and these have been combined with the export values to give what is likely to be a fairer representation of average prices.[17]

For the remaining coalfields there is no easy method of estimating average pitmouth prices prior to 1882, though there are district figures for North Wales and for Somerset in the 1870s and 80s, which were published by the Royal Commission on Mining Royalties,[18] and these may be used to extend the series for these districts for a few years. The same can be done for South Staffordshire, for Leicestershire, and for Warwickshire on the basis of the records of one major company in each district, provided these can be regarded as typical. In some coalfields this would be a totally unwarrantable assumption because of the variety of seams found in different parts. This may well make it unjustifiable for South Staffordshire, since the Cannock Chase coalfield produced a rather different mixture of types of coal from that produced in the Black Country. Despite this we have used the Cannock Chase Colliery Company's average receipts here, though it must be borne in mind that the results are more speculative than most. For Leicestershire and Warwickshire, where the Oakthorpe Colliery and the Baddesley Trustees' records have been similarly used, the homogeneity of the coalfield makes the process somewhat less unreliable.

Before leaving the subject of coal prices, it is worth looking briefly at the question of variations from the averages with which we have been mainly concerned. As was pointed out at the beginning of this section, the prices of different kinds of coal varied, and there were also differences between the qualities of the same basic kind of coal. These things, in turn, had a lot to do with deviations between different collieries in their average receipts per ton sold, and at times these could be very large. Most of our information about individual collieries comes from the later part of our period when, following the creation of a very dense network of railway lines in the coalfield areas and the intensive location of industries on or close to most of them, variations in transport costs between pits and markets were pretty small for most pits in most coalfields. In 1900, for example, of the thirty or so collieries in Scotland, the North-East, Lancashire, Yorkshire, and various parts of the Midlands, whose records have been examined, the biggest difference from the average pitmouth price for the coalfield concerned is the average receipts per ton of Seaton Delaval in Northumberland, which were 28% higher. But this was exceptional, and only one other colliery's receipts were not within 12% of its district's average. In 1883, of the fifteen collieries for which this comparison has been made, one was 25% above average, another 14% above, and the rest less than 10% above or below. Such records as we have for earlier periods suggest that a similar degree of uniformity would be found in the 1860s and 70s, but that in the 1840s and 50s there may have been more divergence, and that in the 1830s and earlier this was certainly the case. One extreme example of this may be quoted from the year 1837, when the implied average pitmouth price in table 9.3 for Northumberland and Durham is 5s 4d per ton. According to accounts in George Johnson's Minute Book, the average receipts in the late summer of that year at South Hetton colliery was 11s 6d per ton, and in explanation of this it was stated that very little 'inferior' coal was sold. The reason for this was not only, or even mainly, the acknowledged high quality of its coal, but the distance of this newly opened part of the Durham coalfield from any existing markets for small coal. Inferior coal from Hetton simply could not be sold to the industries of Newcastle or Sunderland at prices which would cover its transport cost at that time. Small coal was therefore stowed in the goaf so far as possible and used in the engines, or 'burnt in the heap' when it had been raised.

Putting the matter generally, one may say that individual collieries' receipts per ton diverged from average according to the quality of the seams worked and to the amount of their transport costs to market. Of these two, the latter was by far the more important until a cheap means of transportation was available to all collieries. Except, perhaps, for the heavily canalised West Midlands, the latter development had to await the wide diffusion of railway lines within the coalfields. Though this process of diffusion began as early as the 1830s, it could not be described as reasonably complete until the end of the 1860s.[19] By that time, quality of seams worked had probably become at least as important a cause of divergence from average receipts as differences in transport costs; but the extent of the divergences was also by then much more limited than it had been formerly.

Costs

It is extremely difficult to generalise in any precise way about colliery costs of production, for, whilst in every market there was, broadly speaking, one price for coal of the same type no matter whence it came, the costs of each colliery were peculiar to itself. This is true of every industry, of course; but the range of costs per unit of output was probably wider in coal than in any other important economic activity in the country, with the possible exception of agriculture. Even in that sector, the range between individual enterprises in the same area is likely to have been much narrower than in mining, for the peculiarities of each seam in each pit surpass even those of each farm. Unfortunately, the number of colliery records or contemporary analyses of colliery accounts which has survived is only sufficient to indicate the size of the range of different costs. It is not large enough to constitute an adequate sample, and there is no way of establishing how representative it might be. Most of the records concern colliery companies, rather than individuals' or partnerships' enterprises, which may make them untypical; whilst they none of them relate to really small collieries, which were still numerically important even as late as the 1900s, and had contributed a significant fraction of output as late as the 1870s. Nevertheless, this is the only material which is available about costs of production, and it is necessary to see how far it can get us, whilst at the same time making it quite clear that no really authoritative analysis of the subject is possible. Despite this, it would be a mistake to paint too gloomy a picture of our ignorance. Both the structure and the movements of costs do show considerable similarities in all our sources, and this does make some generalisations possible.

Perhaps the best way to begin the study of costs is to show the main series that we have for total costs per ton, for this will bring out both the scale of variation between pits and the considerable degree of uniformity that there was in the way in which costs changed. Before doing this, however, it is worth pointing out that, although we have stressed the wide range of costs at different collieries, there were limits on this range. At the one end there was the limit provided by the royalty owner's ability to appropriate at least part of the benefit of an especially valuable or easy-to-get seam, though if he or his agent were not perspicacious, or if the district was in the pioneering phase when a lease was made, he might have to wait a long time before he could do so.[20] At the other end there was the limit provided by the inability or unwillingness of any coalowner to sustain losses for very long. Very good quality seams and proximity to markets could make relatively high costs – resulting from (say) antiquated pit arrangements, long underground hauls, or irruptions of water in an old colliery, or the need to leave large pillars in sub-urban workings – bearable up to a point; but only up to the point where they led to sustained losses.

Figure 9.1, which presents the main continuous series of total costs per ton which we have found, is in two parts simply for convenience of presentation. A good deal more material is available from 1883 onwards than earlier, and in order to reduce the confusion it is shown on an enlarged scale. The two parts are continuous, however, with no break.

The considerable difference between costs per ton at some collieries at all dates for which more than one series is available is readily apparent, and whilst some of the difference may, in certain cases, result from differences in definition, this does little to weaken the general point. Taking some of the more isolated figures which could not be included in the diagram, one could pick out a much wider range than this. In 1829, for example, R.W. Brandling reckoned the costs at the highest-cost colliery in the North-East were about 23s per Newcastle chaldron f.o.b., whilst the lowest cost would be 15s to 16s per chaldron.[21] Again, in 1842, cost per ton sold in the Five Quarter seam at Wingate Grange colliery in Northumberland was 11.38s (or 9.91s excluding 'leading' – that is, carriage to the staithes), and at the Cowden pit of Dalkeith colliery in Midlothian it was 11.98s per ton.[22] These contrast strongly with the 5.1s per ton at Heaton colliery. Or, to take other examples from single owners in 1862, the cost per ton sold from the Marchioness of Londonderry's collieries ranged from 4.39s at Seaham to 8.56s at Pittington; whilst the Monkland Iron Company's costs per ton varied from 3.88s at Chapelhall Colliery to 5.63s at Crewswood.[23] Reverting to the diagram, one can see that broad differences continued at the height of the boom of the early 1870s, when costs at Walbottle colliery were 17.24s per ton (in 1873).[24] At the Cannock Chase Colliery Company's pits in the same year they were 9.86s,[25] and at Oakthorpe in Leicestershire they were 10.1s.[26] The width of the range declined subsequently, but even as late as 1910, costs at Barber Walker

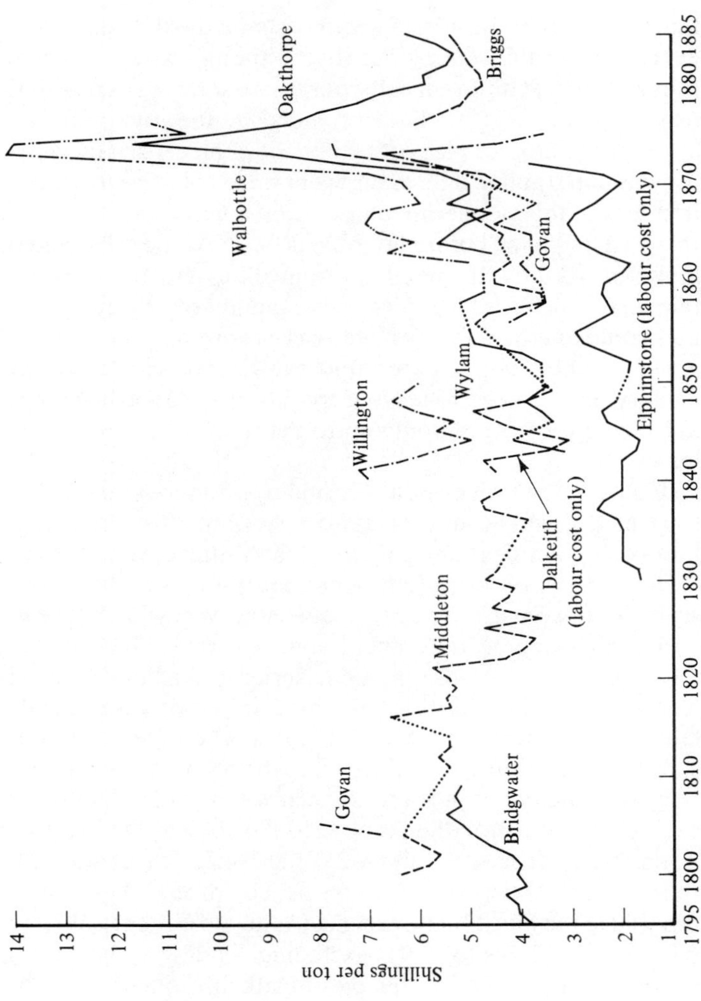
Shillings per ton
14
13
12
11
10
9
8
7
6
5
4
3
2
1
1795
1800
1810
1820
1830
1840
1850
1860
1870
1880
1885
Walbottle
Oakthorpe
Briggs
Willington
Wylam
Govan
Middleton
Dalkeith
(labour cost only)
Elphinstone (labour cost only)
Govan
Bridgwater

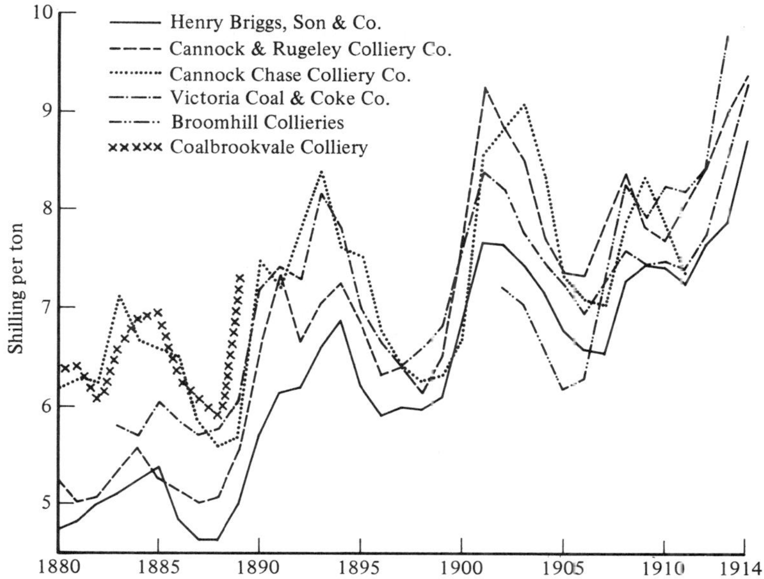

Figure 9.1 Costs per ton for various firms, 1795–1914

and Company's pits in Nottinghamshire could vary from 6.83s to 10.37s per ton, the pits in both cases being neither very new nor approaching exhaustion.[27]

Despite the wide differences in costs between different collieries, it is quite clear from figure 9.1 that, at least from the 1860s, the general movement of costs over the long run was very similar in all cases. One cannot make such a statement about any earlier period simply because of the lack of adequate continuous data; but there is no reason to suppose that things had become different. As well as moving together in the long run, there was also very close coincidence in the occurrence of *sharp* changes in costs.

One outstanding general feature revealed by the diagram is the long-run rise in costs which began for certain in the 1880s, but quite probably may have been going on from the 1850s, and possibly earlier still. Unfortunately there is only one continuous series for costs which extends for a sufficient period on either side of the great disruption of the 1870s, that of Oakthorpe colliery. But for one other firm, at least, Henry Briggs, Son and Company, which was very typical of its district,[23] there was a quite substantial rise in costs per ton between 1860 and the mid-80s.[29] The visual impression conveyed by figure 9.1 that the rise did not occur gradually, but took the form of discontinuous leaps followed by a failure of costs to return fully to their old level, may not be misleading, despite the fact that there are few series which are continuous over the periods of the main leaps. Data which cover periods too short to include in the diagram, or are too broken, do nevertheless support this interpretation. The jump in costs in 1853–4 at Wylam colliery is matched by the Heworth colliery figures, which averaged 3.87s per ton for 1849–53 and 4.98s per ton from 1854–7.[30] And partial cost figures for labour on its own at Dalkeith and at Elphinstone collieries also provide corroboration.[31] It must be noted, however, that the rise in costs at this period is more or less in line with the rise in average wholesale prices, and does not indicate a rise in real costs. This does not apply to any rise which may have occurred between the 1860s and the 80s, when wholesale price indices fell. There is no similar comparison that can be made with the failure of the Oakthorpe figures to return by the 1880s to the average levels of the 60s, other than the Briggs case just mentioned; but there is indirect confirmation in the long series for labour cost at the coalface at Govan colliery.[32] These *do* return to the level of the late 1850s and early 60s in 1877–9 and again in 1885–8; but given what must have been happening in an ageing colliery to labour costs per ton away from the face, this suggests a rise in total labour costs over the period of about the same amount as Briggs's.

It is really unsafe to generalise about costs before 1850, for the data are too thin. The impression they give is that there was a rise during the French Wars, a sharper decline up to the early 1820s, and then little change through to 1853–4. This behaviour conforms to what we know of the behaviour of coal prices and of wholesale prices generally, and thus seems quite plausible. But one cannot say more than that.

Having stated the position about total costs so far as can be done, the mechanism by which they were influenced will be examined, by breaking down the aggregate figures into their chief constituent items. Broadly speaking these are wages, stores and other materials, rents and royalties, rates and taxes, and miscellaneous other charges. In the shipping coalfields there were also quite substantial 'leading' costs, shipping dues, and so on, for it was the custom to include in colliery accounts receipts for coal sold f.o.b., and therefore costs between pitmouth and ship had also to be included. One further cost – depreciation and redemption of capital – has not been included here, though in any complete statement of the costs of the industry it clearly requires a place. In most nineteenth-century colliery accounts, however, the treatment of this item was confused, ambivalent, or simply non-existent, despite the clear recognition of the necessity for covering it by such viewers as John Buddle, Nicholas Wood, or George Johnson, early in the period.[33] It was not generally included in colliery working accounts, though; and if it was put into the overall profit and loss accounts of an enterprise – when, indeed, such were kept, which was not common until well into the second half of the nineteenth century – it was either in the totally conventional fashion of adding 2d per ton to costs, of else it took the form of writing off part of the book capital as and when profits were considered sufficient to allow it.[34] In many, perhaps most, cases this did not matter very much from the point of view of the financial well-being of the enterprise, since, apart possibly for the initial capitalisation, investment on new openings, machinery, and other equipment was met out of revenue, and not ascribed to a separate capital account. Indeed, the influence of the methods of estate accountancy was paramount until the 1860s, and separation of capital and revenue accounts was probably confined to the few limited companies in the industry until the 1870s or 80s.

Since they constituted by far the largest item in costs, wages per ton almost invariably moved in a similar way to total costs, so that there is little point in showing them in a separate diagram. The changing position of wages in total costs is better illustrated by discussing the one as a proportion of the other. The data are insufficient both in amount and in precision to warrant detailed statistical display or analysis, but there are some quite clear general impressions to be gained from them. The earliest figure we have is Rimmer's statement that the labour costs at Middleton colliery near Leeds in the period 1810–15 constituted 76% of total costs.[35] But this seems to be aberrantly high, and probably includes the cost of delivering the coal to Leeds. Buddle – not one to underplay the burden of wages – reckoned that they were 60% of costs (excluding leading) in the late 1820s,[36] and this conforms well with the data which we have about individual collieries for the following forty years. Of seventy-five such individual figures as we have for the period 1834–72, all but seven fall into the range between 50% and 64%. Of those seven, the lowest was 45.6%, and the highest 66.7%. The continuous series which we possess for this period are, admittedly, scanty, but such as they are, they show no tendency for the wages proportion to change secularly until the last few years. Of 142 individual figures for the period 1873–89, 118 show a wages proportion in the range from 60% to 69%, with only eleven under that and thirteen over it. And of 283 figures for the period 1890–1914, 253 are in the range from 65% to 80%, with twelve below and eighteen above it. In other words, our admittedly inadequate data do lend support to the hypothesis that wages as a proportion of total costs were stable up to the boom of the mid-1860s; that they shifted upwards between then and the peak of 1873, and stayed at the new level until the late 1880s, when there was again a rather discontinuous upward movement over the course of five or six years. The first of these shifts took the average proportion from somewhere around 55% to 60% to about 65% to 70%, and the second shift took it to about 70%. In 1898–1900 there was another small rise of two or three percentage points. In fact, according to the Home Office statistics prepared for the Sankey Commission, wages over the whole country averaged 77% of total costs in 1889–93, 78% in 1898–1903, and 77% in 1913, though not counting royalties as a cost.[37]

There are two things here about labour costs which require explanation. First, what caused them, along with total costs, to display a rising trend from the mid-1850s onwards? And second, what lay behind the rise in the relative importance of labour to other costs from the 1870s onwards? The two questions are, of course, connected, though the answer to the second requires an examination of the other kinds of cost than labour. The two determinants of wage cost per ton are the average wage rate per shift and the output per man-shift, and the reasons for changes in wage costs lie in what happened to these two things. Wage rates were discussed in a previous chapter, and productivity is the subject of chapter 10, though, as we shall see, there is little direct evidence on overall output per man-shift, either in the country as a whole or in the individual coalfields, and the more plentiful data on output per man-year is of most use as a guide when it can be taken in conjunction with data on the regularity of working. However, since there are no aggregate cost data, this scarcely matters for the present purpose, and we must perforce look for guidance to the records of individual companies. Only three of these are complete enough to give a reasonably long run of statistics of wages cost per ton, average wages per shift, and output per shift. For 1870–1914, the period covered by these companies, 44% of the annual change in wages cost per ton was accounted for by the change in wage rates, and the remaining 56% by the change in productivity.[38] If, however, the annual changes are cumulated over the whole period, the rise in wage cost sums to 108.71 percentage points, 94.90 of which result from rises in wage rates, leaving only 13.81 percentage points due to the net fall in productivity. This latter is the better measure of the relative importance over the long run of the rise in wage costs compared with the fall in productivity. The former measure merely tells us that productivity was even more volatile than wage rates in the short run – a rather surprising result, perhaps, but of little importance in the present connection. However, although over the whole period 1870 to 1914 changes in wage rates appear to be a much more significant influence on wage costs per ton than do changes in productivity, this conclusion has to be modified if the period is split into two at the time when the trend of productivity changed from being upward to being downward, namely 1888–9. In the first subperiod, 1870–88, of the cumulated fall in wage costs per ton of 80.59 percentage points, 20.99 only was due to the fall in wage rates, with rising productivity roughly three times as important a factor. In the second subperiod, wage rate changes *were* rather more significant than the fall in productivity – 106.75 out of 189.31 percentage points – but not by nearly so much as appeared when the forty-four years were taken together.

Our tentative conclusion, then, about the reasons for the long-run rise in labour costs from roughly the 1850s onwards is that so long as the trend of wage rates was rising, this was probably the most important causative influence. Up to 1873, it was probably the sole influence, for it seems likely that productivity was rising,[39] and so served partially to offset the movement of wage rates. After 1888, with productivity falling, the effect of the rise of wage rates was exacerbated, though wage rates were still the leading factor in what happened to labour costs in the long run.[40] In the interlude, for 1873 to 1888, when, with productivity still on a rising trend, money wage rates were generally falling, then the picture is very different, and labour costs fell a good deal more than the fall in wage rates alone would have brought about. In fact, rising productivity was the more important influence, though it is unlikely that this was so by anything like the three-to-one margin suggested by our very inadequate sample; perhaps one might guess that a proportion nearer to 60:40 would be appropriate.[41]

The second major item of costs which we distinguished was stores and other materials, a composite item with many constituents – timber, iron and steel, horses and fodder, ropes, oil and grease, lamps and candles, gunpowder and fuses, and mechanical and electrical repairs, to name only the more important, together with the collieries' own consumption of part of their own output in their pit engines.[42] Naturally, the cost of these items did not vary in exactly the same way; but there was a tendency for most of them to move together, especially in the bigger short-run fluctuations, largely because the demand for most stores ebbed and flowed together. Over the long term, however, there were changes in the relative importance of various items, mainly consequent upon technical developments, though these changes appear to have been remarkably small between the 1830s and the 1870s, despite the undoubted improvements in winding and hauling especially. Between the later 1870s and 1914, however, there was a significant fall in the relative importance of collieries' own consumption of coal, and of horses and fodder. The former would probably have taken place earlier, when the switch from furnace to fan ventilation first became widespread, had it not been that the period was still witnessing the spread of mechanical haulage. After the 1880s, as subsidiary mechanical haulage became more and more common, this brought about a reduction in the relative importance of horses, but it did not offset the economising of coal brought about both by the virtual elimination of furnace ventilation and by improvements in engine and boiler design. Other items which became proportionately less important from the later 1870s onwards were timber, rope, and lamps. The increased use of timber occasioned by the Mines

Regulation Acts and by greater depths and underground distances was just about offset by the longer life of tapered timber and by the increased use of steel girders, and also, of course, by the reduction in timber prices. Ropes also became cheaper at this time, as well as longer-lasting, and safety lamps, too, declined in price.

In contrast to these, there was a rise in the relative importance of the other major stores items, particularly of repairs, as the use of machinery increased. This also required more oil and grease. Iron and steel, too, loomed larger, mainly because new uses were found for them.

On balance, the result of these various increases and decreases was to keep stores and materials costs as a whole steady, ignoring year-to-year movements, for long periods. In the 1840s and up to 1853, a reasonable average figure to take would be 7d to 8d per ton. With the rise in prices of many materials in the mid-50s, stores costs rose to 9d or 10d a ton on average, and stayed there until the 1900s, except for short periods during booms, and a longer one following 1872–3. In this last upswing, stores costs rose to well over double their normal level, principally because of the rise in the price of timber, springing from the short-run inelasticity of supply coupled with the sudden increase in demand brought about by the rush to expand capacity, and perhaps also by the Mines Regulation Act of 1872.[43] The fall from these high levels was fairly gradual, and it was not until 1878 or 1879 that stores costs had returned to normal, for it took time for supply to adapt to the increased demand. From then until 1900 there was no clear trend; but in that year there was another sudden jump, though not to the levels of the 1870s, and the return to normal was much swifter. But the normal level to which they returned was about 2d per ton higher than before – i.e. roughly 1s per ton. The next sharp rise in stores costs was in 1911–14, and was mainly the result of the steep increase in the price of most materials, assisted (as had been the case earlier in the decade) by falling productivity.[44]

Practically the only important costs which fell in the later part of our period, and then not in all coalfields, were rents and royalties, which were discussed in the last chapter. It will be recalled that the average cost of this item seems to have changed very little in the nineteenth century up to the 1890s,[45] though with some tendency to fall a little in some districts, and an opposite tendency, having a small effect on the average, in the boom of the early 70s. Of course, since royalties were fixed for long periods, their relative importance changed quite considerably during the course of each trade cycle; but in the long run there were only minor changes prior to the 1890s. From that time onwards, royalties tended gently downwards in most of the inland coalfields, whilst moving in the opposite direction in the major exporting districts of South Wales, North-East England, and Eastern Scotland. This has been discussed and the figures have been given in chapter 8, and it is not proposed to say any more about it here.

It is doubtful how far the item rates and taxes should be interpreted as a cost of production. Rates certainly are a cost, but most economists would say that income tax was not, even though coalowners treated it as such in their books. But before 1914, this item formed such a small proportion of total costs that it is of little significance and can be ignored. Rates per ton certainly rose in the last few years before 1914, but it is doubtful if they ever much exceeded 1d per ton for the great majority of colliery concerns. Taxes were still less until the mid-1900s, even for very profitable concerns, and not much more until 1914.

The major constituents of the remaining, catchall, item, 'other charges', before 1899 were salaries, the expenses of agents and other sales costs, upkeep of offices, and so on. These did not rise in anything like the same proportion as output for any individual concern, and as the size of business units (and of pits) gradually rose, the benefits of scale economies were felt in the slow downward trend in this item. There were, of course, discontinuities, and the Mines Regulation Acts, especially that of 1872, which introduced (and required) managers' certificates, and called for official statistical returns, were one of the main causes of these. As with so many other costs, the return to normal cost levels after the great rise of the early 70s was a return to a normal level which was a bit higher than it had been before the boom – and, in this case, before the Act. The Workmen's Compensation Act introduced another charge into this item, and since it increased with the years (as well as with falling productivity), this tended to retard the previous downward trend. The arrival of National Insurance in 1911 definitely reversed it.[46]

In conclusion, it can be said that the accelerated increase in costs per ton which took place in the last few years before the outbreak of war was but a continuation of a trend which had been maintained since the 1880s, certainly, and, if we ignore the great disruption of the 70s, since the mid-1850s. During the period from the 80s onwards, this trend was either contrary to or more extreme than the behaviour of general price indices, indicating that the real cost position of the industry was deteriorating. There can be little doubt that this trend would have continued in any event, at least until the cost position made more widespread mechanisation attractive to entrepreneurs. The war ensured a continued acceleration of the cost rise, and added an element of monetary inflation to it.

Profits

There are two ways of looking at profits which are of interest here – first, as a rate per ton, and second, as a return on capital. These will be considered in turn, but first it must be stressed that, perhaps even more than in the study of costs, many conclusions are bound to be very tentative, for not only are business records scanty, but also they assess profits in several different ways, and it is not always possible to isolate those figures which we should like.

The variations from firm to firm in profits were naturally much greater than in costs. The Sankey Commission was told by an official witness that the spread in 1913 was from 6s per ton profit to 6s per ton loss,[47] and the range for each colliery could vary very greatly from year to year. We shall come to this later, but for the moment let us begin with aggregates and averages, for, unlike costs, there are some which exist in the case of profits.

As Sir Josiah Stamp pointed out many years ago, 'owing to a happy accident of legal classification in the earliest Income Tax systems, we have always had the statistics of assessment on mines shown separately.'[48] The assessments, however, are far from easy to interpret. They are based, in principle, on the average profits of the five years preceding the year of assessment, and up to 1910/11 it is impossible to distinguish coal from all other kinds of mine. Moreover, on the one hand, income from royalties is included in the assessment, whilst on the other, provision for redemption of capital was not allowable as a cost. Finally, there are the intricacies of the '133rd section' – a provision for relief and reassessment if actual profits turned out to be below the assessment, which had the effect of blurring the precise attribution of the assessment statistics to any five-year period[49] – and the fact that if the individual assessment were based on an average loss, rather than profit, it was treated as nil and not as a negative value. The tax statistics, therefore, are likely to overstate the average profits of such a fluctuating industry as coalmining in depressed periods, especially towards the end of prolonged slumps. However, against this must be set the generally agreed fact that the income tax collectors gradually improved in efficiency, probably throughout the period.[50] An unknown, but decreasing, proportion of successful evasion must therefore be taken into account – or at any rate borne in mind. In consequence of all these considerations, the income tax assessments take a good deal of unravelling, and it may be doubted if they can be successfully unravelled very much further than Stamp succeeded in doing. In fact, what he did was to provide an annual series of gross profits for mining for 1888 to 1915, though still including royalties, and ignoring the existence of mines other than coal mines.[51] His method involved the use of the published profit figures of 'a large unbiased group' of collieries, which may have been feasible for the period he covered, but is scarcely so for any earlier times. Certainly there are some colliery records which can be used as a guide, but they can hardly be described as an unbiased sample. It may seem easy to push Stamp's estimates back beyond 1888, using the four figures for 1888–91 and the five-year profit averages of the tax assessments applicable to 1887–91 to arrive at the missing figure for

1887, and to continue this operation for each earlier year in succession. Unfortunately, this method comes to grief almost immediately, for the resulting estimates for 1885 to 1887 are respectively £2.7 million, £4.5 million, and £0.7 million. There is no warrant in any evidence, quantitative or otherwise, for believing that profits in 1886 were higher than those of 1885, 1887, and 1888 (or, for that matter, 1884, for which the estimate is £1.3 million). On the contrary, the majority of available colliery records indicate that the profit experience of each of the years from 1884 to 1888 was rather similar, and was bad. But not only does the method produce estimates for the mid-1880s which cannot be believed, the result of these false estimates is to set up an echo effect which leads to increasingly absurd estimates, including, by as early as 1882, negative figures, which, by definition, cannot be allowed in a series based on income tax assessments, for by the rules average losses over an assessment period were treated as nil income, not as negative profits. This is almost certainly the main reason why the method is unworkable, and it points to a danger in the five-year assessment statistics themselves. Since the coal industry was one in which fluctuations were wide and variation between individual operating units was large, it is obvious that some concerns were likely to be making losses in nearly every year, and that quite a substantial number was probably making losses in depressed years. The occasional loss-making firm in reasonably prosperous periods would not affect the statistics much, but when several depressed years occurred in succession, there would almost certainly be significant numbers of enterprises with five-year average losses.[52] So far as can be told from examination of surviving colliery records, this situation did not crop up to any significant extent in the period with which Stamp was concerned, and though it must certainly have affected his estimates for 1888 and 1889, he seems to have tried to allow for it by putting 'or less' after the figures. But for some earlier periods, perhaps, above all, for the years from the later 1870s through to 1889, it is pretty safe to say that the tax assessments exaggerate the amount of profits made by the industry in the depressed years and during the earlier stages of cyclical upswings.

In the light of this, it seems wise to conclude not only that there is no way of reliably estimating the aggregate profits of the industry on an annual basis, prior to 1890 or so, but also that the five-year averages themselves must be treated with a certain amount of suspicion.[53] However, since they are the only aggregate data which we have, they must be looked at; and before doing that, it is worth trying to bring them closer to the commercial concept of profits. The first step is to eliminate royalties, which cannot be done in any precise fashion, of course, but which, in the light of the discussion in the last chapter, is probably adequately effected by taking royalty payments to be equal to an average of 6d per ton of output over the country as a whole. It is possible that this is a shade too low in the earlier years for which the tax statistics are available, and it may also result in a degree of understatement during, and for a few years after, the great boom of the early 1870s, when new and renewed leases were at temporarily inflated levels. But these errors are unlikely to be very significant. The profits and royalties of non-coal mines are rather more difficult to exclude. One way to remove them would be to take the late-1900s level per ton of output (or of value produced) of minerals other than coal and extrapolate it backwards for as long as such production records are available – i.e. to 1854. When this was done, however, the resulting estimates of coal profits and royalties were clearly much too low for most years prior to 1872. In the end, the method preferred was to assume that the proportion of total mine profits (excluding royalties) which was attributable to collieries was the same as the proportion of total mineral receipts – or, to be more precise, the receipts of the five major minerals[54] – which was contributed by coal. Using this method, the estimates of profits in coal mines still seem surprisingly low in the 1850s and 60s, and it may well be that the profitability of minerals other than coal was relatively lower than that of coal at this period compared with the 1900s. If this is so – and the virtually uninterrupted decline of copper mining after 1856, coupled with the lack of long-run growth in the output of iron ore, lead, and tin,[55] suggests that it may be – then the colliery profit figures shown in table 9.4 will be significantly underestimated during the years when coal was not so overwhelmingly the major mineral produced in Britain as it became from about 1870. This must be borne in mind when analysing the table.

What, then, do the tax assessment statistics tell us? In the first place, they confirm, if confirmation is needed, the fluctuating character of the industry's profits, which shows up despite the dampening effects of a moving average and of the failure of the statistics to take account of average losses made by individual companies over any five-year period. This is especially apparent in – or, to be more precise, just after – the two periods when prices rose most steeply, the early 1870s and the two or three years around 1900. This prompts a closer examination of the relationship between prices and profits, which is undertaken in figure 9.2. This reveals the curious lag of average prices behind average profits which Stamp pointed to, and explained, for the period from 1888 onwards, through the operation of Section 133 of the Income Tax Act, which had 'the effect of partly advancing the profits a year'.[56] Prior to the mid-1870s, the lag is two years rather than one, which may perhaps have been the result of different practice on the part of tax officials. However, there is another possible explanation of the lag of prices behind profits, which we have already touched on. This is the fact that, in some seams, rising

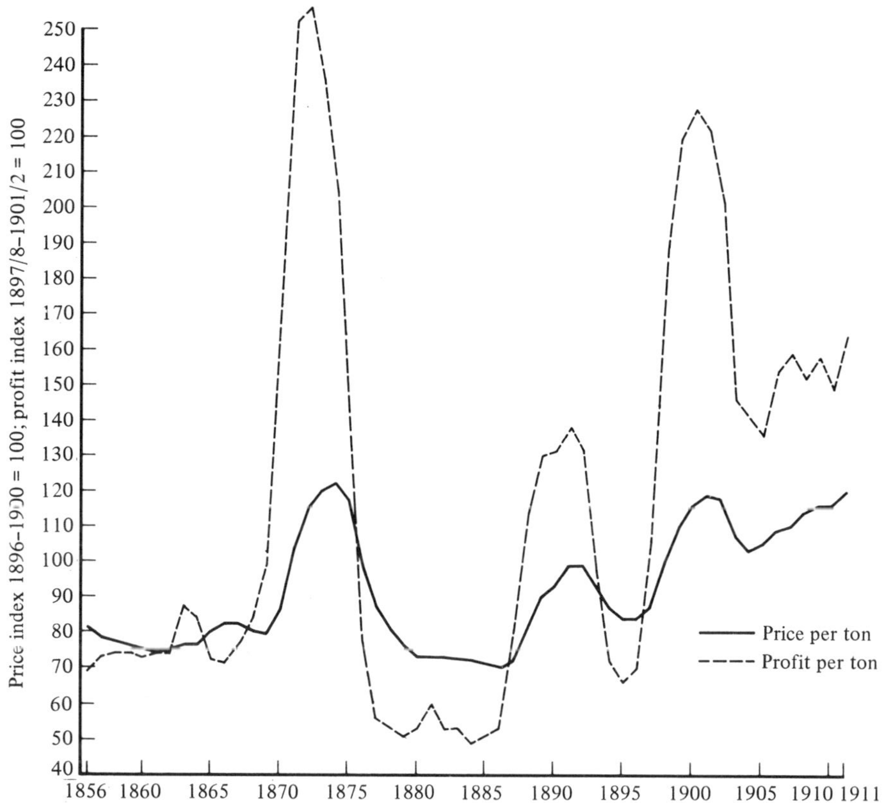

Figure 9.2 UK average prices and profits per ton, 1856–1911 – five-year moving averages (centred on year shown)

Table 9.4 *Tax assessments on gross profits of mines in the UK, tax years ending 1815–1915*

	All mines (£000)	Coal mines (£000)	Coal mines (excl. royalties) (£000)	Coal mines (excl. royalties) (pence per ton)
1814–15	679			
1842–3	2,081			
1845–6	2,213			
1847–8	2,280			
1848–9	2,434			
1849–50	2,410			
1850–1	2,390			
1851–2	2,374			
1852–3	2,340[a]			
1854–4	2,851			
1854–5	2,937			
1855–5	3,036			
1856–7	3,061			
1857–8	3,882			
1858–9	3,950			
1859–60	4,084	2,958	1,327	4.9
1860–1	4,269	3,111	1,443	5.2
1861–2	4,436	3,295	1,550	5.3
1862–3	4,516	3,443	1,613	5.3
1863–4	4,572	3,562	1,651	5.2
1864–5	4,829	3,786	1,768	5.3
1865–6	5,080	4,004	1,883	5.3
1866–7	5,612	4,462	2,249	6.1
1867–8	5,745	4,615	2,312	6.0
1868–9	5,480	4,465	2,043	5.1
1869–70	5,544	4,579	2,078	5.0
1870–1	5,892	4,901	2,327	5.4
1871–2	6,335	5,242	2,607	5.9
1872–3	7,283	5,891	3,177	7.0
1873–4	10,546	8,511	5,702	12.2
1874–5	14,108	11,680	8,743	17.9
1875–6	14,614	12,238	9,205	18.2
1876–7	14,091	11,903	8,756	16.7
1877–8	12,900	11,022	7,791	14.5
1878–9	10,088	8,627	5,343	9.8
1879–80	7,501	6,328	3,024	5.5
1880–1	6,667	5,585	2,245	4.0
1881–2	6,728	5,589	2,181	3.8
1882–3	6,734	5,597	2,089	3.6
1883–4	7,065	5,916	2,296	3.8
1884–5	7,603	6,416	2,640	4.2
1885–6	7,519	6,410	2,499	3.8
1886–7	7,487	6,479	2,506	3.8
1887–8	7,152	6,256	2,267	3.4

Table 9.4 (*cont.*)

	All mines (£000)	Coal mines (£000)	Coal mines (excl. royalties) (£000)	Coal mines (excl. royalties) (pence per ton)
1888–9	7,290	6,438	2,421	3.6
1889–90	7,452	6,615	2,567	3.8
1890–1	8,808	7,893	3,764	5.5
1891–2	10,872	9,881	5,640	8.0
1892–3	12,047	11,083	6,703	9.2
1893–4	12,322	11,431	6,952	9.3
1894–5	12,676	11,833	7,382	9.8
1895–6	12,278	11,515	7,008	9.3
1896–7	10,509	9,872	5,324	7.0
1897–8	9,078	8,522	3,925	5.1
1898–9	8,901	8,356	3,657	4.7
1899–1900	9,492	8,923	4,036	5.0
1900–1	12,031	11,334	6,287	7.5
1901–2	17,642	16,724	11,500	13.2
1902–3	20,259	19,261	13,919	15.6
1903–4	21,194	20,258	14,791	16.2
1904–5	21,236	20,346	14,737	15.8
1905–6	20,000	19,193	13,523	14.3
1906–7	16,372	15,679	9,954	10.4
1907–8	16,400	15,648	9,763	10.0
1908–9	16,614	15,846	9,757	9.6
1909–10	18,460	17,629	11,384	10.9
1910–11	19,343	18,481	12,079	11.3
1911–12	19,681	18,348	11,805	10.8
1912–13	20,299	19,069	12,422	11.2
1913–14	19,611	18,233	11,623	10.6
1914–15	21,525	19,962	13,222	11.7

[a]Figures are for Great Britain only up to 1852–3. This cannot have made a significant change in this series.

Sources: column 1: Stamp (1916), except for the figure for 1914–15, which comes from the *Report* of the Commissioners of Inland Revenue. The latter is also the source of the figures for 1911–12 onwards in column 2, the remainder being arrived at by deducting from column 1 a percentage equal to the percentage which the value of the production of ores of copper, iron, lead, and tin was of the value of those ores plus coal. Column 3 results from the deduction from column 2 of a sum equal to 6d per ton of coal output. For column 4 the average output of the five years to which each assessment nominally relates has been used. If the lag mentioned in the text is taken into account, the average output of the five years up to the one of each assessment should be used. The effect of doing this would generally be to reduce the profit per ton figures by a small amount.

demand made it possible to sell less valuable types of coal, which, in the slump, were not brought to the surface at all. Average receipts per ton might thus fall at the same time as profits were rising. Another reason why profits could rise before prices is that fixed costs were being averaged over higher output. Whatever the explanation of the lags, however, it is readily apparent that profits and prices practically always moved together, at any rate if we ignore years when changes in either direction were small and inconsequential. However, the ratio of the change in profits to the change in prices of each period was far from being precisely regular, though it usually happened that when the change in prices was relatively large the change in this ratio tended to be large. The close connection of profit movements with price changes also suggests that the impact of other influences on profitability was very much less. Whilst this seems almost certainly correct, it is worth noting that after 1900 or thereabouts, the index of profit per ton stayed well above that of prices, whereas in all previous cyclical downswings for which there are reasonably adequate data, profitability had plunged even further than prices. The explanation still probably lies in the behaviour of coal prices, but in this case in their behaviour relative to other prices. The fall in coal prices from the very high levels of 1900–1 was quite sharp, but it did not carry them down to anything like the levels prevailing before the boom, and it left the index of coal prices substantially higher, compared with the mid-1890s, than the index of wholesale prices generally. From the mid-1900s up to the War, this divergence was widened, even in the recession of 1908–11. As we saw in the previous section, this was associated with increasing costs in the coal industry; but it is obvious that it was also associated with rising demand for coal, and with a considerable degree of inelasticity over the relevant range of demand. In the 1900s, unlike in the periods following the previous two major upsurges in prices, such rising demand could not be met by high elasticity in the long-run supply curve: for geological, legislative, and other reasons, supply was becoming more inelastic, and coalowners' profits benefited from this.

So far we have examined profits in terms of the country as a whole, and for all the coalfields except Scotland, and for all years except a few at the very end of our period, that is all that can be done, except in a very speculative fashion. The Scottish tax assessments were separately recorded, and, on the probably reasonably accurate assumption that all concerns operating in Scotland were assessed there, it is possible to reproduce for Scotland alone the calculations which were presented in table 9.4 for the United Kingdom, though prior to the 1880s information about both the level of royalties and the output of other minerals is less reliable, and some royalties must certainly have been assessed in England. Nevertheless, the picture of fluctuations which emerges is so very similar to that for the country as a whole that there is no reason to question it, and little point in showing it in detail. The one difference which does seem to show up is in the level of profit per ton, which was generally some half to three-quarters of the UK average, and was probably even lower in the extended depression of the 1880s.[57] A small part of this difference is likely to have been the result of firms being assessed in England, though deriving some of their profits in Scotland; a rather larger part is likely to result from excessive deduction for royalties, since some of the royalty owners lived and were assessed in England. It is, indeed, possible that all of the apparently lower profitability of Scottish mines is an illusion resulting from assessments being made outside the country; but the chances are that it is not entirely illusory.

When the tax statistics are available by counties, for 1911–12 onwards, there are other coalfields than those of Scotland which apparently had profitability records varying markedly from the national average, and the different coalfields within Scotland also seem to have had widely varying experiences. Again, this may be largely the result of assessments being made outside the county where the profits or royalties were made, but that this is a major factor in Scotland itself seems unlikely in view of the relatively low profit level shown for Midlothian where some Scots who received income from other counties might be expected to have been domiciled and assessed. Perhaps the figures in table 9.5 should not be taken as precisely accurate for our purpose, and in one or two cases one may have reason to suspect they are misleading; but for the most part the impression they give is probably not far wrong.

The main county figures in this table which seem likely to be misleading are those for Derbyshire and Leicestershire and Nottinghamshire, and those for Ayrshire and Lanarkshire. Many of the newer Nottinghamshire collieries were owned by firms based in Derbyshire (with a few from both Yorkshire and Lancashire), and it is probable that their profits raised the Derbyshire and Leicestershire level significantly above what it ought to have been, and depressed the level of the rather smaller Nottinghamshire coalfield still more. Similarly, in Scotland, several Ayrshire collieries were owned by firms based in Lanarkshire or Renfrewshire, and whilst the level of profits in the much larger Lanarkshire district is probably not greatly inflated by this, the Ayrshire level is likely to be understated. For all that, it can be taken as very probable that, Nottinghamshire apart, the general order of magnitude of profits per ton, and the ranking of the coalfields towards the end of our period are pretty well represented by table 9.5, though one further qualification has to be made. That is that the Cumberland figures are probably as low as they are

Table 9.5 *Profits per ton (excluding royalties) by counties, 1911–14*

	1911–12	1912–13	1913–14
Northumberland	11½	10	8½
Durham	13	12	11
Cumberland	4	3½	5
Yorkshire	11½	12	12½
Lancashire	11	11	10½
Denbighshire	...	6½	...
Derbyshire & Leicestershire	11	11½	12
Warwickshire	14	14	14½
Nottinghamshire	6	6	5
Staffordshire	7½	8	8½
Carmarthen, Glamorgan & Monmouthshire	15	15	14
Ayrshire	1½	½	1
Lanarkshire	2	9	6
Stirlingshire	3½	2½	4
Midlothian	5½	6½	4½
Fifeshire	14	13	10½

Note: the statistics of profit relate to assessment years, i.e. to an average of a preceding five-year period.
Sources: based on the *Reports* of the Commissioners of Inland Revenue, the royalty estimates of chapter 7, and the output statistics for the five-year periods after the nominally correct ones.

because of the relatively high royalties there, which, as has been noted, contained an element of payment for capital equipment originally installed by the royalty owner.[58]

As to what things were like in the different districts at any much earlier date, it is really useless to speculate, though the temptation to assume that much the same general pattern would be found is strong. In general, the highest average profits seem to have been found in those districts which had been expanding most in the recent past – South Wales, Warwickshire, Yorkshire, Fifeshire. Perhaps it is not illegitimate to conjecture that the same correlation of relatively rapid expansion and relatively high profit levels would be found at all periods. Economists would surely be surprised if it were not.

We have not so far attempted to say anything about profits in the period before the income tax statistics are available, nor, indeed, about the earliest years of these statistics, when it is virtually impossible to eliminate from them the non-coal component. There is, in fact, little we can say about this period, covering the first half of the nineteenth century, which is not conjectural. A few profit statistics for individual collieries have survived, but only one or two cover a continuous period of reasonable length, and it is impossible to generalise on this slender basis. Later experience, indicating the close correlation between profits and prices, suggests which periods were likely to be the most and the least profitable. But this was not necessarily the case for each individual colliery, and, indeed, the Middleton colliery profits, known (with some gaps) for the period up to 1839,[59] show some surprising fluctuations, as well as expected ones. Nevertheless, from this series, from that for the Bridgwater Trustees' collieries,[60] from scattered figures for Cumberland collieries,[61] and from the evidence to the various Select Committees of the period, we can tentatively accept that the price indices are a reasonable rough guide to profitability. There would seem to have been a period of brisk activity in the industry, bringing high profits, in 1803–5, followed by a few slacker years and a boom in 1810, which brought what Buddle thought were the highest prices of his experience. One can see this reflected in the fact that the Bridgwater collieries' profits rose from 17.6d to 36.3d per ton in that year – fractionally lower, however, than in each of the years 1803, 1804, and 1805. Such a period of high profitability does not seem to have come again until the mid-1820s, and in the Middleton series there were particularly low profits in 1816 and 1821. The late 20s were, again, relatively unrewarding, with a vigorous recovery of profitability not coming until 1836. The coalfields of the manufacturing districts doubtless shared the deep depression of the early 40s, but in the North-East the effects seem to have been delayed, with 1844 still starting depressed.[62]

Apart from such very general comments on the way in which profits probably fluctuated, it is only possible to speculate about the average level of profits. Guessing at the proportion of the tax assessments of the early years of the renewed tax which are attributable to coal mines, and excluding royalties by the method already used, we get some rather low average figures for profit per ton. Even if profits from non-coal mines were nil, the profit on coal was only assessed at a maximum of 7d per ton in any year prior to 1854. This seems to be improbably low, even though there was depression at both the beginning and the end of the period, and the intervening upswing saw the virtual ending of the Limitation of the Vend. One's immediate reaction is to suspect that the assessments were incomplete, improving gradually perhaps as the collectors gathered experience. There is, however, nothing to support this hypothesis other than *a priori* reasoning, and there is indirect evidence which suggests that the tax assessments may not be too far wrong, except possibly in the first two or three years. This comes from an examination of the relationship between wages and prices, such as was undertaken in chapter 7.[63] There we were concerned with the movements of the two series: here it is their relative levels which matter. If the ratio of the price index to the hewers' wage index in each year is taken, and if it can be assumed that the latter can be taken as a reasonable proxy (so far as changes are concerned) for total costs, then changes in the ratio will be a rough indicator of changes in profitability. Of course, this is very imprecise, because the data are not totally reliable, and errors in one series may either offset or compound errors in the other series in a way which we cannot anticipate. But a large change in the ratio over a comparatively short space of time should signify a genuine change in profitability. such a change does occur in the Durham and Northumberland ratio in the mid-1840s. From 1832 to 1844 it was never below 1.47, and averaged 1.61. After 1844 the ratio only exceeded 1.40 four times (one occasion being 1873, another 1900, and the other two being 1848 and 1853), and in the ten years from 1845 to 1854 it averaged 1.30. The subsequent average was about 1.2. The clear implication of this, surely, is that profitability in the major coalfield of the country declined quite sharply just after the time when the income tax was reimposed and the assessment statistics become available (the first assessments themselves being affected by the long and continuing depression of the early 40s). In the light of this, perhaps one's first suspicious reaction that the early assessments were incomplete is largely unjust. If that is so, and if our reasoning from the price:wage ratio is correct, we can go further, and suggest that the average level of coalmining profits in the 1830s, and probably earlier, was higher than in the 1840s and 50s. Maybe the Middleton colliery decade averages of about 1s to 1s 7d per ton (presumably before allowing for redemption) are not unrepresentative.

Turning now to profits as a rate of return on capital, we can use the estimates which we made in chapter 3 along with the tax assessments to arrive at what may perhaps be best described as a rough idea of the rate of return at various periods – though Stamp's annual profit estimates allow us to suggest annual rates of return from 1888 onwards. Taking first periods corresponding to the trade cycles affecting the industry we get figures of average capital and profits as shown in table 9.6.

The low figure for 1848–57 (matched, incidentally, by the last, and usually more prosperous part, of the previous cycle) must be viewed with considerable scepticism, for it included three or four years in the mid-1850s which have a reputation for being very prosperous. However, such a reputation is itself to be taken very cautiously, for it is usually based on a rise in prices. This, as we have seen, normally meant a still bigger rise in profits, but need not necessarily do so if costs rose in proportion. Usually, later in the century, costs did not do this; nor do they seem to have done so in earlier booms, judging by the rate of increase of hewers' wages. But between the bottom of the depression of the early 1850s and the peak of 1855–7, hewers' wages appear to have risen by a substantially

Table 9.6 *Capital and profits in the UK, 1848–1913*

	Average capital (£ million)	Average profits (£ thousand)	Gross rate (%)
1848–57	19.1	798	4.2
1858–67	29.75	1,859	6.2
1868–73	52.45	5,640	13.3
1874–83	54.85	3,533	6.4
1884–91	67.80	4,103	6.1
1892–1900	87.75	7,004	8.0
1901–7	115.95	11,987	10.3
1908–13	140.75	(12,600)	(9)

Sources: see text. The profit figure for 1913 (assessment year 1914–15) has been assumed.

bigger proportion than prices in all coalfields for which we can make estimates, except Durham and Northumberland. Since this was a time when wholesale prices in general rose rapidly, it seems reasonable to suppose that costs as a whole kept up more closely with prices than in most other booms. It is possible, therefore, that the low profit rate indicated by the tax statistics for 1848–57 is not entirely illusory. Perhaps it might be reasonable to reckon that it was little different from that shown for 1858–67, and approximately repeated in the late 1870s and the 1880s.

The extraordinary prosperity of the early 1870s and the rather weaker repetition in the boom peaking in 1900 stand out in table 9.6; but the overall conclusion which must be drawn from it is that coalmining capital received *on average* a rather low rate of return over the period covered by the income tax statistics – low, that is, relative to industrial capital in general. Recall that these statistics do not allow for redemption of capital, which in an industry where most of the equipment had little or no scrap value was a more than usually necessary part of keeping capital intact. If one allows for an average pit life of thirty years – a common practice of colliery viewers and managers throughout the period – and employs the straight-line depreciation method to the average colliery capital in existence over each trade cycle, the rate of return net of redemption was roughly three percentage points less in each case than is shown in table 9.6. An average yield of 3% to 5%, with one short bonanza period when it reached 10%, can hardly be considered anything but low, and was, for example, very little different from that received by UK railway capital, despite the latter's much more stable character.[64]

The highly fluctuating character of the coal industry's profit rate is apparent if we take Stamp's annual profit series, adjusted to exclude non-coal mines and royalties, and turn it into rate of return on capital by

Table 9.7 *Annual gross profits and rates of return on capital in coal mines in the UK, 1888–1914*

	Gross profits (excluding royalties) (£ million)	Estimated capital (£ million)	Rate of profit (%)		Gross profits (excluding royalties) (£ million)	Estimated capital (£ million)	Rate of profit (%)
1888	2.9	69.4	4.2	1902	11.2	109.7	10.2
1889	8.0	71.0	11.3	1903	8.6	113.9	7.6
1890	12.1	72.6	16.7	1904	7.0	118.1	5.9
1891	9.8	74.2	13.2	1905	7.1	122.3	5.8
1892	6.5	77.2	8.4	1906	10.3	126.5	8.1
1893	5.0	80.2	6.2	1907	19.7	130.6	15.1
1894	5.7	83.2	6.9	1908	18.0	134.0	13.4
1895	2.9	86.2	3.4	1909	9.5	137.4	6.9
1896	2.8	89.2	3.1	1910	10.6	140.8	7.5
1897	3.8	92.2	4.1	1911	9.8	144.1	6.8
1898	7.3	95.2	7.7	1912	16.4	147.5	11.1
1899	17.4	98.2	17.7	1913	23.2	150.9	15.4
1900	30.0	101.3	29.6	1914	16.6	154.3	10.8
1901	19.7	105.5	18.7				

Sources: column 1: Stamp (1918), adjusted to exclude non-coal mines and royalties as described in the text. Column 2: based on table 3.1 on p. 43 by regularly interpolating between the peak-year estimates without attempting to allow for differences between current prices and historical prices.

roughly interpolating between the capital stock estimates for peak years presented in chapter 3. These are shown in table 9.7. The range of profit over this twenty-seven-year period was from 3.1% to 29.6%, or, if redemption of capital is allowed for, from nil to about 25%. The mean rate was 10.2% gross, or perhaps something like 7% net of redemption.

In their discussion of profits in the eighteenth and early nineteenth centuries, Ashton and Sykes note that 'industrialists and their friends are notoriously inclined to understate the rewards that come to them', that various expert viewers estimated profits at a low level, and that 'there is reason to believe, however, that the results of mining were by no means so unsatisfactory to the investor as these authorities imply'.[65] They then go on to quote various high rates of return on viewers' valuations. These can be matched by similar examples at many times throughout the nineteenth century; but there is still no reason to disbelieve entirely the low *average* rates indicated by the tax statistics from the 1850s onwards, even if those for the early years of the revived income tax may have been unduly low. For the period of the Napoleonic Wars covered by the income tax the average profit would not seem to have been very high either – if some rather heroic guesswork at output and capital can be accepted – perhaps somewhere between 4% and 7%, it would seem. This accords with what the vastly experienced John Buddle told the 1829 Select Committee of the House of Lords, that 'according to the best of my knowledge, taking a run of years from the time I have been in the trade, I should think that by no means ten per cent has been made as simple interest, without allowing an extra interest for redemption of capital'.[66] He may not have been far from the truth, though against this, and suggestive of a higher average rate of return than in the middle of the nineteenth century, is the wider spread between prices and wages in Durham and Northumberland up to 1844 than subsequently – surely an indication of higher profit rates there than after the effective dissolution of their oligopolistic arrangements.

If, as seems to be firmly established, the average rate of profit on capital invested in coalmining was highly variable from year to year, was subject to periods of two or three years at a time when it was scarcely, if at all, positive in net terms, and was low relative to that from many other activities,[67] the question must arise as to why investment in collieries was forthcoming in the amounts which it was. The answer to this takes us back to a point made at the beginning of this section, and touched on again just now – the great variety of experience in different colliery enterprises, and the possibility of one earning (say) 6s per ton profit at the same time as another was making 6s per ton loss.[68] This might be taken to imply that the industry was one for gamblers, and, indeed, there was always a speculative element in it, and on one or two occasions some people from both inside and outside the existing personnel of the industry behaved in a fashion rather similar to that of investors in railways during the Mania of the 1840s. But these were exceptional and unusual, and the majority of capital drawn into the industry took its risks more clear-sightedly. And in practice these risks were not quite so big as they appear, from the average rates of return, to have been. As Stamp put it: 'For a man considering industry as a whole, and not a particular proposition, it was a very difficult industry to put money into.'[69] But most men who invested in coal mines *were* considering particular propositions. Ashton and Sykes grant the probability of runs of ill-fortune not being unknown, but say that

> the fortunes accumulated by families like the Brandlings, the Tempests, the Newdigates, the Bradshaighs, the Mackworths, the Lowthers, and the Curwens suggest that the compensation for risk-bearing was by no means small. And the frequency with which leases were renewed by the original lessees, when they fell in, implies that the entrepreneurs were not dissatisfied with the share of the earnings of the industry that came to them.[70]

These examples could be matched by many more from later in the nineteenth century – the Dixons, the Bairds, D.A. Thomas (Lord Rhondda), the Markhams, H.D. Pochin, and a host of others. The point that they make is not that coalmining was a fantastically profitable venture for all who invested in it, but that those with knowledge of the industry, of markets and transport capabilities, of costs, and of the fundamental geological basis – that is, the professional coalowners, who were either themselves mining engineers, or who were closely associated with one – these people could, given reasonable luck, reckon to make more than the average rate of profit. What induced them to invest in collieries was not the average rate of return, but their normally well-informed estimate of the likely particular rate of return on the enterprise they had in mind. For example, it was not the distinctly poor average rate of profit of the 1880s which induced Emerson Bainbridge and his colleagues to open up Bolsover colliery in 1890, but the well-founded belief that much more could be made, given the recent proof of the continuation of the coal measures eastwards under the Permian strata, given the ease of connection to railways going to the Humber ports and to both London and the northern and Midland manufacturing districts, and given recently proved methods of sinking.

If some collieries were profitable well above the average, others, of course, had to be the opposite. How did investment in such as these come about? To some extent bad luck on the part of professionals who, though informed, could never be completely informed, especially about the detailed geology of a particular area. Some mines were sunk at various times which seemed reasonable investments *ex ante*, but which proved to be uneconomic to a greater or lesser degree because of faults in the seams, excessive and unexpected water (either raising the capital cost of sinking or the running cost of pumping), or some similar difficulty, which could be anticipated in a proportion of pits but not in any particular individual case. This was the speculative element which always remained in the industry. If such a pit was so uneconomic that it made losses even in years of ordinary trading conditions, it might very likely be closed altogether. In that case, its capital would no longer appear in our estimates of the capital stock of the industry, and to that extent the average profit rate which we have suggested will be somewhat inflated above what it ought to be. But if such a pit was not catastrophically uneconomic, it would probably continue to produce, and in so doing it contributed to reducing the average profit rate of the industry.

Another reason for investment in uneconomic collieries was ignorance on the part of investors. This was not a reason of really major importance, for outsiders did not come into the industry in great numbers at most times. But the major booms always attracted some, most especially the boom of the early 1870s, when talk of vast profits was frequently correct. With the legalisation of joint stock limited liability companies, it became quite easy, at such times, for promoters to attract speculative stock exchange investors into ventures which proved of little or no long-run worth. But so long as they remained in production, they depressed average rates of return.

In addition to unlucky or unwise investments in new pits, average profit rates were also kept down by the continuation in production of elderly collieries, which had at one time been very profitable ventures, but which were declining towards marginality as their easier-got or more valuable coal was worked out, as their equipment became outdated or costly to maintain, or as the weight of water to be pumped out inexorably increased with the expansion of the workings. This problem was dramatically put by Buddle to the Commons Select Committee of 1829 when he said that 'if Wallsend was a virgin mine, I should be better satisfied, and the proprietors better paid at 19 shillings or 20 shillings than at this time they would be at 34 shillings'.[71] Of course, such ageing collieries eventually obsolesced and were closed. But so long as they seemed likely to be capable of making a profit in the next boom which would be greater than the losses in a current slump, there was adequate reason to keep them open, and so depressing the industry's average profit rate, both directly through their own profit and loss record, and indirectly through their effect on the supply of coal and so on the prices that even the more profitable concerns could get.[72]

Finally, there was a depressing effect on average profit rates from the impact of some temporary interruption to the normal output of coal from any colliery. There were many possible causes of such an interruption. Perhaps the most common was strikes, which, if confined to a single firm, might affect its profits drastically, for there would be none of the compensating effects of a rise in prices such as accompanied wider strikes. Other temporary interruptions arose from irruptions of water, explosions, or other less dramatic forms of accident. Another cause of stoppage could be the failure of the management (or of a railway company) to have enough wagons to take away the output, though since this was usually associated with a period of very brisk trade, its effects on profits would tend to be masked. Finally, there was the weather – at any rate in the days before screw steamships and mechanical ventilation, when contrary winds might block up the pipeline, or temperature inversions might incapacitate the ventilation system. Such stoppages, however, tended to be short; or, if extended, as periods bad for shipping by sail could be, they brought some compensation in the form of higher prices, though it was the coal merchants rather than the coalowners who took most of this.

We can perhaps sum up this question of profit rates by saying that the shrewd professional coalowners, though they occasionally made mistakes and burnt their fingers, generally expected to make, and did make, profits well above the industry's average, though their personal average might be reduced if they were involved in ageing, but still viable, pits. Owners of elderly pits who did not extend their interests into newer ones – and who were consequently, in effect, retiring from the industry – made profits which increasingly fell below the average for the industry, quite often, indeed, becoming losses. Speculators from outside the industry, drawn in by the occasional periods of very high profits, were usually disappointed, and formed another group whose yield fell below average.

We have argued here that the rate of profit in the coal industry was relatively rather low, basing this principally on the income tax statistics. There is another approach to the question, which was adopted by a number of contemporary students of the industry, and which usually brought them to a rather different conclusion. One example of this is a book (or, rather, extended pamphlet) by Thomas Richardson and John A. Wallbank, which takes as the basis of its analysis the published balance sheets and profit and loss accounts of public colliery companies. Their conclusion was that total profits (including interest on borrowings and additions to reserves, but not deducting anything for redemption of capital) was 11.8% in 1909 for companies producing some 80 million tons of output; and they argued that this must be lower than the 'true' average, quoting W.J. Ashley that 'there is a widespread impression among businessmen that larger profits are, in fact, made in private unlimited concerns than in public limited ones'.[73] The work is a somewhat polemical one, though not bitterly so, and it is a shade misleading from the historical point of view to look only at the period from 1898 onwards, and not at the less prosperous years which had preceded it. However, so far as it goes, their estimate of profits is probably correct for public companies, though another enemy of the coalowners, A.E. Davies, general manager of the British, Foreign and Colonial Corporation, disagreed with them (and with Ashley) so far as small colliery companies were concerned, telling the Sankey Commission that 'it is the big companies which make big profits. I am quite prepared to admit a lot of small collieries have made losses.'[74] The trouble with the Richardson and Wallbank approach, however, is that whilst published statistics of profits, dividends, additions to reserves, and so on, generally can be trusted, capitalisation figures cannot. After a careful study of both the balance sheets and the profit and loss accounts of some companies, it is possible to formulate reliable figures of the historical cost of capital formation. But this can seldom be done simply from the summary figures of liabilities and assets. On the one hand, the recorded capital of a firm may be unchanged over the years at the historical cost of the initial sinking and equipping, or the historical book value of the company's purchase, taking no account of additional capital expenditure undertaken out of the revenue account, and perhaps not even of that which was formally paid for from profits.[75] Furthermore, even if it did represent the cumulated historical cost, it need not necessarily be anywhere near to the current valuation. Some collieries which were excessively costly to sink were sold for much less when the original owners went into liquidation, and subsequently proved quite adequately profitable at their new, lower capitalisation. On the other hand, this latter point could cut both ways. Moreover, the capital account might be inflated by 'watered' capital – the addition of nominal sums which were never, in fact, paid up or paid out. However, it is doubtful whether this was often of importance in the case of colliery companies, and the establishment of asset and liability entries that were both 'too high' was much more commonly the result of valuing an existing colliery excessively when its ownership was transferred during a boom period. A good example of this is Broomhill Collieries Ltd, formed in 1910 with a nominal capital of £350,000 and debentures amounting to £325,000 to acquire Broomhill Coal Co. and Radcliffe Coal Co., paying out £565,000 in cash together with £20,000 in ordinary shares and £30,000 in debentures. The company's working profit was better than average, being about 1s 6d per ton over its first ten years; but it only managed to pay £63,000 in ordinary dividends over that period (i.e. an annual average of 3.6% on £175,000 of ordinary shares), though always meeting the debenture interest and the preference dividends (except in 1910), and managing to reduce the outstanding debentures to £157,000. The board of directors had, in fact, devoted profits to this repayment of loan capital because, as they said, in commenting on the original formation of the company, 'a reasonable capital would have been half a million'.[76] In other words, the boom-time euphoria of 1900 had encouraged (or, if one prefers, enabled) the promoters to overvalue the assets bought by the new company by approximately a quarter.

Short of detailed examination of the accounts of all public colliery companies, which is not now possible in any case,[77] there is no way of knowing whether capitalisation was generally higher or lower in the books than it 'should have been' – should have been, that is, to reflect accurately the cumulated historical cost or the current valuation of the assets concerned. Because of this, the method of Richardson and Wallbank is unreliable, and cannot be preferred to the profit estimates derived from the income tax statistics, and our estimates of capital stock at current valuation. However, they do perhaps give a very rough idea of the rate of return on capital valued at historical cost, since this is probably the element which enters most strongly into balance sheet capital.

10

PRODUCTIVITY

In the last two or three decades, economists have shown much concern with questions relating to productivity, and have developed techniques for measuring both productivity itself and the sources from which changes in it have sprung. Questions remain, however, both about some of those techniques, and about the meaning of the concept of productivity. At one level, of course, it is easy enough to define productivity. The (partial) productivity of labour, for example, is given by the quantity produced per unit of labour time; and the total factor productivity of an industry is the quantity produced in a given time per composite unit of all the factors of production employed. Assuming that difficulties of measurement can be surmounted, the problem arises in interpreting exactly what they mean.

Total factor productivity is often taken to be an index of the efficiency of an industry,[1] but, as Domar has pointed out, it is not this in the sense of a ratio between the actual and some potential maximum or optimum output.[2] However, it seems to be the behavioural implications of the word 'efficiency' which are in question here, and one can perhaps avoid those by saying that comparisons of total factor productivity over time do provide an indicator of changes in the effectiveness of a group of factors of production. Partial productivity statistics, on the other hand, are not capable of the same relative simplicity of interpretation. Changes in the output per man-hour (or -day, or -year) do not necessarily reflect any change in the productivity, and certainly not in the efficiency, of labour. Any particular change may just as easily have been caused by an alteration in the quantity or quality of other factors of production. So that a change in the labour productivity statistics implies, like total factor productivity, a change in the productivity of a group of factors, both the quantity and quality of all of which were liable to vary in time – and in place, too, if a comparison between different districts is to be made. Given the same regularity of work, such a change implies a change in the effectiveness of a representative group of factors of production, but which of them was responsible for the change cannot be discovered from the partial productivity data alone.

When we move from concepts to actual measurements, we come up against a number of difficulties, especially with total factor productivity.[3]

In the first place, there is the difficulty, which is really one of concept still, over what to measure as capital or as its cost. Should capital be valued at historical cost, at replacement cost, at the discounted value of its expected future stream of income, or at what? Should the *ex post* profit rate be taken as the cost of capital, or should one use a rate of interest plus depreciation, and if the latter, what rate of interest, and depreciation by what method over how long a period? In the particular case of the nineteenth-century coal industry, the decision about capital has already been effectively taken, for the estimates of capital stock which we made in chapter 2 had to be based on specific assumptions about the relation between capital and (full capacity) output. Profit rates are necessarily based on these estimates, so that, short of picking an arbitrary figure for interest plus depreciation, there is nothing we can learn about the productivity of capital that we have not already put into the calculation by assumption. In the second place, it is almost impossible to think of a suitable measure for the resource input of the coalmining industry. The biggest single resource used was coal-bearing land, and some sort of bench-mark estimates might be made of the surface acreage affected at various times. But it was in only one or two districts that payment was made on this basis, royalties more commonly taking the form of a toll on output, so that there are difficulties in the way of costing a unit of input of this particular resource. There are, in addition, other resources used in the industry, such as timber, gunpowder, oil, and so on. The first of these, especially, varied with the nature of the coal seam being mined in a way which cannot be quantified in any precise fashion. All of them are subject to the problem that there are no aggregative data on costs.

Finally, when it comes to the labour input of the industry, there is a certain amount of data, but it is not without drawbacks. Output and employment statistics can be combined from 1864 onwards to give figures of output per man-year (OMY), and isolated estimates can be made for some earlier years. But, as we have seen, there was considerable variation in the regularity of work from year to year, and the effects of this can only be partially allowed for, and then only from the mid-1890s onwards, by using the statistics of the average number of days per week on which the pits were open.[4] Even then, they take no account of absenteeism, which was not constant from year to year. Even though it generally moved in such a way as partially to offset fluctuations in the regularity of opening of the pits, genuine figures of output per man-shift (OMS) would be rather higher, by anything up to 10%, than those derived in the way just indicated. Fortunately, as we saw in chapter 4 there is no reason to suppose that there was any major long-run trend in absenteeism, so that we can take these proxy OMS statistics as a reasonably good guide to the long-term movements of genuine OMS, whilst recognising that the fluctuations in the latter would be less prominent. Fortunately again, the same can be said in turn of the OMY statistics, for leaving aside short-period changes in the latter and looking at periods of three to five years or more, the trend of OMY follows quite closely that of the proxy OMS. This is brought out in figure 10.1, which shows the annual OMY and proxy OMS statistics for those coalfields for which there are separate data for the period 1894–1914. The OMY fluctuations are wider, and their peaks are sometimes a year or two later than those of the OMS series; but there is no question but that the trend is very similar. With certain qualifications, principally about trade cycles which were very much out of the ordinary, therefore, it does seem possible to take the OMY series as a reasonably accurate guide to the *trend* of the productivity of representative groups of factors of production. Let us, then, begin by looking at these statistics, and leave until later any attempt to allocate responsibility for their movements.

The five-year moving averages of OMY for the various coalfields in figure 10.2 show that, for the period from 1874 onwards, when the data are consistent and generally reliable, these fall into three groups. All experienced a very considerable increase, sustained for about a decade, from the levels of the first half of the 1870s, but thereafter they behaved differently. In the most important group, consisting of Durham, Northumberland, Cumberland, Lancashire, South Wales, and both North and South Staffordshire, the falling trend of OMY after the early or mid-1880s was steep, and in some cases only briefly interrupted, and the level which OMY reached by 1910–14 was below that of 1874–8 (or whenever the trough of that period happened to be in each individual coalfield). In a second group, including all the remaining larger coalfields – Yorkshire, the East Midlands, Scotland – as well as North Wales, there was a similar sustained decline in OMY from its peak in the 80s, but it was generally less steep,[5] and the 1910–14 level was above that of the mid-70s. The third group comprises a handful of small coalfields – Leicestershire, Warwickshire, Somerset, and Gloucestershire, with Shropshire's experience putting it closer to this group than to any other. For this group, the trend of OMY was upwards, or at least level, after the peak of the 80s, and the level reached in 1910–14 was sometimes substantially higher than it had been in the mid-70s.

The data prior to 1874 are a good deal less reliable than subsequently, and, at best, offer only the possibility of rough estimates of productivity prior to the mid-1860s. The Inspectors of Mines' returns of numbers employed prior to the 1872 Coal Mines Regulation Act represented only perfunctory guesses in one or two cases, and most were probably too low for accurate comparison with later statistics, in that they seem to have excluded a number of surface workers who were covered by the 1872 Act, and so included in the statutory returns. The same almost certainly applies to the Census returns of numbers of coalminers. In some coalfields, however, the effect of this was more than offset by the fact that the output from coal mines of subsidiary minerals – fireclay, ganister, and ironstone, in the main – was not included in the inspectors' estimates of coal output. These could, of course, be taken out of later figures in order to facilitate comparison, though in doing so one would lose the full picture of the collieries' productivity. In many districts these subsidiary minerals were of little significance, but in South Wales, Scotland, the Black Country, and, above all, Shropshire and North Staffordshire, substantial amounts of ironstone came from the coal measures. Indeed, in North Staffordshire, quite a large number of mines which came under the Coal Mines Regulation Acts were engaged primarily in extracting iron ore.[6]

For all these difficulties, however, it is likely that the inspectors' figures of numbers employed for 1864–72 are consistent with each other, and probably also with the decennial census returns, which provided the inspectors with their bench-marks. Combining them with the output estimates derived from those discussed in chapter 2, it is possible to derive estimates of OMY for various years up to 1864 and annually thereafter up to 1872, which, though not precisely comparable with the later figures, and of doubtful accuracy, especially prior to the 1850s, are at least on the same basis as each other in principle, and should tell us something in broad terms about the movement of productivity.

Of course, it cannot be stressed too much that these statistics must not be taken as precisely accurate, most especially in the first half of the nineteenth century. Both the output figures and those of numbers employed are themselves estimates, and their margins of error, which are considerable, especially for districts other than the North-East, are combined in this table. Moreover, as has already been pointed out, quite wide variations in OMY could occur from year to year, resulting mainly from changes in regularity of employment rather than from anything affecting productivity more narrowly defined. Of the years shown here, both 1830 and 1850, being years of depression in the industry, with short-time working, are likely to be too low to be truly representative of their immediate period. But this may apply also to 1800, 1816, and 1820, all of which were far from boom times, so that the general trend may not be much distorted by these two years. In contrast, 1840 was still a reasonably prosperous year in the coal industry, and with more shifts per man likely to have been worked in that year than in any other shown here up to 1861, the OMY figures for 1840 probably suggest a faster rate of growth in the previous decade and a slower rate in the subsequent decade than is really justified, though this would not apply to OMS, for depressions tend to increase productivity in that sense. Nevertheless, taking the period as a whole, it does

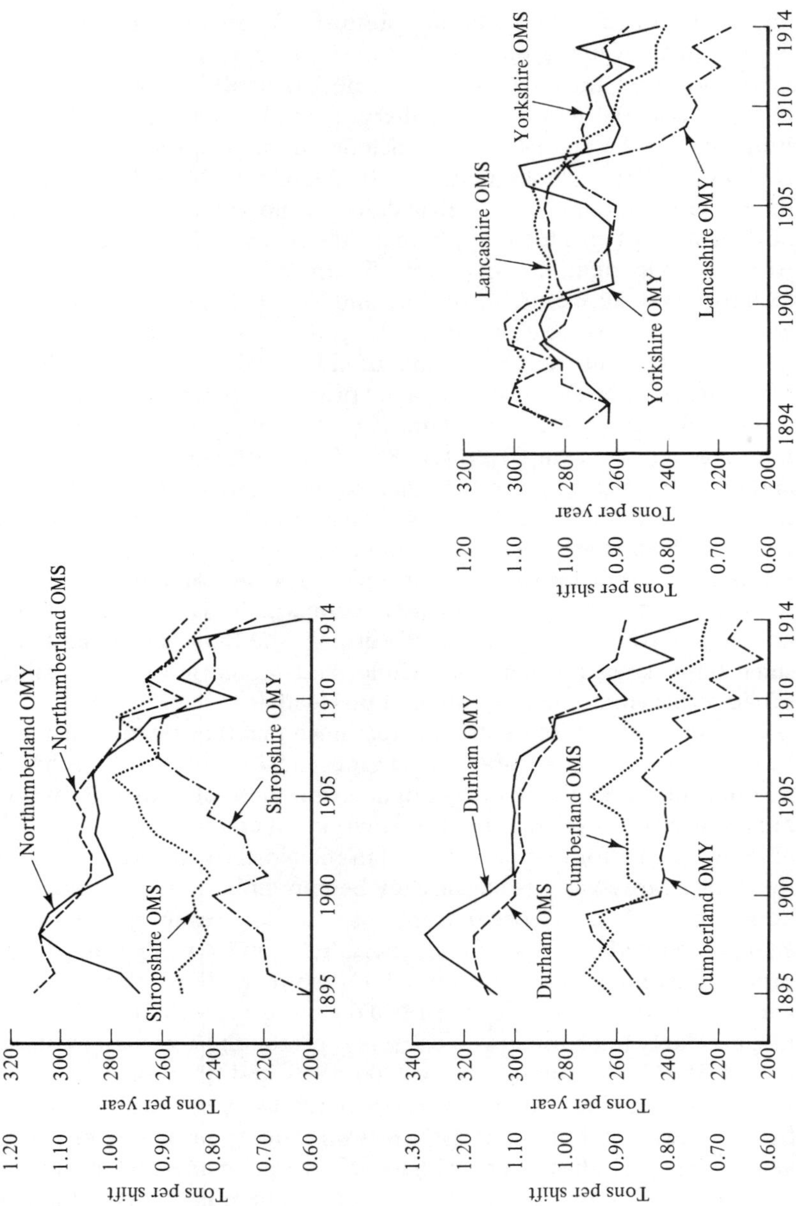
Tons per shift
Tons per year
1.20
1.10
1.00
0.90
0.80
0.70
0.60
1.30
320
300
280
260
240
220
200
340
1895
1900
1905
1910
1914
1894
Northumberland OMY
Northumberland OMS
Shropshire OMS
Shropshire OMY
Durham OMY
Durham OMS
Cumberland OMS
Cumberland OMY
Lancashire OMS
Yorkshire OMS
Yorkshire OMY
Lancashire OMY

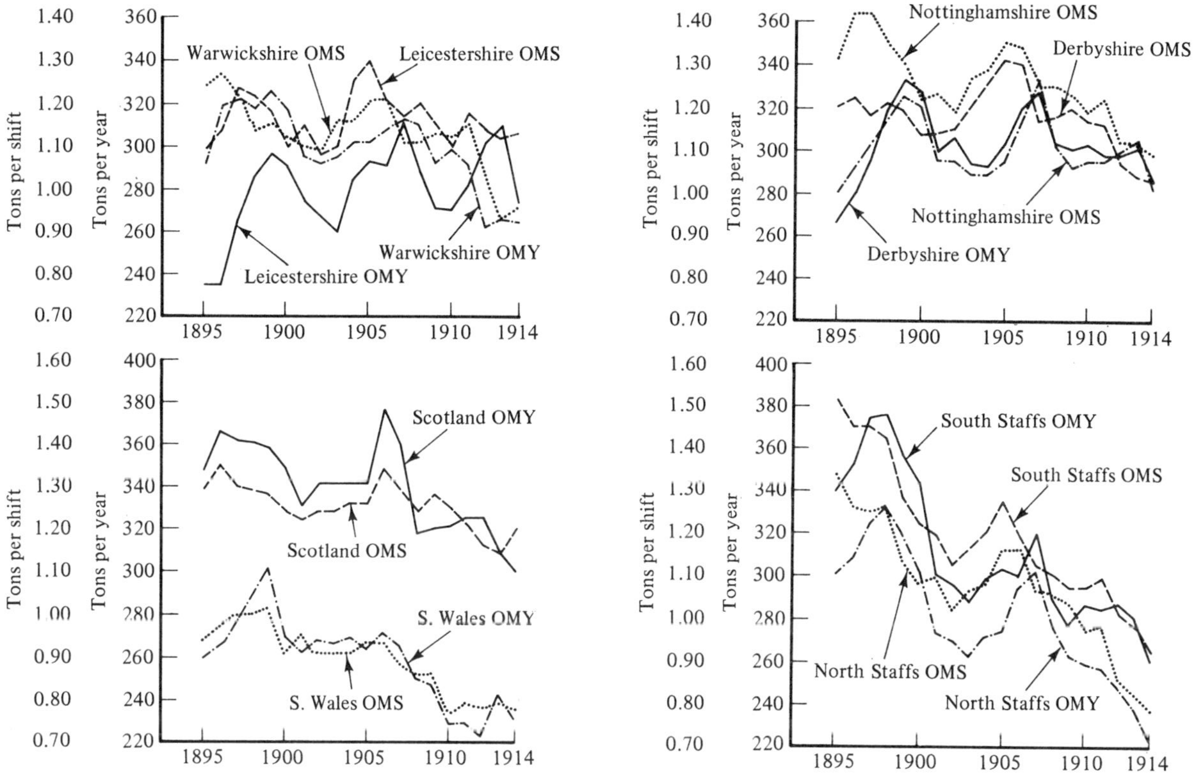

Figure 10.1 Output per man-year and output per man-shift, 1894–1914

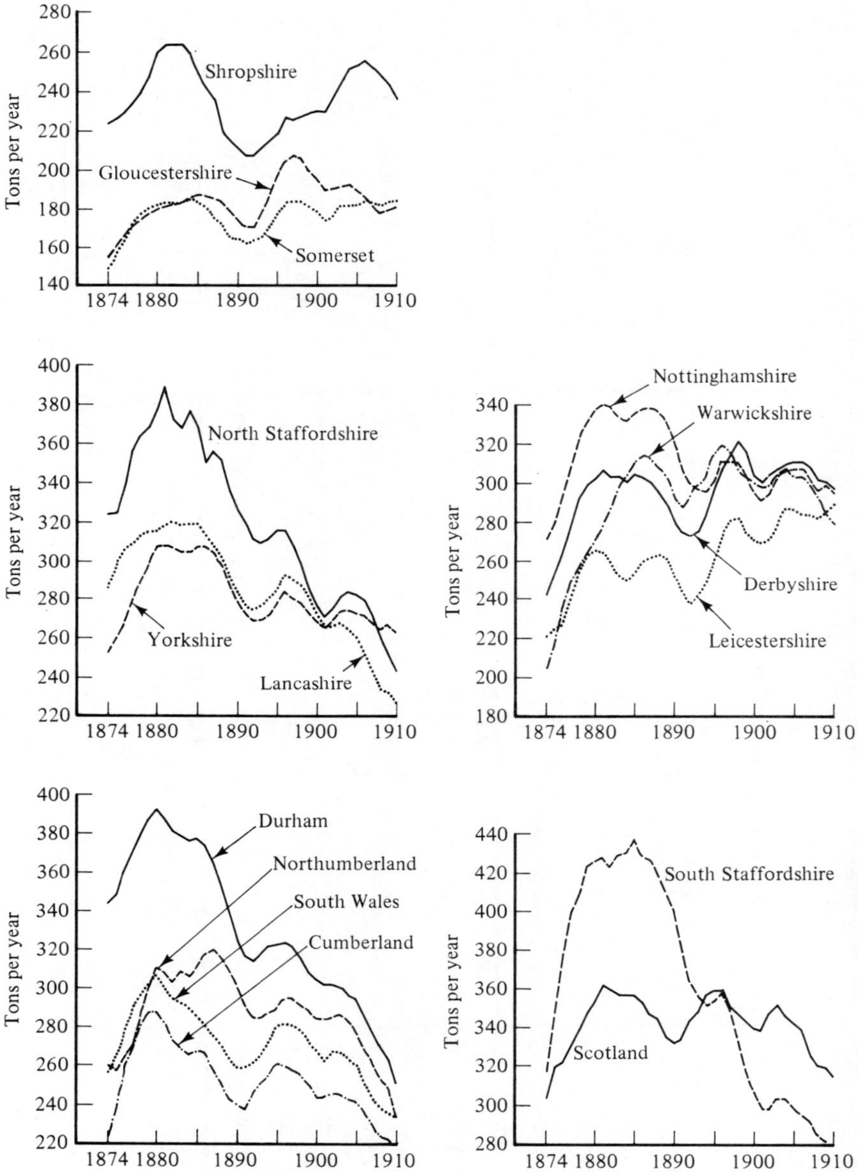

Figure 10.2 Output per man-year, 1874–1910 – five-year moving averages (centred on year shown)

Table 10.1 *Estimated output per man-year in British coalfields, 1800–72 (in tons)*

	North-East	York-shire	Lancashire & Cheshire	Midland Inspection Division	West Midlands	South Wales	Scotland	UK
1800	225	211	158	190	190	155	212	170
1816	265	219	165	229	250	160	219	190
1820	261	244	169	246	248	179	202	194
1830	276	270	189	277	210	194	213	215
1840	343	297	241	300	212	233	225	238
1850	295	229	252	250	231	236	209	231
1861	400[a]	302	296[b]	304	258[b]	272[c]	314	302
1864	428	270	298	274	334	232	316	302
1865	449	266	289	280	329	251	320	311
1866	435	266	276	280	335	279	306	317
1867	417	266	279	281	338	282	282	314
1868	374	262	276	275	324	275	293	297
1869	389	303	291	284	336	231	300	311
1870	390	316	292	290	338	237	318	315
1871	386	332	297	298	326	249	329	317
1872	322	285	293	271	285	242	304	295

[a]Subsequently including Cumberland.
[b]Subsequently Cheshire is included with the West Midlands, not Lancashire.
[c]Subsequently including Gloucestershire and Somerset.
Notes:
1. It is assumed that the figures of numbers employed in the census years are applicable to the previous year. The estimate for 1816 is interpolated on the basis of scattered evidence about the opening of new pits and about trade activity in various districts.
2. The output referred to in the table is of coal only and does not include subsidiary minerals, which are, however, included in figure 10.2.
3. There is probably a major break in the West Midlands series between 1861 and 1864, the cause being the inclusion of many iron mines in the census statistics of the numbers employed.
Sources: see text.

seem to be clear that the trend of productivity was upwards in all coalfields up to the 1860s. It would be unwise to make anything of the different pace of increase in the coalfields at different times, for they were liable to be affected unequally by boom and slump. Perhaps all one can say is that the upward trend of OMY was sustained practically throughout the first sixty and more years of the nineteenth century, with the only exception apparently being the North-East in the 1840s. Even this may be more apparent than real because of the difference in the state of trade in 1840 and in 1850.

As has been said, exact comparability between the statistics up to 1872 and those subsequent to the effective operation of the 1872 Coal Mines Regulation Act cannot be achieved. But unless the number of workers covered by that Act who had not been counted by the inspectors previously was extraordinarily large, far larger than there is any reason to suppose,[7] then there must have been a very considerable fall in OMY from the mid-1860s to the mid-70s. By far the greater part of this must have taken place in 1873 and 1874. The recovery after the mid-70s, which has already been referred to, took OMY to levels which, a decade or so later, were in most cases above those of the 1860s, even after excluding subsidiary minerals from the later statistics. This made the 1880s the peak time for productivity in British coalmining during the whole of our period, with only Northumberland and Durham not quite reaching the levels of the 60s, and three or four small coalfields still to go on to higher OMY figures in the 1890s or later.

We have now established, at least roughly, the statistical dimensions of the movement of productivity in British coalfields, expressed in terms of output per unit of labour input (per year, but reflecting the same trend per shift), but indicating the productivity of each unit of labour together with its proportionate share of the other inputs to the industry. It remains to attempt explanations of what we have found. It is quite easy to suggest lists of plausible reasons for the rising trends of productivity to the 1860s or 80s, and for the subsequent fall. It is a different matter to try to indicate the relative importance of these reasons, and, indeed, it is impossible to do this with any precision. Nevertheless, some attempt must be made.

Taking first the long period of rising productivity, it is hard to avoid the conclusion that the main cause of this was technical progress, although in the first thirty or more years of the century, the greater regularity of working associated with improved transport, especially in winter, must have been at least as important in areas not served by good water transport, and in districts where much of the output at the start of the century still came from shallow mines operated on a virtually part-time basis.[8] In this connection, the much higher OMY in Durham and Northumberland than anywhere else up to the 1860s, and especially towards the beginning of our period, must have owed something to the superior organisation of distribution there, and also to the very small proportion of output drawn from surface workings and tiny pits on the outcrop. (Of course, their superiority could not have been sustained without relatively thick seams, and the district had other natural advantages.) Technical change, with the single exception of the use of gunpowder to break down the coal, did not directly raise the productivity of the hewer with his pick, though better ventilation must have had a generalised impact on the energy of everyone down a mine. The main thrust of technical change was, rather, in the improvement of the haulage of the coal from the face to the pitmouth. To some degree this was merely a means of offsetting the need for more haulage imposed by deeper mines; but, especially, though not solely, in the half-century from the 1830s to the 1880s, actual economies in haulage labour were achieved, thus directly increasing average productivity. This, along with shotfiring and better ventilation, must surely be the main reasons for the rising productivity of this fifty-year period. Of course, with the exception, once again, of shotfiring, most technical change had to be embodied in increased capital, and this led to an increase in the amount of capital available per man in the industry from less than £50 in 1816 (at constant 1869 prices) to well over £100 in 1883.[9] It is possible to argue that this increase in capital per man is what lay behind rising productivity. But this is surely too superficial a view. The increase in capital per man was to a very large extent the result of technological response to what amounted to a decline in the quality of the natural resources available to the industry, though one must not exaggerate this – the major decline in the quality of coal seams until almost the end of the century was simply the increase in the depth at which they had to be sought, not a change in average thickness. Still, deeper pits cost more to open; deeper pits required bigger areas for their economic operation and this implied more haulage; deeper (and gassier) pits required more ventilation. Of course there were technical changes requiring increased capital which were undertaken for the sake of their productivity-improving effects, but in most of those which affected haulage or ventilation the main motive for innovation was to offset the threat of productivity reduction.[10]

The first major pause in the long upward trend of productivity came in the frenzied boom of the early 1870s, and was one of the major topics considered by the Select Committee of 1873 appointed to examine 'the Dearness and Scarcity of Coal'. Many different reasons for falling productivity were adduced by witnesses before that committee, but the one which HM Inspectors and representatives of both the unions and of the owners agreed was that the miners were working less – either less energetically, or less regularly, or both, and it is worth repeating here some of the statements made earlier. John Normansell, secretary of the South Yorkshire Miners' Association, put this quite bluntly when he said: 'They do not work as hard as they used to do, but they work, in my opinion, quite hard enough. They worked more like fiends, in my opinion, in former days, and they intend taking it a little easier.'[11] Lionel Brough, the Inspector for South Wales, made the same point rather more primly: 'I cannot believe in statements that the men are doing their work badly, but they are doing it in less degree.'[12] Most of his colleagues agreed that the miners, especially hewers, were working less regularly,[13] although two inspectors, from the Midlands, did not accept this.[14] The owners were virtually unanimous in making the point, though a Yorkshire owner, Robert Tennant, thought that 'they have worked better during the time that they have worked', a point with which John Dixon of the West Yorkshire Association of Miners agreed.[15] This refutes the suggestion that work was less energetic, and perhaps the contrariness of the testimony can be taken as an indication that this was an unimportant influence.

More of what later came to be called absenteeism, therefore, may certainly be accepted as one reason for falling OMY in the early 1870s, though how important it was cannot be precisely determined. The few statistics that can be adduced suggest that it may have been less significant than the weight of the evidence to the 1873 Committee would indicate. Managers of important collieries in the North-East gave their own experience. G.B. Forster, of Northumberland, had had a slight increase in the average number of days worked per fortnight in 1872, and a fall of about 10% only in the first quarter of 1873, whilst Lindsay Wood from Durham had seen his hewers' absence percentage rise from 14.5% to 15.5% in 1872, and to 16.5% in the first quarter of 1873.[16] At the Lochgelly Iron and Coal Company's pits in Fifeshire, however, the average number of days worked by the men fell by much more – from 251 in 1872 to 212 in 1872,[17] which, if typical for Scotland, would more than account for the fall in OMY between these two years. However, since there were certainly other forces at work depressing OMY, perhaps too much should not be made of this.

Of these other forces, two stand out as having both *a priori* plausibility and some support from contemporary testimony. One is the influx of unskilled labour accompanying the great boom, which clearly must have tended to reduce average productivity whilst the newcomers were picking up the rudiments of the job. This was later noted by Joseph Dickinson, one of the Lancashire Mines Inspectors, as having been an important factor in falling OMY.[18] Moreover, such a sudden increase in the labour force implies a fall in the amount of capital per man, which must be expected to reduce productivity *ceteris paribus*. The second influence, which was itself partly responsible for the inflow of labour, was the Coal Mines Regulation Act of 1872. This obviously could not have contributed to the decline in OMY in 1872, but may have played a part in the continuing fall of the next two years. As we saw earlier,[19] hours of work were effectively cut by this Act in many districts, especially those of byeworkers, whose numbers therefore had to be augmented if they were to service the same output from the hewers. The need for improved safety measures may also have increased the proportion of offhand workers. And temporarily, at least, there was probably some effect from the disruption of working schedules, especially from the difficulty now experienced in employing boys under twelve years old.[20]

When prices, wages, and profits all fell in the middle and later 70s, many of the forces which had operated to reduce OMY in the boom went into reverse. Absenteeism fell, hours were raised so far as possible, and no doubt the intensity of work increased once again, since the pieceworker had to cut more coal in the attempt to maintain his earnings. For managements, the prevalence of low profits, and in some cases losses, had continued long enough by the end of the decade to ensure special concentration on efficiency, and the elimination of as much deadwork as possible. Bankruptcies helped to weed out the least productive pits, and the outflow of labour probably took a disproportionate number of the least efficient and most troublesome workers. Moreover, the early 70s had seen a great deal of new sinking undertaken, and by the 1880s much of this was at the most productive stage of its life – that is, fairly fully opened out, but still working relatively easily-got coal, not very far from the shaft. Thus each coalfield, with the likely exception of Durham and Northumberland, contained an unusually high proportion of mines in their prime. In addition to this, district-wide exhaustion of the best seams had not yet begun to affect any coalfield very seriously, though it was to be found in the older-exploited areas near the River Tyne and in parts of the Black Country.

Natural conditions and the circumstances of the labour market were thus exceptionally favourable to the attainment of high productivity. They were helped by the recent relatively large accession of capital to the industry, which raised the capital per man (at constant 1886 prices) to £118, a figure not subsequently exceeded until after the First World War. Moreover, technical progress in haulage and winding was still taking place, and the 1870s was the period of most rapid adoption of mechanical ventilation, which must have had a considerable, though incalculable, impact on work effort.

After the 1880s there was, as we have seen, a general decline in physical productivity, with only a few of the smaller coalfields enjoying any different experience. Again, it is easy to list plausible reasons – the resumption of new sinking, a renewed influx of untrained labour, increasingly stringent safety legislation, a deterioration in labour relations (or, at any rate, increasing trade union power), and so on. All may have played some part: none looks to have been nearly so important as the increasing average age of mines, the cause which was accepted by both the Sankey and the Samuel Commissions as by far the most significant. The increase is strikingly illustrated in table 10.2, which shows the proportion of men employed in collieries that had been opened for less than thirty years in 1894 – the first year for which the data are available – and again in 1900 and 1913. The figures, of course, cannot give an entirely accurate picture of the ageing of mines, for some collieries sank new pits and closed old ones without any trace of this appearing in the *List of Mines* that the Home Office published annually. This seems to have been especially the case in some of the small coalfields, above all in Leicestershire, though there is no very obvious reason why this should have been so. Moreover, the further sinking of existing shafts to new, deeper seams did sometimes take place, effectively renewing old collieries, if not making them precisely as good as they had originally been. But the broad position pre-

Table 10.2 *Percentages of men employed in collieries with over fifty men which were thirty years old or less*

	1894	1900	1913
Durham & Northumberland	44	30	21
Cumberland	57	34	16
Yorkshire	62	56	36
Lancashire & Cheshire	56	32	23
North Derbyshire	59	49	30
Nottinghamshire	78	61	41
Leicestershire & South Derbyshire	41	19	10
Warwickshire	41	50	24
North Wales	68	45	19
North Staffordshire	46	26	9
Cannock Chase	78	51	6
Bristol & Somerset	28	18	37
Forest of Dean	24	28	24
South Wales	86	63	40
Scotland	55	50	37

Note: owing to the large number of small pits, many with divided ownership, and to the fact that many were closed and reopened, often more than once, to work out as much as possible of the pillars left in the original workings in the Ten Yard seam, it is impossible to make similar calculations for the Black Country.
Source: HOLM.

sented by table 10.2 is accurate enough. Of the major coalfields, only the North-East did not have more than half of its men in relatively young collieries in 1894. By 1900, the North-East and Lancashire both had less than a third in such mines; and by 1913 all had well below 50%.

The principal way in which this increase in the average age of mines made itself felt was in the greater distances which both men and coal had to travel underground. This cut the effective working time of face-workers, especially when their hours were limited. According to the Samuel Commission it also led to an increase in the proportion of haulage workers needed, which we found, in chapter 5, was consistent with the evidence.[21] Certainly in individual cases the expansion of the proportion of haulage and maintenance workers as pits grew older was very marked – as at Govan colliery.

But though the effects of the ageing of mines were the most important reasons for falling productivity, it is worth noting that these effects had long been felt in individual collieries, in quite large numbers in the North-East by as early as the 1820s and 30s. They had, in fact, been one of the main incentives to technical progress in underground haulage, such as had been a feature of the middle decades of the nineteenth century. By the 1890s and 1900s, however, there seemed to be an inability to respond sufficiently to the increasing need for technology to offset the effects of deteriorating natural conditions. Some aspects of this were discussed in chapter 4, and the general conclusion then – which applies also to the mechanical cutter as a means of offsetting the declining productivity of faceworkers – was that the failure to apply more mechanical (and electrical) technology was mostly a matter of relative costs. Put rather crudely, demand for coal was relatively inelastic over the relevant range of prices, so that in the long run prices could be raised as costs went up without any major impact on the size of the market.

Whether this would have been true if prices had gone up even further cannot be known, but that they did not go up even further owed a lot to the relatively elastic supply of labour at wage rates which, though rising relative to most other kinds of labour, were still not rising steeply. The important point about this so far as technical innovation is concerned is that wage rates still remained low enough in the 1900s to make newish, but proven, techniques, like mechanical cutting and face-conveying, uneconomic in the majority of natural conditions still met with in British pits. In consequence, physical productivity continued to decline, but the profitability of the industry was maintained – improved, indeed, compared to the period from the mid-1870s to the late 90s – by its ability to put up prices and so keep up average revenue-productivity. The latter, for the United Kingdom as a whole, averaged £83 per man for 1882–98 and £117 for 1899–1913, whilst the averages of the Board of Trade wholesale price index for the two periods are the same. However, when the period 1882–1913 is disaggregated further, it is noticeable that the average revenue for the trade cycle beginning in 1907 was rather below that of the cycle beginning in 1900, the figures for complete cycles being as follows: £85 per man for 1883–9, £98 per man for 1890–9, £117 per man for 1900–6, and £114 per man for 1907–12.

So in the last few years of our period, the situation was beginning to change. Average revenue productivity did not recover to its previous all-time peak of 1900 in either 1907 or 1913. After 1900, in fact, its trend was steady and flat in current prices, but with the wholesale price index rising from 100 in 1900 to 117 in 1914 this implied a fall at constant prices. Three influences may be identified as having had a major effect in bringing this about. On the supply side there was, first of all, the increase in the number of bye-workers resulting from the Eight Hours Day Act. Then there was the continuing ageing of mines, which was probably a more important factor.[22] Finally, on the demand side, there was the beginning of the emergence of competing fuels, and coupled with this was the greater economy in the use of coal which was effected by progress in the electricity supply industry. It is unlikely that elasticity of demand had changed very much before 1914, but it may have been enough, in conjunction with continued rapid expansion of coal output elsewhere in the world, to prevent a sufficient price increase during the upswing of 1912–13 to offset the fall in physical productivity that had occurred since the previous cyclical peak. Between 1907 and 1913 average OMY fell by 11% and average pitmouth prices rose by 11.5%, leaving revenue product per head at the same level; but wholesale prices in general had risen by 10% in the meantime. Undoubtedly, the progress which was made in installing such labour-saving devices as coal-cutters and conveyors in the last few years before the outbreak of the First World War owed something to the incentive provided by this constellation of stationary revenue productivity, rising general prices, and the rising wage costs we found in chapter 7. The war exaggerated all these trends; and after delays due to various causes,[23] the British coal industry did eventually mechanise more rapidly and completely. By the late 1920s the trend of physical productivity was at last reversed, and went upwards again until the Second World War. That this had not happened before 1914 may be put down to the framework in which the industry operated – labour was still reasonably plentiful and not rising rapidly in cost, demand was still relatively inelastic, and foreign competitors were either too far from sea transport (like the USA collieries) or still operating at levels of productivity that were below that of the British, even though these were mostly still rising and so catching up.

The discussion so far has been in terms of the productivity of the labour force as a whole, but it is possible to look separately at that of underground workers from 1874 onwards. However, to do so merely shows that they moved in very much the same way in all districts except South Staffordshire, where productivity underground fell rather faster than overall. This development, which might seem surprising in view of the increasing attention paid to coal preparation of one sort or another, is explained by the closing of most of the small Black Country pits, which had a disproportionately large number of surface workers.

The more interesting information as to whether productivity at the face changed differently from that of general underground productivity cannot be told from any direct aggregative evidence. However, as we have seen,[24] there was some increase in the proportion of men employed at the face during the 1870s and 80, which implies a relatively greater improvement in the productivity of non-faceworkers during that final period when OMY was rising to its (national) peak. In view of the spread of mechanical haulage, and the absence of any equivalent technical progress applicable to the faceworkers alone, this is not unexpected. The Eight Hours Day Act, and possibly the spread of mechanical cutters and face-conveyors, produced a change in the opposite direction, in some districts at any rate. Unfortunately, it is impossible to be other than rather vague about the extent of either of these movements.

For the reasons given at the beginning of this chapter, it is impossible to make a formal analysis of total factor productivity in the nineteenth-century coal industry. But the necessarily imprecise examination of productivity which we have undertaken in terms of output per man is, in effect, an examination of the productivity of all the factors involved in the industry. To sum up what we have found, it seems reasonably certain that it rose without major interruption from the 1820s (and probably earlier) to the 1860s largely because of technical progress and the increase in capital per man in which most of that progress had to be embodied. The quality of the labour force probably underwent no significant change: both the quality and the quantity (per ton); of the materials component of the resources input probably remained much the same; and whilst the average depth at which coal was mined certainly increased, and by implication so did the average distance from coal face to pit bottom, the quality of the coal seams probably did not change very much in other respects – apart, perhaps, from the mining of more fiery, but more productive, seams which followed the invention of the safety lamp.

In the early 1870s there was a considerable fall in productivity, resulting from a mixture of influences: (i) a massive influx of new labour, necessarily of lower quality for the time being; (ii) a lowering of the average amount of capital per man, consequent partly on cramming more men into the same space, partly on the employment of more men to offset the lesser regularity of attendance, and partly on the opening (and reopening) of small, ill-equipped collieries on the outcrop, as everyone who could, strove to make something out of the massive price rise; (iii) for the same reason, some decline in the quality of coal-bearing land; (iv) some disorganisation of the industry's routines consequent upon legislation; and (v) perhaps some slackening in the workers' efforts, and certainly a falling off in the regularity with which they worked, which, whilst it would not have reduced the individual hewer's output per shift, may well have cut the average for other workers, who could not easily be employed in exactly the optimum proportion.

With the end of the great boom, many of these influences were felt in reverse, and productivity climbed to new peaks in the 1880s Decline thereafter was affected primarily by the falling quality of coal-bearing land, partly the result of the exhaustion of thicker seams[25] and the continuing increase in the average depth at which coal was won, but still more springing from the increasing average age of mines – though this was itself partly a consequence of the greater depth of workings and the desire to get the most out of the heavy investment in deep shafts. Age had several deleterious effects, which an American writer of the 1930s, Harold M.

Watkins, noted as follows, in a comparison with collieries in the United States:

> The greater age of the British workings, in addition to imposing the handicaps of greater length of haulage, maintenance of air ways and travelling distance from shaft bottom to face, makes for conservatism in mining practice. A new mine starting with a clean slate may introduce changes in methods without serious resistance and may take advantage of the latest developments in engineering practice much more readily.[26]

Legislative interference with the length of the working day may have aggravated this, especially where underground travelling times were long, in the last few years of our period – though the worst effect was almost certainly the purely temporary disruption of routines. However, where there was a need to increase the haulage labour force, there was also, *ceteris paribus*, a decline in the amount of capital per man. In fact, it seems true to say, in general, that there was little, if any, increase in the capital:labour ratio after the 1880s, so that the industry was certainly not becoming a more capital-intensive one in any significant measure. Finally, there is the possibility that the average quality of effort by miners fell as wage rates rose after 1887, and even more after 1897, and did not go back to the old levels in the downswings. This was discussed earlier,[27] and there does not seem to be any good reason to suppose that this was a significant influence.

Whilst a total factor productivity study for the industry as a whole cannot be undertaken, it can be done after a fashion for two of the firms whose records we have examined. As has been pointed out by various critics,[28] the total factor productivity formulae imply that the firms are in a perfectly competitive industry and are in a state of long-run equilibrium. Neither of these implications is true in any real-life situation, but it is probably fair to say that British coalmining in the sixty or seventy years up to the outbreak of the First World War was as close to perfect competition as it is possible to be. It was, in fact, a good example of Chamberlin's concept of 'pure competition'.[29] Both the firms for which we have sufficient data were, however, growing, at least for part of the period which we can cover. As a result there is likely to be a little upward bias in our measure of the rate of change of productivity.

The data we have only allow the use of the particular formula which relates total factor productivity to the rates of change of wages, resource costs, profits, and prices, namely $\overset{*}{A} = \alpha\overset{*}{w} + \beta\overset{*}{r} + \gamma\overset{*}{i} - \overset{*}{P}$, where $\overset{*}{A}$ is the percentage rate of change of total factor productivity, $\overset{*}{w}$ is the rate of change of wages per ton, $\overset{*}{r}$ is the rate of change of resource costs per ton, $\overset{*}{i}$ is the rate of change of profits per ton, $\overset{*}{P}$ is the rate of change of average receipts per ton, and α, β, and γ are the respective shares in total proceeds of wages, resources, and profits (averaged between the years covered). It is, perhaps, more than usually questionable in the case of the coal industry whether the highly fluctuating profits can really be accepted as the cost of capital, as is implied by the formula. However, this can be partly, though not entirely, got round by using average profits for groups of years, though since few records extend over any lengthy period it is this more than anything else which cuts down the number of firms which can be usefully analysed in this way. The data and calculations for the two firms for which there is adequate coverage – Henry Briggs, Son & Company and the Cannock Chase Colliery Company – are as follows, for periods as near as possible to the trade cycles affecting the industry.

	w	r	i	P	α	β	γ
Briggs							
1876–83	41.49	21.04	3.57	66.11	.63	.32	.05
1884–90	40.80	19.81	4.61	65.22	.63	.30	.07
1891–1900	53.53	19.99	11.88	87.33	.63	.23	.14
1901–7	63.69	22.47	12.71	96.75	.64	.23	.13
1908–14	67.58	24.50	12.81	104.89	.64	.23	.12
Cannock Chase Co							
1874–83	59.34	35.88	10.40	105.62	.56	.34	.10
1884–90	48.37	27.33	5.82	81.53	.59	.34	.07
1891–1900	59.80	25.34	9.94	95.09	.63	.27	.10
1901–7	67.77	29.01	9.14	105.92	.64	.27	.09

Briggs		
1876/83–1884–90	$\overset{*}{\dot{A}} = .63 \times (-1.7) + .31 \times 5.8 + .06 \times 29.1 - (-1.3)$	$= 3.8$
1884/90–1891/1900	$\overset{*}{\dot{A}} = .63 \times 31.2 + .26 \times 0.9 + .105 \times 157.7 - 33.9$	$= 2.6$
1891/1900–1907/7	$\overset{*}{\dot{A}} = .635 \times 19.0 + .23 \times 12.4 + .135 \times 7.0 - 10.8$	$= 5.1$
1901/7–1908/14	$\overset{*}{\dot{A}} = .64 \times 6.1 + .23 \times 9.0 + .125 \times 0.8 - 8.4$	$= -2.33$
Cannock Chase Co		
1874/83–1884/90	$\overset{*}{\dot{A}} = .57 \times (-18.5) + .34 \times 23.8 + .085 \times (-44.0) - (-22.8)$	$= 0.5$
1884/90–1891/1900	$\overset{*}{\dot{A}} = .61 \times 23.6 + .30 \times (-7.3) - .085 \times 70.8 - (-6.6)$	$= 1.6$
1891/1900–1901/7	$\overset{*}{\dot{A}} = .63 \times 13.3 + .27 \times 14.5 + .095 \times (-8.0) - 14.4$	$= -2.9$

One cannot, of course, draw any conclusion for the industry as a whole, or even for the two coalfields involved, from two individual firms. Nevertheless, it is interesting to see that the falling productivity of the end of our period is reflected in both companies, setting in a few years earlier in the Cannock Chase Colliery Company than in Briggs's, or than our earlier discussion would suggest was the general case. It is also noteworthy that, since productivity is being measured in value terms here, the increase between the 1870s and the 1880s was very small, the rise in the 1890s being considerably larger. In fact, one may say that after the 1880s the main reason for rising revenue productivity was the rising price of coal. There is no reason to suspect, in this matter, that the experience of these two companies was at all atypical. And, of course, it carries implications for the performance of the British economy as a whole, for whilst the coal industry itself might achieve prosperity by raising its prices by more than its costs were going up, this was a subtraction from any productivity increase achieved in other sectors of the economy which used coal as a resource input. In the later 1900s even this prosperity for coal was declining, though the industry was still more profitable (and paid its workers better) than twenty years earlier, when its physical productivity was at its height. In the light of this it does not seem to be exaggerating to say that the efficiency of British coalmining was declining, even though one hastens to add, before any behavioural conclusions are drawn, that the main reason had to do with the falling quality of the resource base – something which is not brought out by the study of the individual firms since their resource inputs were a heterogeneous mixture of stores and payments for coal-bearing land, and the latter was measured by a payment which, as a toll fixed for long periods at a time, probably did not fully reflect the falling quality of what was being paid for.

11

CONCLUSION

The coal industry of Durham and Northumberland, and the scattered local coal industries of the other British coalfields, were vastly expanded, and transformed in many ways, in the course of the nineteenth century. The number of uses to which coal was put increased, though there was no change so fundamental as those which had taken place before 1800, with the invention of steam engines and with the first applications of coal to the manufacture of pig and bar iron. However, the diffusion of these innovations, and especially the diffusion of steam engines, was mainly effected in the nineteenth century. Because of this, including the effects on transport costs of the application of the steam engine to the railway, and because, too, of the growth of population and the improvement of living standards, the production of coal was increased by roughly twenty-five times. To achieve this, the size of the industry's labour force rose from perhaps 70,000 to well over a million, and the capital invested grew from about £4½ million to somewhere around £124 million at the constant prices of 1886.

Among the major changes, apart from those of sheer size, was the much greater integration of the marketing set-up of the industry, mainly brought about by the spread of railways. This had the effect of bringing the different coalfields closer together in other ways, such as in mining practices and in wages. However, transport costs for more than a few miles still constituted a substantial proportion of final total costs, so that the coalfields did remain distinct in many ways. Each retained substantial local markets free (or virtually so) of outside competition, and this enabled local differences in prices, costs, customs, and practices to survive. Parochial attitudes on the part of owners, workmen, and their organisations were also slow to disappear.

Competition within local coalfield markets was nearly always strong, for the number of firms was seldom small enough to constitute an oligopoly, and new entry was comparatively easy until quite late in the nineteenth century. It was only in selling to the London market in the days when transport costs ensured that it could only be supplied economically from a relatively small number of collieries near to the shipping ports of the North-East Coast, that there was effective limitation of competition. Even then it was intermittent, mainly because new entrants to the industry fought the oligopolists until they were accommodated within the organisation. When railways brought effective competition in the London market from other coalfields, that was the end of limitation. Competition between coalfields thereafter ruled in all national markets – in London and other areas away from the coal mines, in export sales, and to a large extent in the supply of railways and shipping. Competition also ruled between firms within local markets.

The organisation of the business units which directed the capital and labour force of the industry was also altered during the course of the nineteenth century, though much of the change in this respect was in form and appearance only. The old form of individual ownership or partnership continued to be extremely important, though by the end of our period it most commonly took the guise of a private limited company, a form available since the company legislation of 1907. In effect, though, there was little change, with decisions being taken by owners (or, at any rate, owning families) just as they had been in the eighteenth century. This applied even to some of the larger, public companies, where control of the ordinary voting shares had been retained by the company's founders. Perhaps the main change within this group of owner-managers during the century was the gradual departure from it of most, though certainly not all, of the landowners, who had formed a large part of it in 1800. In their place, the dominant constituent of this group came to be professional coalowners, some originally from outside the industry, some from the growing body of mining engineers, some direct from the ranks of working miners, at any rate up to the 1860s or 70s. But if the old-style owner-manager still provided the main direction of the majority of firms in the industry, larger companies with at least partial divorce of ownership and control, and attracting capital from outside, had begun to come on the scene in the 1860s, and more especially from about 1890 onwards, and they had a fair deal of importance by 1914, notably in South Wales.

Almost as a corollary of the continued importance of the private firm and of the professional coalowner was the fact that ploughed-back profits, in some form or other, constituted the principal source of the increased capital of the coal industry. There was always some which came from outside in every spurt of investment, notably in the early 1870s and around 1900; but it was never a majority. In the light of the unexciting long-term profit record of the industry as a whole this is perhaps surprising, though, as we have argued, it probably owes a lot to the unusually great variance between the profits of different collieries, and to the probability that most professional coalowners did rather better than average. Nevertheless, when all is said and done, one would have expected to find average net profit rates that were distinctly higher than the yields on the ordinary shares of the main home railway companies, which surely cannot have been regarded as equally risky as collieries, even in the 1900s. All in all, there would seem to be some justice in the view of both miners and coalowners that those who received virtually risk-free royalties were the main beneficiaries of the industry. The movement of landlords out of entrepreneurship would seem to have had a lot to commend it from the point of view of the individuals concerned.

On the side of labour there was also a good deal of change during the nineteenth century, though there, too, some of it was rather superficial. The more than tenfold increase in the size of the labour force was mainly the result of the natural increase of the coalmining community itself, with which was associated a fair amount of migration between coalfields in the second half of the century, as growth of capacity slowed down or stopped in some districts. But that community was reinforced from time to time in periods of high demand by inflows from other occupations – mainly from agriculture, including especially Irish agriculture in the middle twenty or thirty years of the century. Metal miners, whose jobs were vanishing, and others from declining trades – framework knitters, nailers, and so on – were another source of recruits. Except in the later 1870s, the inflows were usually absorbed by the existing mining community without difficulty, though the Irish tended to remain a separate part of it for a long time in those districts where they formed a sizeable group, particularly in western and central Scotland and in Durham.

By 1914, collective bargaining with trade unions was established in every coalfield. But even at the beginning of the nineteenth century it is a mistake to think that the supposed individual bargain between manager and workman was a practical reality. Informal collective bargaining at the workplace level was what actually took place. In the course of the century the locus of bargaining shifted to larger units – from the particular face to the pit; from the pit to the group of pits in a small area; from that to the whole of a district; and eventually, after 1908, to the country as a whole, though in an incomplete and ambivalent fashion, which still left most of the real power of decision-taking at the district level. As the locus of bargaining shifted, so paid trade union officials came on the scene and began to take an increasing part in the process of negotiation, though never being free of rank-and-file control, especially in determining the basis rates of each seam and pit. The power of the trade union was, in fact, still severely restricted right up to 1914 by the existence of competition in the markets for coal, and despite the declared opposition of the Miners' Federation, wage levels largely followed prices. Given the elasticity of supply in the short run up to capacity working, this meant that it was the short-run inelasticity of demand for coal which was the main influence on wages. This, in turn, meant that the miner, especially the hewer, though one of the very highest-paid, on average, of manual workers, was also subject to greater fluctuations in his wage rates than practically any other workers. This volatility was transferred to his earnings to quite a substantial extent, despite variations in the intensity of work and in voluntary irregularity of attendance, because of the industry's practice of short-time working in depressions.

Whilst the technology of the industry was another aspect which experienced a fair amount of change in the nineteenth century, it was mostly on the way from the coalface to the pitmouth that this occurred, rather than at the face itself. Except in the thin seams where coal-cutting machinery was used, the hewer in 1914 was not doing a job that was very much, if at all, different from that which his predecessor had done in 1814 – or, perhaps, even in 1714. His way to the working place might be a bit easier, if also a good deal longer on average; the dangers he faced might be just a little less; and his physical condition would probably be better because of the improved ventilation, though greater depth could mean greater discomfort from heat. But once at the face, the only major change in the typical hewer's job had been brought about by the use of powder for shotfiring. In the winding and haulage of coal, however, there had been a great, if not revolutionary, change as the result of steam power, and at the end of our period electricity. And at the surface, the treatment and grading of coal had become increasingly extensive and sophisticated, enabling a bigger proportion of the output to be sold, and a bigger price to be got for the poorer grades.

The effects on the production of coal of technical change in shifting coal, of shotfiring, and so on, had been at least partially hidden by the gradual deterioration in the natural conditions which were the framework in which the industry operated. Indeed, to a large extent it had been the need to win coal at greater depths which had been the main incentive for technical progress. Greater depth required a larger area to be tapped from each of the now-costly shafts. This meant longer hauls underground, and, without mechanisation, such operations would have required so many oncostmen as to make them uneconomic at the price and wage levels which ruled – probably at any wage level which could conceivably have ruled. As we earlier quoted A.J. Taylor as saying,[1] 'invention . . . has been and is the indispensable condition of the industry's continued existence'. In fact, innovation was more than just that for a large part of the nineteenth century, and it was at least partly responsible for the rising trend of productivity, which was most notable in the period from the 1830s to the 1860s. Greater regularity of working the pits, not so much in the North-East, but practically everywhere else, had earlier done most to raise annual productivity, and other factors were also at work; but technical progress was certainly the main one up to the 60s. After that time, there was a confused period of twenty years, in which miners' psychology, coupled with the effects on wages of the great boom of the early 70s and the subsequent prolonged depression, played a major part. Thereafter, the trend of productivity was downwards, and the continued application of technical progress was insufficient to do more than slow it down.

That this was the case must be attributed at least partly to the failure to mechanise the hewer's job to any major extent. This was not essentially a matter of a failure of technology itself or of entrepreneurship, despite a long tradition of blaming that. Rather, it was a matter of relative costs and of the nature of markets – principally the large premium price received for round coal. The main element in the cost calculation was the wage level for hand-hewing. Despite the fact that hewers were relatively highly paid for manual workers, their wage rates were still low enough to make it unprofitable to mechanise their job in the majority of working conditions met with in British coal mines in 1914. As these conditions changed, and especially as thinner seams had to be exploited, and as wage rates rose, then the calculations must change; but they had not yet done so by the end of our period on a sufficient scale to induce enough mechanisation to reverse the trend in productivity. That did not happen until the 1920s and later.

As has been implied, coalmining was in many ways a fairly self-contained industry, with most of both its entrepreneurs and its labour force coming from the ranks of those already involved in it. What, then was its place in the wider economy? Mainly, it would seem, it was a responder. Its role was to meet the increasing demand for fuel from other industries, from transport enterprises, from households, and increasingly, after 1870 or so, from overseas customers as well as British ones. Except to some degree with exports, there was no aggressive seeking after new markets, as opposed to competition between firms for existing ones, which could often be aggressive enough in cyclical downswings. It might be argued that, in a sense, there was no need for this, since markets grew rapidly enough in the booms in response to the extension of the use of the steam engine and of iron, and in slumps there was rarely a massive fall in sales. But this did mean that the main thrust of both investment and technical progress was rather to seek to expand output at something like constant costs than to strive to cut costs in order to tap new markets. In so far as the latter did take place, it was mostly done by aiming to cut transport costs between the pits and the consumers – by investing in canals and later railways. And there is certainly plenty to connect coalowners with both these transport innovations, though more with short, local routes than with the main components of the water and rail networks – with the Duke of Bridgwater's original canal, with the Stockton and Darlington Railway, or, much later, the Barry Docks and Railway and the Hull and

Barnsley Railway, not with the Grand Trunk Canal or the London and Birmingham Railway.

The coal industry's response to rising demand was, as we have seen, largely successful until towards the very end of our period. Costs were held and even cut (in real terms) right through to the last decade of the nineteenth century, despite the need to exploit less favourable seams and areas. But eventually this need, coupled firstly with the slowing down of the expansion of demand, and secondly with the consequence of both these factors, the increasing average age of collieries, led to falling productivity and rising costs. The rise in the relative price of coal which accompanied this kept the coalmining industry itself comparatively prosperous – more so than it had been when its physical productivity was at its peak, in the 1880s; but it was a deleterious influence on the economic performance of the country as a whole, and it presaged considerable difficulties in the future, even if demand had not become less inelastic.

NOTES

1. Introduction

1 Jevons (1864).
2 Note, however, that Wrigley in effect supports the older view in his two penetrating articles on modern industrialisation: Wrigley (1962) and (1972).
3 See Peter Temin's contribution in Rosovsky (1966).
4 Allen (1935), p. 32.
5 Note that Feinstein's estimates of GDP per head in 1900 prices show an average rise of 1.3% per year between the trade cycle of 1891–9 and that of 1900–7, but only of 0.3% per year between 1900–7 and 1908–13. (Based on Feinstein (1972), T5, and the estimates of mid-year population in Mitchell & Deane (1962), pp. 9–10).
6 Rostow (1948), p. 33.

2. Markets

1 There were some c.i.f. sales, mostly when they were made direct to foreign consumers (such as railway companies) rather than to foreign merchants.
2 Jevons (1915), p. 303.
3 Daunton (1977), pp. 57–8. See also HOLM.
4 Jevons (1915), p. 293.
5 Sweezy (1938).
6 Mushet (1840), p. 311.
7 Scrivenor (1854), p. 260.
8 The North British Railway ran into Northumberland, but this was the only exception.
9 Porter (1847), p. 250.
10 SC 1873.
11 *CG*, 12 April 1878, p. 586. See also issues of 5 January 1883, p. 18, and 22 January 1886, p. 138.
12 *CG*, 2 February 1883, p. 178.
13 *CG*, 3 January 1908, 'Review' of 1907; and RC 1903–5, Q. 15455.
14 RC 1871, Report of Committee E, pp. 196–7.
15 *Ibid.*, pp. 185–7.
16 See p. 31 below.
17 Rimmer (1960), p. 229.
18 *CG*, 23 March 1888, p. 411; 29 March 1889, p. 448; 30 August 1889, p. 304; 5 August 1890, p. 241; 1 January 1909, 'Leeds Notes'; and 5 January 1912, 'Barnsley Notes' and 'Leeds Notes'.
19 See p. 24 above.
20 RC 1871, Report of Committee E, pp. 98–106; *CG*, 10 September 1880,

p. 411. The Chesterfield correspondent wrote: 'South of the Midlands there is a vast area and population whose requirements for general and domestic purposes have been yearly increasing.'

21 *CG*, 2 August 1907, 'London Notes'; 5 January 1912, 'London Notes'.

22 *CG*, 1907–13, 'Hull Notes', *passim*.

23 *MS* annually.

24 RC 1871, Report of Committee E, pp. 186–7.

25 *CG*, 13 May 1881, p. 742.

26 Note in this connection the passage in Raistrick (1953), pp. 266–70.

27 RC 1871 put total consumption in potteries at 730,000 tons, of which the great majority must have been used in North Staffordshire.

28 RC 1871, Report of Committee E. It must be remembered that Yorkshire and the East Midlands supplied part of the Lancashire market. Judging by evidence to the Select Committee of 1873 and by reports in the *Colliery Guardian* in the course of 1873, the amount was in the region of one million tons.

29 RC 1903–5, Q. 15465; and *CG*, 5 June 1908, 'Barnsley Notes'.

30 Based on *MS* and on Meade (1882), pp. 817–19.

31 Thomas (1903), p. 46. The reference to 'steam coal for navigational purposes' is to coal exported to bunkering stations overseas, not to coal used for bunkering at home ports, which was distinguished separately.

32 See, for example, the essays by R.B. Zevin and by Robert W. Fogel and Stanley L. Engerman in Fogel & Engerman (1971), Part III.

33 In this connection, note the disarming way in which Fogel and Engerman, having used their model to find that 'most of the rise in the supply of anthracite was due to the growth of capital', a supply-side factor, then effectively repudiate this by saying that 'while K appears in the supply term of equation 29, part of its value may be attributable to changes in demand'. Surely they are right – but where has the precise division between demand- and supply-side factors in their explanation of industrial expansion gone? (Fogel & Engerman (1971), p. 161.)

34 Sweezy (1938), p. 41.

35 Mitchell & Deane (1962), pp. 8–10. The results are the same whether or not Ireland is included. It has been suggested that it would be more meaningful to take consumption per head of *urban* population. This is debatable, and the decision to include rural population here has been made rather arbitrarily. One small piece of evidence came the author's way by chance and may have been given undue weight. According to the books of the Overseers of the Poor for the village of Toft (now in the Cambridge County Record Office), the final change from 'turf' (i.e. peat) to coal in the fuel distributed to the poor came between 1813 and 1819, though some coal had been bought regularly in the 1780s and again after 1800. Toft is seven miles from the head of navigation of the River Cam, and this suggests that it is reasonable to suppose that, even at such an early date, more than half of the rural population of the country was favourably placed to receive coal. Moreover, only a few heavily wooded areas were so well situated for a competing fuel as a village only seven miles from a waterway serving the peat fens.

36 This is completely consistent with the findings of Hawke (1970), ch. 6.

37 von Tunzelman (1978). This is the central theme of a large part of the book and page references are really superfluous, but see especially ch. 6.

38 Mitchell & Deane (1962), pp. 185, 198, 203, and 210.

39 von Tunzelman (1978), ch. 9. See also his earlier remarks on Watt and on compounding in ch. 4, especially pp. 67–70 and 86–8.
40 See the summary in Dyos & Aldcroft (1969), pp. 238 *et seq*.
41 Mitchell & Deane (1962), p. 218.
42 SC on Railways (*SP* 1844 XI), appendix 2.
43 RC on Railways (*SP* 1867 XXXVIII), p. lxv.
44 Mather (1970). The statistics are given in appendices A and B. There is extensive discussion of competition with railways, notably in ch. 6. The quotation comes from p. 137.
45 An index of the price of common bar iron, based on Mitchell & Deane (1962), p. 493, usually has the same peaks and troughs as an unweighted average of the coal price indices for the ironmaking districts shown in table 9.3 below. The iron price index occasionally leads by one year. Prior to 1871–3, and again in the early 1880s, the iron price fluctuations were considerably more ample than those of coal prices. In the early 70s this was reversed. After 1886 their amplitude was very similar.
46 The amount of coal used in producing 1 ton of pig iron fell between 1870 and 1880 from 3 tons to less than 2 tons, according to *MS*.
47 See the estimates of both Deane (1968) and Feinstein (1972).
48 The amount of coal used per million cubic feet of gas scarcely changed from 1887 to 1903; and rough estimates for earlier years, made on the basis of the author's unpublished estimates of capital formation in the industry, suggest similar constancy. (The basis of estimates is *Returns from Gas Companies* in *SP* 1847 XLIV, *SP* 1850 XLIX, *SP* 1865 L, and *SP* 1866 LXVI, and *Annual Returns of Gas Undertakings* from 1881 in *SP*.)

3. Capital

1 Morris & Williams (1958), p. 77.
2 The only figures we have both for the individual collieries and for the whole industry are those of numbers of men employed, and these exist only from 1894 onwards (except for East Scotland, where they go back into the 1870s). To rely on these to relate the capital figures of a sample to those of the whole industry would be unwise, since the validity of doing so would depend either on the wholly unwarranted assumption that capital per miner was the same for all collieries, or on the unlikely and unverifiable assumption that one's sample was by chance a statistically representative one.
3 See the next chapter for discussion of technical innovation.
4 Records of the colliery are in the Northumberland County Record Office.
5 See the correspondence between the Cramlington and Seaton Delaval Companies in the Cramlington files at the Northumberland County Record Office.
6 They were standard in Durham and Northumberland and in a few parts of Scotland, but not in any other coalfield – though several individual enterprises did build them.
7 CIC Q. 771; *CG*, 19 and 27 May 1910. J.C. Stamp adopted this figure, as did Lord Aberconway and H.F. Bulman, a leading mining engineer and co-author of one of the foremost mining textbooks.
8 MSS. of Barber, Walker & Co., placed at my disposal in 1953 by the kindness of Mr H. Horspool, secretary to the company.
9 Merivale (1903).

10 RC 1903–5, Q. 10976.
11 See p. 314 below.
12 *TFIME*, V, pp. 3–4.
13 Figures derived from the registration particulars and other returns filed at the Companies Registration Office (in the 1950s), and from HOLM.
14 See chapter 4 below.
15 *Report of the Select Committee of the House of Lords on the State of the Coal Trade* (*SP* 1830 VIII), p. 52.
16 Thomas Wood, one of the coalowners who was hostile to the Limitation, gave evidence to the 1836 Select Committee on the State of the Coal Trade (*SP* 1836 XI), which suggested that a colliery with a capacity of over 250,000 tons a year would require only £50,000 to open out, exclusive of working capital. At the opposite extreme, the Midland Mining Commission in 1843 reproduced a paper on the Tyne and Wear collieries 'furnished by a gentleman well acquainted with the subject', showing the annual output and numbers employed at seventy-nine collieries, with the 'probable amount of capital' at forty-four of them (*SP* 1843 XIII, pp. 105–8). The capital–output ratio of these forty-four comes to over 12s per ton. However, some of the figures for capital seem to be wild guesses – Shield Row and South Moor Collieries, producing 56,703 tons, for instance, are credited with £200,000 capital, whilst Elvet Colliery's 99,000 tons is supposed to have required only £20,000 capital. On the whole it seems safest to ignore these estimates.
17 See chapter 10 below.
18 Feinstein (1972), T51–2.
19 The sources are as follows: hewers' wages – chapter 7 below; engineers' wages to 1850, Bowley & Wood (1905–6), and thereafter Spicer (1928); wood prices to 1837, from Tooke (1838–58), and then first Memel fir to 1853 from *The Economist* (four quotations a year) and subsequently the average values of imports of sawn fir; brick prices to 1862, *Laxton's Builders' Price Book* (with extrapolations prior to 1830; 1863–1900, *Report on Wholesale Prices* (*SP* 1903 LXVIII), and thereafter the *Builder* (four quotations per year); bar iron prices to 1845, *London Mercantile Price Current*, thereafter A. Sauerbeck in the *Journal of the* [*Royal*] *Statistical Society* (1886) and annually subsequently); steel prices – average export values of steel bars.
20 Using those of Deane to 1855 and of Feinstein subsequently, both as reproduced in Mitchell (1980), pp. 818–19 and 838.
21 HOLM. See References, below, for details of publication.
22 In *SP* annually for the years from 1856 onwards.
23 The term is used here in the sense which was normal prior to nationalisaition of the coal mines; that is, to describe the owners of the collieries, which were established to exploit the coal seams which were owned by the landlords, who were also known as royalty owners.
24 J.T. Ward in Ward & Wilson (1971).
25 See also David Spring in Ward & Wilson (1971), especially the example of Lord Ravensworth and Partners referred to on pp. 37–8; and Thompson (1963), pp. 264–6. Duckham (1970), p. 159, also has a nice comment: 'The risks of mining very early persuaded some landowners that it was a game most safely played from the touchline.'
26 An example is the Duke of Northumberland's involvement after the flooding of Percy Main in 1838, and on other occasions. (See Spring (1963), p. 127.)
27 Williams (1962), pp. 42–4.

28 *Ibid.*, pp. 39–40.
29 MacGeorge (1875), pp. 16–20 and 46–52.
30 Morris & Williams (1958), pp. 127 and 140–1; Walters (1977), pp. 51 *et seq.* Griffin (1969) says of the Leicestershire and South Derbyshire coalfield: 'Mining engineers who had accumulated both capital and mining experience during their years of practice were certainly in demand as partners' (p. 254).
31 Very many examples could be quoted to add to those cited by Morris and Williams. A few noted mining engineers who became coalowners must suffice – William Hedley, viewer to the Blackett family in the North-East, became founder of Hedley Collieries Ltd; H.A. Allport, manager of the Bestwood Company in Nottinghamshire in the later 1870s, later became an owner in Yorkshire, and Cooper, Earl Fitzwilliam's manager in the 1860s, became senior partner in the Roundwood concern in 1871. The Blackett conveyor, incidentally, was invented by a member of the old landed family who became a mining engineer.
32 A great many registers were examined in the early 1950s, when many of the companies had not been wound up, and their documents were still kept at the Companies Registration Office. The majority of the records of defunct companies, transferred to the Public Record Office, were destroyed, only about 10% being retained for each company. Some of these latter did, however, prove of use.
33 See p. 44 above.
34 Morris & Williams (1958), pp. 150 *et seq.*
35 One Warwickshire coal company was involved in iron *working*, and the Scottish coal and iron company, Merry & Cuninghame Ltd, did develop a colliery in the same district after 1900, but this was wholly unconnected with its involvement in the iron industry.
36 Morris & Williams (1958), p. 147.
37 Griffin (1971), pp. 106–7.
38 Information supplied by Mr H. Horspool.
39 Made available by the kindness of the owners before liquidation.
40 Griffin (1969), p. 257.
41 The accounts were inspected either at the Companies Registration Office or in the Public Record Office, and gaps were filled from the *Stock Exchange Year Book* and the *Stock Exchange Official Intelligence*.
42 Sometimes, especially before 1900 or so, this was not carried through completely, only debentures or preference shares being issued to the public. Where this happened, only the types of stock which were issued publicly have been included in public companies' capital, so far as it was possible to separate them.
43 A similar survey for the Midland Inspection Division, covering the public iron companies as well as the purely coalowning ones, showed a figure of nearly 70% of new capital from public subscriptions.
44 His daughter, Miss Clare Warburton, then in her eighties, kindly gave me copies of his professional cards and of some of his pamphlets in 1953.

4. Technology

1 Taylor (1961), p. 51.
2 Ashton & Sykes (1929), chs. 2, 3, and 4.
3 It is not intended to imply that innovations which improved the safety of

mines and miners had no obvious economic importance. This is certainly not the case, as HM Inspector Morton pointed out when he said in his *Report* for 1851: 'Happily, the requirements of humanity and of sound economy are to a certain extent identical as regards improved ventilation and mechanical contrivances in mining' (p. 7). Unhappily they were not always identical, most especially in the short run. Unhappily, too, there were various avoidable accidents, including major explosions, which showed that there were owners and managers who were prepared to risk heavy financial loss and heavy loss of miners' lives in order to keep down immediate costs. This emerges clearly from Duckham (1973).

4 Also known as stoop and room, bord and pillar, etc.

5 Known as longwall retreating. It has become increasingly favoured recently.

6 Nef (1932), I, p. 365. The system may have been devised independently in other districts. Nef has an early reference to Somerset.

7 Griffin (1977), p. 108.

8 Ashton & Sykes (1929), ch. 2.

9 Ashton & Sykes (1929), p. 30.

10 Parnely (1904), p. 242.

11 The forcefully managed Dowlais Iron Works began to change to longwall in the 1850s, and experienced much labour trouble during the process. The manager, M. Truram, admitted this, but also said that owners as well as miners were prejudiced against longwall. (RC 1871, Q. 374–9.)

12 *Ibid.*, Q. 1339.

13 *Ibid.*, Q. 362–4.

14 *SP* 1842 XVI and XVII.

15 Griffin (1977), p. 109.

16 RC 1871, Q. 888.

17 DC 1907, Q. 11419 *et seq.*

18 Ashton & Sykes (1929), p. 31.

19 RC 1871, evidence of G.C. Greenwell, *passim*.

20 *Colliery Guardian* articles on individual collieries during the 1890s.

21 Transactions of the Manchester Geological Society, IX, part II, p. 44.

22 Galloway (1898–1904), I, p. 357.

23 RC 1871, Q. 1338 and 1351–2.

24 *Colliery Guardian* articles on individual collieries in the 1890s.

25 RC 1871, evidence of M. Truram and T. Evans, *passim*.

26 Galloway (1898–1904), I, p. 357.

27 Griffin (1977), pp. 108–9, implies that the changeover occurred simultaneously in Yorkshire and the East Midlands, beginning in the period 1840–60. His examples are all drawn from the East Midlands (including Leicestershire) however, and there is too much evidence from Yorkshire of the later vigorous survival of the bank system to accept his dating so far as that district is concerned.

28 RC 1871, Q. 2834–5.

29 RC 1903–5, Q. 5520.

30 RC 1871, Q. 1042 *et seq.*, and 2476.

31 Griffin (1977), p. 108, suggests a reason why the East Midlands bank system gave way before that of Yorkshire, namely that 'the much shorter multiple banks of Yorkshire allowed transport to flow more smoothly'.

32 RC 1903–5, Q. 5520.

33 Welbourne (1923), p. 11.

34 Galloway (1898–1904), I, pp. 422–35, is perhaps the standard account, but see also Hardwick & O'Shee (1916), p. 548.
35 Ashton & Sykes (1929), p. 53.
36 Galloway (1898–1904), I, p. 485.
37 Ashton & Sykes (1929), p. 15.
38 Bulman & Redmayne (1906), p. 112. There is a specific reference to the costs of opening a drift in Durham in 1881–2.
39 This can be shown by an example. Suppose forty hewers got 40 tons per shift before shotfiring was introduced, and this required sixty other workers to do the haulage. Overall productivity was therefore 0.4 tons. After shotfiring, the forty hewers got 160 tons per shift, requiring, *ceteris paribus*, 280 other workers for transport. Average overall productivity was therefore 0.57 tons – a rise of 42.5%. Of course, in practice the proportion of the labour force engaged in hewing was never reduced to this extent, since the productivity of other grades was raised by other innovations being introduced at the same time.
40 Hughes (1904), p. 101.
41 *SP* 1854 IX, Q. 1094.
42 Taylor (1961), p. 57.
43 Dunn (1852), p. 15. See also Ashton & Sykes (1929), p. 68, for a reference to this happening in Scotland.
44 SC on Accidents in Mines (*SP* 1835 V), evidence of George Johnson and George Stephenson.
45 *SP* 1842 XVI, p. 392.
46 *Ibid.*, pp. 37 and 187, and *SP* 1842 XVII, p. 156.
47 *SP* 1842 XVII, pp. 67–8. The use of the term corf is misleading, implying that the old type of basket made of hazel rods was used. In fact, as the illustrations in the *Report* show, the cast iron tub was the vehicle implied.
48 E.g. in North Staffordshire (*ibid.*, p. 132), in Derbyshire (*ibid.*, p. 254), in South Wales (*ibid.*, p. 476), and in Cumberland (*SP* 1842 XVI, p. 307).
49 Dunn (1852), p. 122. See also Galloway (1898–1904), II, p. 344.
50 *SP* 1830 VII, p. 316 – evidence of John Buddle to the Select Committee of the House of Commons on the State of the Coal Trade.
51 Dunn (1852), p. 214.
52 Morris & Williams (1958), pp. 68–9.
53 Galloway (1882), p. 261.
54 *Report* of HM Inspector of Mines, H. Hall, for 1901.
55 *Transactions of the South Wales Institute of Engineers*, IV, pp. 174 and 183.
56 *TFIME*, VII, p. 363.
57 In the fifth edition of his *Textbook of Coal Mining*, published in 1904, H.W. Hughes was still writing that 'where limited quantities . . . horse haulage compares most favourably with mechanical means' (p. 262). Ten years earlier, E.B. Wain, a prominent North Staffordshire mining engineer, had thought mechanical haulage needed at least 500 yards straight run to be clearly superior. (*TFIME*, VII, p. 367.)
58 CIC, Q. 2535.
59 *TFIME*, LLL, p. 488.
60 CIC, Q. 2543 *et seq.*, 9117, and 16450.
61 Certainly this was the view (accepted in the subsequent discussion) of a paper given to the North of England Institute by T.E. Forster and F.R. Simpson in 1898 (*TFIME*, XV, p. 139). They were provoked into this paper by one given

earlier by W. Galloway, in which he suggested that, from his experience, there was about a 25% saving in costs to be achieved in South Wales by mechanisation. This has been corroborated by a Derbyshire engineer. (*TFIME*, XII, pp. 267 and 271.)

62 *Reports* of HM Inspectors of Mines.

63 See pp. 92 *et seq*. below.

64 Water balances continued in use in the rather special conditions of South Wales, where many pits had free drainage, until the 1870s or later. (Morris & Williams (1958), pp. 70–1.)

65 Electric horsepower used for winding, according to the Inspectors' *Reports* for 1913, were as follows: South Wales, 22,952; Durham, 6,742; Scotland, 2,208; Northumberland, 425; North Staffordshire, 150.

66 Galloway (1882), pp. 214–16. Galloway attributes the invention to T.Y. Hall at South Hetton, but George Stephenson, in his evidence to the SC on Accidents in Mines, implies that the idea had been applied earlier in 'the South' – i.e. in Derbyshire. (*SP* 1835 V, p. 119.)

67 Galloway (1882), p. 219.

68 Galloway (1898–1904), I, pp. 481–4, and II, pp. 33–4.

69 Galloway (1882), pp. 147 *et seq*.

70 SC on Accidents in Mines (*SP* 1835 V), Q. 500.

71 Morris & Williams (1958), p. 63.

72 *Reports* of HM Inspectors of Mines; *CG* and *TFIME* articles on individual collieries; Lones (1898), p. 109.

73 Kerr (1900), p. 329.

74 *Ibid.*, p. 334.

75 Galloway (1882), p. 262.

76 *Reports* of HM Inspectors of Mines and HOLM.

77 *The Economist*, 16 March 1872.

78 See below pp. 149–55.

79 *TFIME*, LXXI, p. 415.

80 *Reports* of HM Inspectors of Mines.

81 To this can be added the western part of Yorkshire, where almost 18% of coal was machine-cut, compared with 6% in South Yorkshire, despite the much larger proportion of new collieries in the latter area. (RC on the Coal Industry, *SP* 1926 XIV, vol. III, pp. 155–6.)

82 *Historical Statistics of the United States* (Washington, 1975), pp. 589–90.

83 *TFIME*, XXV, p. 109.

84 SC 1873, Q. 822–4; *TFIME*, I, p. 14.

85 *TFIME*, XXXI, pp. 397–8.

86 *TFIME*, XV, pp. 187–8, and XXXI, p. 395.

87 *TFIME*, XXIV, p. 141.

88 RC 1903–5, Q. 1077.

89 RC 1903–5, Q. 670.

90 RC 1903–5, Q. 10787.

91 Taylor (1961), p. 62.

92 See, for example, RC 1903–5, Q. 2765, or Kerr (1900), pp. 95–6. Typical of the bigger claims for thin seams were those made by G.B. Walker of Wharncliffe Silkstone. In very hard conditions in a seam only 18 inches thick, he reckoned the machines saved 10½d per ton on costs and added as much again to the value of the coal, a total of 1s 9d. In favourable conditions in a 2½ feet thick seam he achieved 1s per ton 'saving'. (*CG*, 11 March 1892, p. 463.)

Twelve years later, it was reported that machines saved only 3d per ton directly in a seam 1 foot 10 inches thick at Stringer & Jagger's West Yorkshire pits, but they produced 29% more large coal, which itself meant at least 1s 6d extra. (*CG*, 29 April 1904, p. 909.)

93 Taylor (1961), p. 61, refers to this as 'a relatively minor attraction in the twenty years before 1914 when the price differential between large and small coal was steadily shrinking'. This is a mistaken emphasis. No matter whether the differential was decreasing or not, it was at least 6s per ton between slack and best house-coal at any time before 1914. Even allowing for intermediate grades, 10% more large coal in a best house seam must have meant an extra 3d or 4d per ton on average. Of course, the differential was lower in steam coal seams; but against this, the extra proportion of large coal could be over 20% in very hard thin seams.

94 RC 1903–5, Appendix VII, p. 44.

95 Parnely (1904), p. 1003.

96 The books of Barber, Walker & Co., kindly made available to me in 1953 by Mr H. Horspool, were not designed to answer questions about cost differences made by the introduction of machinery; but so far as it was possible to make out, about 3d per ton was directly saved, and about as much again added to the average value of coal got.

97 The figure usually mentioned by thin-seam managers when speaking loosely in discussions after Institute papers, or in a casual way in evidence: e.g. F.G. Meachem of Cannock Chase to RC 1903–5, Q. 6887 (27).

98 Though the *Colliery Guardian* thought that electrical cutting had been delayed because 'in too many cases manufacturers have designed apparatus in accordance with the conditions which in their opinion ought to have existed but which very rarely do exist at collieries' (*CG*, 4 January 1901, p. 20).

99 Bulman & Redmayne (1906), p. 133; Parnely (1904), p. 1003.

100 Parnely (1904), p. 1003.

101 *TFIME*, XV, p. 384.

102 This can be shown by two hypothetical examples. Suppose the labour cost to be 2s per ton in each case. Then in the first example, the cost of 3 tons will be 72d. If productivity rises by a third, 4 tons can be got for 72d, or 1s 6d per ton, representing a gross saving of 6d per ton, or 3d per ton for each shilling of original labour cost. In the second example, one ton costs 24d, therefore, if productivity doubles, 2 tons will be got for 24d, or 1s per ton. This represents a gross saving of 1s per ton, or 6d per ton for each shilling of original labour cost.

103 Parnely (1904), p. 947.

104 Perhaps it is not surprising to find Parnely saying that cutters 'are not likely to give economical results where coal can be worked by hand at 2s per ton or under as the cutting price' (Parnely (1904), p. 946). The same figure was quoted by J.T. Todd in *CG*, 5 September 1903, p. 663.

105 Account books of Barber, Walker & Co.

106 It will be seen from table 4.2 that, at 2s per ton hand labour, one needs a productivity rise of two-thirds plus an increase in average value of 2s 4d per ton to achieve a total 6d per ton 'saving' and so produce £500 annual profit. In a three- to three-and-a-half-feet seam, 2s per ton was over the odds for hand labour, and only if it was a top-grade house-coal would one expect anything like 2s 4d per ton extra in sales value from the higher proportion of large coal.

Since top-grade house-coal was, by definition, soft rather than hard, the £500 figure can certainly be said to be a maximum. Barber, Walker & Co. did come fairly close to achieving it in the last few years before 1914, but this was mainly because they managed a productivity increase of around 80%, part of which was attributable to a *general* reorganisation of the pits which accompanied a large-scale switch to mechanical cutting.

107 They reckoned to get 30% more, increasing their average sales value by 1s 6d per ton, whilst direct cost saving only came to 3d per ton (*CG*, 29 April 1904, p. 909).

108 *CG*, 11 March 1892, p. 463.

109 An argument much – probably too much – favoured by the eminent owner and mining engineer, Emerson Bainbridge. (See *CG*, 6 May 1904, p. 959.)

110 Carter & Williams (1957) and (1958).

111 *CG*, 4 January 1901, p. 20.

112 Cf. the evidence of W.E. Garforth, a protagonist of the machines, to RC 1903–5, Q. 2862.

113 *Report* of the Coal Conservation Committee (*SP* 1918 VII).

114 *TFIME*, XXVI, p. 388.

115 *TFIME*, XXVII, p. 270; *CG*, 6 November 1903, p. 973, and 21 April 1911, article on Mont Cenis colliery.

116 But note that this only came about to any significant extent during the First World War. (*TFIME*, LXXXVI, pp. 156–7.)

117 *TFIME*, XIII, p. 491, XIX, p. 3, and XXVI, p. 325; *CG*, 6 November 1903, p. 993.

118 See the statement of J. Gibson in *TFIME*, XLVII, p. 250: 'In the case of a coal-cutter, the manager who instals one in a section where the draw is over 100 yards long is courting disaster, because the cost of drawing more than swallows the prospective economy; rapid clearing of the faces cannot be done.'

119 *TFIME*, XLI, p. 212, and XLII, pp. 391 *et seq*. Kerr (1914), pp. 424–5, seems to have been alone in taking the view that few roofs were good enough to be left sufficiently wide to contain both cutters and conveyors without excessive expenditure on timber.

120 *TFIME*, XLII, p. 407.

121 30 tons an hour was regarded as the capacity of the conveyor of the late 1900s, which meant a theoretical maximum of 57,600 tons from a year of 240 eight-hour shifts. (*TFIME*, XLI, p. 221.)

122 *TFIME*, XXIX, p. 460, and XLI, p. 221.

123 Bulman & Redmayne (1906), pp. 34 and 134A.

124 The long continuation of hand holing in South Wales was almost wholly the result of the friable nature of most of the coal, which made it exceptionally easy to work, but which also demanded a great deal of bye-work on maintenance.

125 Galloway (1882), p. 264, referred to the experimental use of electricity at Pleasley Colliery in Nottinghamshire and at one colliery each in South Wales and Scotland. Trafalgar Colliery, Forest of Dean, has often been credited with being the first to use electricity, also in 1882 (e.g. *TFIME*, LXXIX).

126 *TFIME*, VI, p. 366.

127 Departmental Committee on Electricity in Mines (*SP* 1911 XXXVII, Q. 189).

128 *Ibid.*, Q. 4269 *et seq*. See also discussions following various papers on electricity in *TFIME*.

129 CG, 4 January 1901, p. 20.
130 Departmental Committee on Electricity in Mines (*SP* 1911 XXXVII, Q. 4269 *et seq.*).
131 *Ibid.*, *Report*, p. 6.
132 Averages of quotations in the *Colliery Guardian*.
133 *TFIME*, XLVII, p. 241.
134 *TFIME*, XLI, pp. 104 *et seq.*
135 *TFIME*, XL, pp. 358–9. In reckoning the yield, depreciation was charged at 10%, which may not have been sufficiently generous.
136 *TFIME* and *CG* articles on individual collieries.
137 Griffin (1977), p. 116, notes that even as late as the 1920s and 30s, fillers in the East Midlands had to use forks to load the coal. Presumably this was only in non-coking seams.
138 *TFIME*, XV, p. 119.
139 *TFIME*, XXXIII, p. 336 *et seq.*, and XL, pp. 121–2; *CG*, 8 January 1886, p. 58.
140 *TFIME*, I, p. 83.
141 David (1975), ch. 1.
142 Bulman & Redmayne (1906), p. 30A.
143 *TFIME*, XXXII, pp. 374–5.
144 Taylor (1961), p. 64. The reasons for the prevalence of small units were dealt with by a number of commentators, especially in the interwar years. One of the best treatments is in Neuman (1934), ch. 2.
145 Taylor quotes Emerson Bainbridge in support of his view, but Bainbridge's companies were, in fact, on the large side, especially if one thinks of them as a group, which seems reasonable in view of their interlocking directorships. In general it seems to be the case that companies which tried to attract funds from portfolio investors distributed a high proportion of their profits, since this was necessary in order to keep up the standing of the company on the stock market. Very high proportionate distribution was, of course, the practice of the railway companies, which set the standards in this respect virtually throughout the second half of the nineteenth century.
146 Taylor (1961), p. 65.
147 This seems to be the obvious way to interpret very many discussions following papers to the various engineers' institutes. One of the best examples of these is in *TFIME*, XLI, following a paper on conveyors. The attitude of the President of the Midland Counties Institution of Engineers, J.P. Houfton, seems to justify the description 'conservative'; but other participants are best described as 'cautious'. The pioneer's attitude is well exemplified in Greenwell (1912), p. 236.

5. Labour

1 Brand (1789), II, p. 681.
2 RC 1871, p. 327.
3 DC 1907, Q. 11729.
4 This can be deduced as happening to some extent from various colliery paybooks that have been examined.
5 A modern work, based partly on personal experience and partly on folk tradition, Douglass (1972), puts this pithily: 'The deputy is in reality over a barrel. Management may have very clear lines of operation marked out for him, but the deputy himself is placed in the arena without quite knowing

whether he is supposed to be a safety official, a shot-firer, a general help, or a gaffer. Deputies don't take long to discover that what little authority they have is neutralised by the men's refusal to accept supervision' (p. 6).

6 Principally Western Scotland, some pits in Lancashire, and West (as opposed to South) Yorkshire.

7 See p. 100.

8 Ashton & Sykes (1929), p. 20.

9 Cf. *ibid.*, p. 171, which notes the continuation of the practice in Wigan 'to our own day'.

10 Burgess (1975) says that 'the extent of job control exercised by the hewers especially is unrivalled in capitalist enterprise' (p. 105).

11 Ashton & Sykes (1929), p. 148.

12 In 1841–71 undefined 'miners' in coalmining counties are included. The few coal miners in non-coalmining counties in these years have been excluded here since they were presumably retired miners. It is hoped that by excluding miners aged sixty-five and over the problem of the retired has been finessed. The numbers aged sixty-five and over in 1841 have been estimated on the basis of the 1851 figures.

13 It is impossible to exclude miners aged sixty-five and over in 1841. However, the enumerators were not following a definite policy of including retired people in their former occupation at this census, so the discrepancy is unlikely to be great.

14 *SP* 1874 XIII, p. 316.

15 Morris & Williams (1958), p. 74.

16 Dunn (1844), p. 29.

17 *SP* 1830 VIII, p. 54.

18 *SP* 1847 XVI.

19 If our estimate of the labour force in 1801 is correct, it had grown by something over sixteen times.

20 See p. 194 *et seq*.

21 Cairncross (1953), ch. 4.

22 See Mitchell (1980), table C6, for comparison with other European countries for which there are data.

23 Cairncross (1953), p. 83.

24 It is fair to point out that the Glamorgan group includes the port cities and does not include the stagnant coalfield districts of Carmarthenshire and Pembrokeshire.

25 In 1851 it was everywhere over 50%, and in the vast majority of districts it was over 75%. The proportions had changed little by 1871, and unfortunately it is impossible to continue the comparison for later years, since the census reports cease to give occupation data for registration districts. It is likely that the proportion of miners did go up above 50% in some registration districts by 1911 in South Wales and Durham, for the proportion in the aggregate of all rural districts in each of Glamorgan and Durham was a third (and in Rhondda Urban District it was 71%).

26 Ashton & Sykes (1929), p. 157, say: 'That generation followed generation in work underground is demonstrated by a glance at the records of any colliery concern with a long history. Colonel Blackett informed the writers that his colliery in the 1920s had employees descended from the men who set their marks to an indenture in 1767.'

27 Burgess (1975), pp. 159–60 and 165.

28 Griffin (1971), p. 112.
29 Griffin (1971), p. 42, is one of the few authors to comment on this, and he says of the East Midlands that 'prior to 1850, mining communities . . . produced their own increase'. Duckham (1970), p. 309, appears to suggest that there was a large inflow of Irish to western Scotland in the 1820s, but from the estimates in table 5.7 it appears that this was unlikely before 1830s, and not all that significant until the 1840s. Youngson Brown (1953), p. 38, confirms this movement as taking place mostly after 1830.
30 *SP* 1842 XVI, p. 150.
31 *Ibid.*, p. 527.
32 Welbourne (1923) speaks of an influx of new labour at this period as mining spread north and south from the Tyne (p. 46). During discussion with the author in 1955, Welbourne made it clear that he was referring mainly to local men from other jobs.
33 Morris & Williams (1958), ch. 1, and Mitchell & Deane (1962), p. 131.
34 Hodges (1947), reprinted in Minchinton (1969). The reference to the Irish is on p. 64 of the original. There is another in Evans (1961), p. 21. However, Morris & Williams (1958), p. 236, say the Irish were concentrated in the ironworks rather than the collieries, though mentioning Irish in collieries in the Rhondda in 1857 and near Pontypridd in 1866. Hodges apparently thought that the local rural communities were not a major source of recruit miners, on the ground that their population continued to expand until 1841 or 1851. But their growth was a good deal less than their rate of natural increase, and for the Welsh, at least, there was no other source of attraction within reach than the mines, metal trades, and docks.
35 Arnot (1955), p. 17, refers to both Irish peasants and handloom weavers being brought into the mines in 1837.
36 *SP* 1844 XVI, p. 38.
37 *SP* 1845 XXVII, p. 207. (See the glossary for the definition of darg.)
38 *SP* 1846 XXIV, p. 397.
39 Moore (1974), p. 75.
40 Welbourne (1923), p. 74.
41 Welbourne (1923), pp. 130–1.
42 Fynes (1971), pp. 71–2 and 108–11.
43 See, for example, *SP* 1844 XVI, p. 38.
44 Challinor (1972), p. 41, says that 'better conditions in Lancashire than in surrounding coalfields had a magnetic attraction'.
45 Evans (1961), p. 62.
46 Parliamentary ReturnNo. 142, 1873.
47 Morris & Williams (1958), pp. 235–6; SC 1873, Q. 1533.
48 Mitchell & Deane (1962), pp. 7 and 52. Note that when civil registration began in Ireland in 1864, deaths exceeded births, so that there was a natural decrease of total population as well as emigration.
49 Miller (1864), p. 187.
50 Youngson Brown (1953), p. 38.
51 SC 1873, Q. 1853 and 4596, where the Irish were said to be insignificant for both Lancashire and Scotland.
52 SC 1873, Q. 3562.
53 E.g. SC 1873, Q. 904, 913–18, 2920, 3745, 5085–6, and 7369.
54 See pp. 194–5.
55 SC 1873, Q. 2018.

56 *Ibid.*, Q. 2920.
57 *Ibid.*, Q. 961.
58 Moore (1974), p. 66.
59 Evans (1961), p. 96, refers to 'a notable increase in the number of English workmen employed' in Monmouthshire. 'Many of these had come from Staffordshire in the wake of colliery officials . . . who had been engaged by the local coal proprietors.'
60 SC 1873, Q. 4637.
61 Cf. Broehl (1964), pp. 164–5. Parliamentary Return No. 142 of 1873 gives annual figures of emigration of miners and quarrymen in 1862–72. They totalled 60,000 over the eleven years, but probably the great majority were metal miners.
62 *CG*, 30 November 1877, p. 868, and 24 July 1885, p. 140.
63 A South Wales owner, in evidence to the Royal Commission on Labour in 1892, when asked if there were not more newcomers from the agricultural districts than from other coalfields, answered: 'Oh dear no; not within the last ten years; nothing like it' (*SP* 1892 XXXVI, Part I, Q. 11988).
64 Evans (1961), p. 157. Storm-Clark (1971), p. 65, quotes a man who had been a South Wales miner from 1877 as saying that 'Llwynpia Colliery was all Bristol people.'
65 Jevons (1915), pp. 286–7.
66 In addition to the source cited in n63, see also *Reports* of HM Inspectors of Mines, W.H. Pickering and W. Walker, in 1909 and 1910 respectively.
67 DC 1907, Q. 4344–8.
68 *SP* 1892 XXXVI, Part I, Q. 13903. For an analysis of the Irish influence in Lanarkshire in the mid-century period see Alan Campbell's contribution to Harrison (1978).
69 *The Iron and Coal Trades Review*, 4 January 1901, p. 17, noted that of 545 men set on during 1873 at Pease and Partners' ten main collieries in Durham, 381 had been colliers, 71 lead miners, 52 surface workers around pits, 28 farm labourers, and the rest from miscellaneous occupations.
70 *SP* 1854 IX, Q. 1044.
71 All figures are from the *Reports* of HM Inspectors of Mines.
72 There, an underground proportion of 74% in 1874 rose, twenty years later, to 80%, where it stayed.
73 *Report of the Commissioner to Inquire into the Operation of 5 & 6 Vict. c. 99 and into the State of the Population in the Mining Districts* (*SP* 1844 XVI, p. 423).
74 See p. 77.
75 Welbourne (1923), p. 100.
76 SC 1873, Q. 3022–3.
77 *Ibid.*, Q. 1978.
78 *SP* 1892 XXXIV, Q. 1969–70.
79 *SP* 1890 LXVIII.
80 When weighted by the numbers employed in each pit.
81 DC 1907, *Final Report*, appendix 23.
82 See Watkins (1934) for a discussion of how this affected British, as opposed to American, practice.
83 Slaven (1966), appendix A, table 26.
84 In 1890, for example, 48% of pits in Durham and Northumberland were thirty or more years old, compared with a UK average of 34% (*MS*).

85 These dates are chosen because 1869 is the first for which it is possible to make the calculation for fifteen-year-old pits, and it was thought that this was the lowest age which would be at all meaningful in the context; and the proportion for 1889 is not affected, as would be that for 1890, by temporary, boom-time reopening of closed pits of uncertain age. Sources are *MS* and HOLM.
86 SC 1873, Q. 1184.
87 Welbourne (1923), p. 16.
88 Bienefeld (1972).
89 Bienefeld (1972), p. 55.
90 See, for example, the *Report of the Commissioner* . . . (*SP* 1844 XVI, p. 33). Note also the statement of William Small, of the Blantyre Miners' Association to the Royal Commission on Labour: 'I would say further that the miners of Lanarkshire have learned the true secret of the limitation of production' (*SP* 1891 XXXVI, Part I, Q. 10165). However, Anthony Slaven, in Payne (1967), p. 220, notes that, though widely applied, the Scottish darg was not very successful in attaining any of its objectives.
91 Ashton & Sykes (1929), pp. 163–5, on which this paragraph is largely based.
92 Ashton & Sykes (1929), p. 166, quoting p. 481 of the *Report* of the Children's Employment Commissioners.
93 Ashton & Sykes (1929), p. 165, and Welbourne (1923), pp. 32–7.
94 SC on Accidents in Mines (*SP* 1835 V, Q. 2814–15).
95 *SP* 1842 XV, pp. 118–22.
96 See *Return on Trades* (*Hours of Work*) (*SP* 1890 LXVIII). This return is difficult to interpret, because it is drawn from very divergent sources But only in the Rhondda are more than seventy-two hours a week indicated; and only in one return for Aberdare, another for Shropshire, and the trade union report for Lancashire is a twelve-hour day reported (together with the putter boys of the North-East). All other returns, including some for Lancashire and South Wales, and all for Scotland, record an eleven-hour day or shorter in 1850.
97 For example, Welbourne (1923), p. 16, and the evidence of Ralph Young to the Royal Commission on Labour (*SP* 1892 XXXIV, Q. 2239).
98 Bienefeld (1972), p. 184.
99 Bienefeld (1972), p. 186.
100 Bienefeld (1972), p. 159.
101 Evidence of H. Bramwell, by then a colliery manager in Glamorgan, to the DC 1907, Q. 8733.
102 Evidence of J.G. Weeks of the Northumberland Coalowners' Association and of W.H. Patterson, Secretary of the Durham Miners' Association, to the Royal Commission on Labour (*SP* 1892 XXXIV, Q. 170–84 and 3040–58).
103 DC 1907, Q. 7389–90 and 13328–36.
104 Rowe (1923), p. 115.
105 In *SP* 1890 LXVIII.
106 MFGB (1890).
107 Rowe (1923), p. 112.
108 In this case the very low union figure of eight hours is directly contradicted by the only evidence from Warwickshire to the Royal Commission on Labour, that of a coalowner, who said plainly that hours were nine from bank to bank (*SP* 1892 XXXIV, Q. 5749).
109 McCormick & Williams (1959), pp. 228–9.

110 Rowe (1923), p. 115. The discussion which follows, except for that on 'hours drift', relies to a considerable extent on Rowe's work.
111 McCormick & Williams (1959), p. 229.
112 RC on Labour (*SP* XXXIV, Q. 6449, 6742, 10967, 11143, and 11189, and *SP* 1892 XXXVI Part III), replies to questionnaire to employers.
113 CIC, *Report*, stage i, and Q. 5134–7.
114 CIC, appendix 24.
115 Evans (1961), pp. 194–5.
116 In North-East Lancashire hours on a full day were eight and a half for coal winding, implying about nine bank to bank, according to a witness in 1906 (DC 1907, Q. 13336). The replies to the questionnaire to owners in *SP* 1892 XXXVI Part III suggest that these, or half an hour less, were the hours in Manchester in 1890.
117 Eight hours winding coal plus half an hour each day for winding the men was the rule there (DC 1907, Q. 2486–9).
118 For example, SC 1873, Q. 2143–4 and 5111.
119 CIC, Q. 5137.
120 According to Slaven (1966), p. 66, the taking of Glasgow Fair as a holiday was not common until after 1867. It was certainly established by 1874 (Pay Books of Govan Colliery, UGD/1/19/2 in Glasgow University Library). At Dalkeith Colliery 1872 is the first year when the pits closed for a holiday period in July (Pay Books of Dalkeith Colliery in the Scottish Record Office).
121 Slaven (1966), table B.27, reproduces an extract from the Govan No. 1 pit splint main seam paybook for the fortnight ending 11 April 1857. This shows that only three out of twenty-two colliers worked twelve shifts, whereas twelve worked eleven shifts, and the rest either nine or ten shifts. From my examination of various paybooks of Scottish, Yorkshire, East Midlands, and Cannock Chase collieries, I am confident that this is entirely typical for all except periods of extreme prosperity or depression (including seasonal depression). Slaven does not show the equivalent pay sheet for the oncost men, but a similar extract which I took from the paybooks of the Duke of Buccleuch's Cowden pit in Midlothian, showed twenty-nine out of thirty-two colliers working twelve or thirteen shifts in a fortnight in August (records in Scottish Records Office). This, too, is typical of such paybooks as I have seen.
122 The evidence on this is confusing, and the accounts by historians of the miners' unions of the North-East do not help to clear up the matter. One witness in 1873, Lindsay Wood, a Durham coalowner, said that some pits still worked a six-day week and that this had been universal 'ten years ago' (SC 1873, Q. 3564–83). Yet G.B. Forster, a Northumberland owner and manager, told the Royal Commission on Depression in 1886 that 'before 1873 we were accustomed to work eleven days a fortnight' (*SP* 1886 XXII, Q. 11688). He did not say how long it had been the custom, however. It may have been a very long time, for Nicholas Wood of Killingworth Colliery told the 1835 Select Committee on Accidents that 'the pits were not worked more than eleven days, and very often a fewer number of days, in a fortnight' (Q. 1200). In view of the position in Durham, however, this seems more likely to refer to the number of days the pits had been open recently rather than to the full normal working week.
123 SC 1873, Q. 3032–3, and *SP* 1886 XII, Q. 11688.

124 SC 1873, Q. 3568, and DC 1907, Q. 13870.
125 Welbourne (1923), p. 137.
126 SC 1873, Q. 3770.
127 SC 1873, Q. 204.
128 SC 1873, Q. 5944–8; *SP* 1886 XII, Q. 3161; DC 1907, Q. 3925 and 6881–4.
129 The evidence of G.W. Macalpine, owner of Altham Collieries, to DC 1907, Q. 13328–36, brings out the variety of Lancashire practices, whilst the dating of the change to a five-day week is given by H. Bramall, managing director of Andrew Knowles & Sons Ltd, who said that they had had no Saturday working since 1892 (Q. 7389–90).
130 DC 1907, Q. 4444–5.
131 DC 1907, Q. 651; Anthony Slaven in Payne (1967), p. 225.
132 An interestingly different reason for a slack winter season in the seventeenth century is given by G.G. Hopkinson's contribution in Benson & Neville (1976), p. 8. At Henpit Leyes Colliery at Barlow in 1632, the lease only permitted both the two pits to work in the three summer months when the roads would be dry enough to bear the traffic.
133 A. Sopwith, manager of Cannock Chase Colliery, told the Eight Hours Day Committee that 'even in exceptional times we have never worked full time in the summer' (*SP* 1907 XV, Q. 2475). Similar evidence was given for Warwickshire by W.G. Phillips (Q. 4274) and for the Forest of Dean by F. Brain, managing director of Trafalgar Colliery, who attested to the low level of absenteeism throughout the year 'because they get so much play time in summer in consequence of there being no demand for house coals' (Q. 6877). See also the monthly issues of the *Labour Gazette* from the middle of 1894 onwards.
134 Evans (1961), p. 61.
135 *SP* 1907 XV, p. 226 (Appendix 7 to the Eight Hours Day Committee).
136 Ashton & Sykes (1929), pp. 166–70. Most of the references they cite relate to absences by the men rather than to closure of the pits, but some indicate the latter.
137 See, for example, n122 and the references in n139.
138 October 1895, p. 308.
139 *SP* 1842 XV, pp. 93, 175, and 390.
140 See sources cited in n. 159.
141 Slaven (1966), p. 100.
142 Slaven (1966), pp. 94–6, and Youngson Brown (1953), pp. 45–65.
143 See the evidence of Robert Baird of the Lanarkshire Coal Masters' Association to the Royal Commission on Labour (*SP* 1892 XXXVI Part I, Q. 13456).
144 In assessing these figures one must remember that a five-day week operated in Northumberland from 1873 to 1877 or 1878, and that there was a lengthy strike in 1877.
145 Rowe (1923), p. 169.
146 This system is described by numerous witnesses before various Commissions and Committees. One of the best descriptions was given by J. Greensmith of Eckington in North Derbyshire to DC 1907, Q. 4692–4708. One involving a 'back' shift at Denaby Main in Yorkshire was described by J. Nevin (DC 1907, Q. 5965).
147 See, for instance, the evidence of Alexander Macdonald (SC 1873, Q. 4619), of David Morgan of the Aberdare and Merthyr Miners' Association and of

Isaac Evans of the Neath, Swansea and Llanelly Miners' Association (*SP* 1892 XXXIV, Q. 3971 and 4917); and of E.B. Wain, a North Staffordshire owner, and J.B. Smith, a Nottinghamshire owner (DC 1907, Q. 3635–40, 4522, and 4546). See also Evans (1961), pp. 194–9, on the unions' largely unsuccessful resistance to double-shifts in South Wales in 1910.

148 To the Departmental Committee on the Coal Mining Industry (*SP* 1914–16 XXVIII, Q. 3357).

149 CIC, Q. 5690.

150 Evans (1961), p. 199, and (1959), pp. 83–5.

151 Handy (1968).

152 Liddell (1954).

153 *Report of the Commissioner* . . . (*SP* 1850 XXIII, p. 581).

154 Ashton & Sykes (1929), pp. 169–70. In Earl Fitzwilliam's collieries, apparently, bonuses for good attendance were tried in the 1830s (Mee (1975), p. 161).

155 *SP* 1892 XXXIV, Q. 6089.

156 For instance, the manager of Ebbw Vale Coal and Iron Company said that his haulage capacity would not have been enough if everyone turned up (DC 1907, Q. 9583–5).

157 DC 1907, Q. 641.

158 DC 1907, Q. 840.

159 See, for instance, the evidence of R.E. Moore, J.P. Baker, and J. Normansell (SC 1873, Q. 1826–8, 660, and 7409). Also *SP* 1892 XXXVI, Part I, Q. 12450.

160 Appendix to the *Report* of the Departmental Committee on the Coal Mining Industry (*SP* 1914–16 XXVIII, p. 300).

161 SC 1873, Q. 651.

162 Moore (1974), p. 153.

163 Slaven (1966), appendix A, table 30. I have eliminated official holidays and two days of strike from the possible shifts and from the absences.

164 Duckham (1970), p. 266.

165 For example, Arthur Sopwith, manager of Cannock Chase Colliery, who said: 'There is a large proportion of young men who have got no responsibilities who, as soon as they get what they think is sufficient money, are satisfied with that' (DC 1907, Q. 2720).

166 SC 1873, Q. 1972, 2013, 6501, and 6835; *SP* 1892 XXXVI, Part III, p. 997; and DC 1907, Q. 6045, 8734, and 18253–5.

167 This, of course, tends to understate the percentage absent compared with the measure which assumes that all men on the books are available for work. It does not, however, represent the percentage voluntarily absent, as was claimed for it, since amongst those not present on the day of maximum attendance there must have been some who were voluntarily absent.

168 DC 1907, Q. 3644 and 16376.

169 DC 1907, Q. 6047. The quotation is from John Nevin, a West Yorkshire Coalowners' Association spokesman, but a similar explanation was offered by others.

170 This may not have been so everywhere. A Lancashire owner said in 1854 that 'there is such a scarcity of officers for the pits . . . that I am obliged to send to the North of England' (*SP* 1854 IX, Q. 680). He was probably referring to managers, however. The same owner, J. Darlington, produced some interesting figures on literacy at his own colliery and three others in Lanca-

shire, North Wales, and Warwickshire, which are reproduced in percentage form here:

Percentage of miners able to read and write, 1854

	Males underground		Surface	
	Aged 15–20	21 or over	Aged 15–20	21 or over
Able to read				
Ince Hall	46	52	93	84
Railway, Chorley	40	50	...	...
Hawkesbury, Warwicks.	41	55	40	54
Brynmally, Wrexham	55	54	...	...
Able to make a signature				
Ince Hall	25	29	67	63
Railway, Chorley	8	31	...	...
Hawkesbury, Warwicks.	32	35	40	50
Brynmally, Wrexham	32	28	...	
Able to write a letter				
Ince Hall	11	20	33	30
Railway, Chorley	8	24	...	...
Hawkesbury, Warwicks.	18	25	20	41
Brynmally, Wrexham	26	23	...	...

Of 49 females employed on the surface at Ince Hall, 12 could read, but only one, a girl under 15, could even sign her name.

171 So at least one owner, T.P.Martin of Cumberland, thought in 1906: 'The class of man who 40 or 50 years ago used to be off drinking for a week or more at a stretch is practically extinct, but his place has been taken by a man who has as many or more holidays, but who spends them, generally speaking, in a more rational and respectable way.' The consequence, he thought, was that men were more proficient than forty years before (DC 1907, Q. 8355–8 and 18365).

172 See the evidence of Alfred Hewlett to SC 1873 for both the argument and the confusion (Q. 1972).

173 SC 1873, Q. 434.

174 SC 1873, Q. 2718 and 5111.

175 SC 1873, Q. 7409.

176 *TFIME*, XVI, p. 68.

177 It would appear to be part of the theme of Colls (1977) that, whilst challenges to hewing matches died out after the 1840s, the achievements at work of the 'big hewer' were still a matter of folk admiration in the colliery community in the decade before the First World War. (See especially pp. 65–6 and 164–5.)

6. Wage bargaining

1 Ashton & Sykes (1929), ch. 6.

2 Ashton and Sykes quote a document in the Watson Collection of the North of England Institute of Mining Engineers Library showing that at Heaton Colliery in October 1806, 39 out of 227 men were unbound (*ibid.*, p. 93).

3 Slaven (1966), pp. 50–1.
4 Challinor (1972), pp. 19–20.
5 Ashton & Sykes (1929), p. 85n. Some of the Dowlais Iron Company letters printed in Elsas (1960) suggest that the majority of men in their employment, and perhaps all but apprentices, were on monthly terms by the early part of the nineteenth century (see, e.g., pp. 37–9 and 67).
6 Douglass (1972), pp. 81–2.
7 Jevons (1915), pp. 347–50.
8 The notorious Cambrian Combine strike of 1910–11 began over an attempt, which failed, to get a cutting price of 2s 6d per ton in a new seam at the Ely pit. After the eventual acceptance of a lower price, the Secretary of the MFGB wrote to the pit lodge enquiring if anyone was getting less than 6s 9d per shift. The reply was negative, and said that 'most of the men since they resumed work had earned big wages at the tonnage rates' (quoted in Arnot (1967), p. 267). This suggests that, on that occasion, the owners' contention that the men had been going too slowly in the trial period was correct. The whole episode is an unusually extreme illustration of the problems involved in fixing a new price list. (See in addition to Arnot, Evans (1961), p. 211.)
9 Douglass (1972), pp. 81–2.
10 Jevons (1915), p. 362. On pp. 348–9 he refers to cases of owners in South Wales selling their collieries rather than face the losses of a prolonged strike to reduce the price list.
11 *SP* 1892 XXXVI, Part I, Q. 10868.
12 *Ibid.*, Q. 6747 and 10883.
13 *SP* 1835 V, Q. 2730.
14 Scott (1949), p. 2.
15 *SP* XXXIV, Q. 8612.
16 *Ibid.*, Q. 7196–200.
17 SC 1873, Q. 5807. This is also clear from D.H. Lawrence's novels, *Sons and Lovers* and *The Lost Girl*.
18 Ashton & Sykes (1929), ch. 7. The quotation is from p. 114.
19 Raybould (1973), p. 193.
20 *SP* 1850 XXIII, p. 628.
21 *SP* 1842 XV, *Reports* of Commissioners Fellows, Mitchell, Scriven, and Symons.
22 SC 1873, Q. 540, 1063, and 2461, and annual *Reports* of HM Inspectors of Mines.
23 Evidence of Col. J.B. Cochrane to the Royal Commission on Labour (*SP* 1892 XXXVI, Part I *passim*). One owner, John Beardsley of Langley Mill in Derbyshire, wrote to the *Ilkeston Pioneer* (22 May 1879) saying 'can we wonder that these men boast in the public street that they can make the colliery pay on five half days per week? . . . The men who do the hardest work can, under this vexatious system, scarcely get enough to exist upon, while the butty is feasting upon his ill-gotten gains' (quoted in Williams (1962), p. 639).
24 Griffin (1971), p. 117.
25 *SP* 1892 XXXVI, Part I, Q. 10878–80. 'Broken' mines were those being worked for the second or third time to extract as much as possible of the pillars left in the original working of the Ten Yard seam.
26 *SP* 1892 XXXVI, Part I, Q. 12645.
27 *SP* 1892 XXXIV, Q. 3742–4.
28 Major works are Arnot (1949–53), (1955), and (1967); Challinor (1972);

Challinor & Ripley (1968); Edwards (1926); Evans (1961); Fynes (1971); Griffin (1962); Machin (1958); Webb (1921); Welbourne (1923), and Williams (1962).

29 Storm-Clark (1971), p. 51. His own work and that of Moore (1974) do something to remedy the neglect. Some of the autobiographies of miners' leaders are also useful in this respect, notably Smillie (1924).

30 Sidney & Beatrice Webb (1920), p. 181.

31 Welbourne's treatment of Roberts is more dispassionate than that of most authors, and of the Webbs in particular. He sums up as follows: 'It is to be feared that the boldness of his fight and the excellence of his cause do not entirely clear this champion from the charge of being a rogue' (Welbourne (1923), p. 67).

32 Sidney & Beatrice Webb (1920), p. 300.

33 Clegg, Fox, & Thompson (1964), p. 15.

34 Machin (1958), p. 278.

35 Welbourne (1923), p. 114.

36 Clegg, Fox, & Thompson (1964), p. 15.

37 Ashton & Sykes (1929), chs. 8 and 10. The quotation is from p. 133.

38 This covers collieries in Lancashire, Yorkshire, Warwickshire, Staffordshire, Nottinghamshire, Northumberland, and Scotland. Slaven (1966), p. 165, also comments on this.

39 On this see Ashton & Sykes (1929), ch. 7. They conclude that some time between the early eighteenth century and the mid-nineteenth century a co-partnership system, whereby the collective contract was with the leaders of the 'pit's company', was transformed into the butty system, whereby the contract was with the chartermaster, who then *employed* his fellow workers. So long as the former system was in operation, or traditions arising out of it were influential, collective action by men at different pits was highly unlikely. See above, pp. 170–1, for the prevalence of the butty system. See also Challinor (1972), pp. 18–19.

40 Duckham (1970), pp. 261–2, and (1969).

41 Welbourne (1923), pp. 26–7.

42 E.g. Sidney & Beatrice Webb (1920), p. 124, and Sweezy (1938), pp. 43 and 96 *et seq.* For good accounts of the disputes see Welbourne (1923), pp. 27–44, and Fynes (1971), chs. 4–6.

43 Welbourne (1923), p. 32.

44 See, for example, the pamphlet of 1831 by W. Scott and of 1832 by 'Scrutator', reprinted in Scott *et al.* (1972).

45 See Challinor & Ripley (1968) for an account of this union.

46 Williams (1962), pp. 89–95,

47 See Welbourne (1923), pp. 66–79, for an account of Jude's union.

48 Challinor & Ripley (1968), p. 126, are surely wrong to say that 'in Northumberland and Durham no time could have been less opportune for a strike than 1844. Trade was suffering from a severe depression . . . Sales had declined by 65,957 tons in 1843 compared with the previous year.' In 1843, maybe. But in view of the upturn of trade in other industries in 1844, and the very vigorous growth shown in all coal and iron statistics in 1845 – the 1844 figures, of course, are distorted by the prolonged strike – it seems highly likely that 1844 would have seen the early stages of the cyclical upswing which is associated with the great Railway Mania. Of course, it must be granted that the ability to get plenty of blacklegs in 1844 probably owed something to the collapse of

employment in building railways after the promotions of the first major boom were completed in 1840–1.

49 Anthony Slaven in Payne (1967), p. 219. He is surely wrong to say that enforcement of the darg began soon after the repeal of the Combination Laws in 1824, however, for, as Duckham says, 'the concept of the "darg" . . . was deeply ingrained in Scottish mining history' (Duckham (1970), p. 266). See also Slaven (1966), pp. 94–8.

50 Youngson Brown (1953), p. 47.

51 Arnot (1955), p. 15; Sidney & Beatrice Webb (1920), p. 193.

52 *SP* 1844 XVI, p. 37; MacGeorge (1875), pp. 67–9.

53 Challinor (1972), p. 17.

54 Challinor (1972), pp. 24–5.

55 Challinor (1972), p. 28.

56 Machin (1958), pp. 32–6.

57 Edwards (1926), pp. 7–9.

58 Edwards (1926), p. 14.

59 In addition to Edwards (1926), see also Evans (1961), pp. 9, 19, and 23–4, and on the 'Scotch Cattle', pp. 51 *et seq.*

60 See pp. 238 *et seq.*

61 Hobsbawm (1964), ch. 8.

62 The reference turning points in Gayer, Rostow, & Schwartz (1953), I, p. 347, are employed here.

63 Mitchell & Deane (1962), p. 111.

64 Ashton & Sykes (1929), ch. 8; Elsas (1960), pp. 39–40.

65 MacGeorge (1875), p. 69.

66 Sidney & Beatrice Webb (1920), p. 124n; Welbourne (1923), pp. 23 and 38–42.

67 Arnot (1949–53), I, 51, puts the foundation of the West Yorkshire Miners' Association in May 1863, but Machin (1958), p. 81, says 1858, though admitting (p. 114) that it virtually did not exist in 1859–62.

68 Strictly speaking, a Northumberland and Durham union was founded in 1862, but the Northumberland men seceded when those of Durham accepted the bond, and the Durham part of the union collapsed.

69 Challinor (1972), p. 57.

70 Evans (1961), p. 93; Arnot (1955), p. 65.

71 Williams (1962), p. 105.

72 Burgess (1975), p. 180.

73 For an account and analysis of its ineffectiveness in South Wales see Morris & Williams (1958), pp. 196–201.

74 Sidney & Beatrice Webb (1920), pp. 305–6.

75 Burgess (1975), p. 167.

76 Dennis, Henriques, & Slaughter (1956).

77 Douglass (1972), p. 47, makes a revealing comment: 'No fighter for the working-class can ever win acclaim from his enemies; the degree to which he does so is a fair measure of his role.'

78 Challinor (1972), p. 92; Arnot (1949–53), I, p. 54.

79 Challinor (1972), p. 97.

80 Evans (1959), p. 15, admits that he 'apparently failed to see its major weakness, namely that it invited employers to underselling and overproduction', but he claims that 'although he always had to negotiate from weakness, since the miners lacked an effective organisation until the end of the century,

through his consummate ability he preserved peace in the coalfield for nearly a generation and . . . gained many victories'.

81 Sidney & Beatrice Webb (1920), pp. 338–9.
82 Welbourne (1923), pp. 177 and 196.
83 For example, if an eight-week strike were successful in keeping wages up by 3d per day above the level indicated by a sliding scale (an appropriate magnitude for the mid-1870s), it would take over three years to make up for earnings foregone.
84 Machin (1958), pp. 260–1 and 469–73.
85 See below ch. 7.
86 Clegg, Fox, & Thompson (1964), pp. 336, 362, and 461; Rowe (1923), p. 41.
87 *SP* 1892 XXXVI, Part I, Q. 13265.
88 See, for example, two pamphlets by well-known economists of the day – William Smart (1894) and J.E.C. Munro (1885). The latter was brought up to date in Munro (1890).
89 Arnot (1949–53), I, p. 61.
90 *Ibid.*, p. 62.
91 *Ibid.*, p. 75.
92 Technically these districts included Northumberland, where the sliding scale had been ended in 1887 on the owners' initiative. But the Northumberland union leaders had retained their belief in the principle, and held aloof from Pickard's group.
93 Arnot (1949–53), I, pp. 177 and 218–31.
94 Clegg, Fox, & Thompson (1964), p. 107. The dispute is usually referred to as the strike of 1893, but there is no doubt that the union contention that the men were locked out for not accepting the wage cut is technically correct.
95 Arnot (1949–53), I, p. 253.
96 Williams (1962), p. 349.
97 Griffin (1971), p. 154.
98 *Ibid.*, p. 155.
99 Challinor (1972), p. 197.
100 Clegg, Fox, & Thompson (1964), pp. 110–11.
101 Evans (1961), pp. 186–7.
102 Arnot (1967), p. 25. Cf. Clegg, Fox, & Thompson (1964), p. 434.
103 Evans (1959), pp. 24 and 52. See also Morris & Williams (1960), reprinted in Minchinton (1969).
104 Evans (1961), p. 220.
105 Arnot (1955), p. 65.
106 See, for example, Arnot (1967), p. 26, and Clegg, Fox, & Thompson (1964), pp. 105–6.
107 Evans (1959), p. 57.
108 Evans says of the latter: 'Its policy had little to commend it, since a conciliation board was only a sliding scale committee in effect, except that it left the employers free to press wage reductions at any time' (Evans (1959), p. 57). Of course, it also left the union free to press for wage increases at any time.
109 Arnot (1955), pp. 75–86.
110 See table 5.4 above.
111 Welbourne (1923), p. 13.
112 Jevons (1915), p. 526. The discussion which follows relies mainly on Jevons's account of the origins of the abnormal places agitation, and on that of Evans (1961) supplemented by Phelps Brown (1960), pp. 318–21.

113 A.R. Griffin, an author generally sympathetic to the labour side, implies that it was mainly unwillingness: 'The men do not appear to have exerted themselves to keep up their output' (Griffin (1971), p. 205).
114 Sweezy (1938).
115 Taylor (1953).
116 There is a reference to a West Yorkshire Coalowners' Association in existence in 1830, but it, and four or five later associations, did not last long, and seem to have represented 'less than a fifth' of owners (Goodchild (1970), pp. 5–6).
117 Taylor (1953), p. 39.
118 SC 1873, Q. 3862.

7. Wages

1 Storm-Clark (1971), p. 56.
2 Rowe (1923), p. 50.
3 This is not to say that it generally was wide. One's impression, in fact, is that there was concentrated bunching close to the mean, median, or mode in nearly all cases.
4 Storm-Clark (1971), p. 55.
5 Anthony Slaven in Payne (1967), p. 221. Our index for Scotland differs considerably from Slaven's, especially in some of the slump years of the 1870s and 80s. It seems probable that he divided figures of weekly earnings by the *normal* number of days worked to arrive at estimates of the daily wage rate, whereas the number of shifts *actually* worked was much less. Contrast, for example, the 1886 Wage Census figure (which we use) with his estimate of 2s 6d per shift, which is surely far too low.
6 E.g. Williams (1962), p. 131.
7 Cf. Bowley (1895).
8 Scott (1949), pp. 4–6.
9 Wood (1954).
10 *SP* 1842 XVI, *Report* of E. Waring; Rowe (1923), pp. 162–5.
11 *SP* 1842 XVII.
12 *Returns of Wages Published between 1830 and 1886* (*SP* 1887 LXXXIX, p. 42).
13 E.g. Griffin (1969), pp. 650 *et seq.*
14 A publisher's reader made the point that this only applied to those in employment. This is correct but unimportant, since unemployment in the coal industry normally took the form of closing the pits for a number of days each week, which restricted the working time of both pieceworkers and dayworkers. Since some of the latter had to work in order to maintain the long-run viability of the pit even when it was closed, the point made here about convergence of earnings in slumps is reinforced.
15 Slaven (1966), pp. 163–70, has a valuable discussion of the earnings of oncostmen at Govan Colliery from 1856 to 1913, the central conclusion of which is the greater stability of their earnings than those of the hewers.
16 SC 1873, Q. 283–93.
17 The main sources for this paragraph are *Returns of Wages . . .* (*SP* 1887 LXXXIX), *Reports* of the Children's Employment Commissioners (*SP* 1842 XV–XVII) and the paybooks of certain collieries.

18 In this connection it is worth quoting again Welbourne's statement that 'strong hereditary influence in the coal trade, the completeness of the monopoly which the Newcastle Hostmen enjoyed in the London market, the absence of internal competition ensured by the "Limitation of the Vend" . . . had made it the rule rather to accept labour costs as inevitable than to seek to diminish them except by engineering skill' (Welbourne (1923), p. 26). See also Sweezy (1938), p. 41.

19 Ashton & Sykes (1929), pp. 138–40.

20 Evidence to the Select Committee of the House of Lords on the Coal Trade (*SP* 1830 VIII, p. 36). Ashton & Sykes (1929), pp. 94–5, analyse the background to the strike, and Welbourne (1923), pp. 23–4, has a more colourful account of it.

21 See Duckham (1970), pp. 261–2, for a discussion of the reasons

22 Machin (1958), pp. 32–6. Machin is rightly cautious about comparing wage figures from different sources, and it is unlikely that these rates are exactly comparable. But the difference is so striking as to argue a considerable fluctuation from slump to boom.

23 On this point see Taylor (1953), pp. 23, 26, and 34.

24 This does not contradict the analysis of chapter 4, which argued that there was no *long-term* scarcity of labour for coalmining. Part of the mechanism for ensuring this, especially in the period from the 1840s to the early 70s, was the periodic *short-run* shortage of miners, which raised wages enough to attract workers from other occupations.

25 See pp. 185–6.

26 Burgess (1975), p. 191, would appear to believe that union leaders were mainly anxious for their own position when he writes: 'Disputes were a drain on the funds which underlay the miners' organization and the officials' privileged position in it. Adequate reserves were also necessary to pay friendly society benefits that tied members more closely to the union . . . A reputation for settling disputes amicably also made union leaders more acceptable to "Society", opening the way to new contacts and career opportunities.' This seems far too extreme a view. Even granting that nearly everyone's motives are mixed, and that some union officials were governed by their own personal interests, it seems fairly certain that the major leaders were primarily concerned with (a) the long-term interests of the mining community as they saw them (which required the maintenance of the union, and not the sort of spectacular smash which happened after 1844 or which overtook the AAM in 1875); and (b) the suffering caused by long strikes, which were just the ones which seldom produced a satisfactory outcome for the men. If it took ten years and more to work off the debts incurred in even a successful strike, was it worth it? One should also grant that union leaders were usually better placed than their critics, both contemporary and later, to assess the relative bargaining power of their union at any time. Perhaps the fairest assessment of the motives and actions of one of the 'moderate' leaders whom Burgess criticises is contained in Evans (1959).

27 On this point see especially Morris & Williams (1960), reprinted in Minchinton (1969).

28 Smart (1894), p. 6.

29 Evans (1961), p. 148.

30 Arnot (1967), p. 25.

31 See pp. 186–7

32 See p. 210.
33 Thick Coal wages were 6d to 1s a day higher than those in the thinner seams of South Staffordshire at this period and later (e.g. *SP* 1890 LXXXIX). The reason, according to Raybould (1973), p. 187, was the greater danger.
34 Somerset and North Wales were both rather special cases, however, where the upward trend of wages was muted until the effects of large increases in and after 1900 made themselves felt.
35 See pp. 271 *et seq.* for data on prices.
36 The upswing of the early 1880s, in which the rise in demand for coal was relatively weak, and the existing unused capacity, following sinkings in the previous great boom, was large.
37 The account of the 1831 strike in Sweezy (1938), p. 43, is particularly valuable in this respect. Amongst the references in trade union histories which may be cited are Challinor (1972), pp. 37–40 and 89–90; Machin (1958), p. 160; Griffin (1962), pp. 72 and 77; Williams (1962), pp. 128 and 131–2; Arnot (1955), pp. 40 and 53; and Evans (1961), pp. 19–20, 91, and 143–4. Note especially Evans's statement (p. 20) that 'even without the support of a permanent organisation they could hope to exploit every period of prosperity'.
38 Cf. Smart (1894), p. 7: 'When there is a rise in the price of coal the miners have no hesitation in demanding that wages follow it. Nor do the masters deny the men's right to share in it; the only question raised is whether prices, as distinguished from quotations, have risen.'
39 Arnot (1955), p. 53; Niddrie & Benhar Coal Company records at the Scottish Record Office (CB 18/5–6). The latter source has a number of other similar examples from different years. One particularly good one, which also brings out the fact that ideas of 'fairness' did enter into managers' calculations, including those of long-term self-interest, is the following from 8 February 1887, which was very early in an upswing: 'I am quite alive to the awkwardness of advancing wages whilst the majority of the employers are against it, but . . . I think we must give the advance on the condition you mention, viz: that if it does not become general in the Hamilton district it must come off again as soon as the question is settled at Hamilton. Our men have, as you say, given us full work and the big darg, and as they have acted thus far fairly with us we will stretch a point with them.' He went on to add that 'broken time is worse for us than an advance'.
40 Hooker (1894), p. 636.
41 Challinor (1972), pp. 45 and 126.
42 Elsas (1960), pp. 39–40.
43 Challinor (1972), pp. 59–64 and 77–9; Machin (1958), pp. 193–207, and 399–402.
44 Quoted in Machin (1958), pp. 399–400.
45 Burgess (1975), pp. 191 *et seq.*
46 See pp. 185–6.
47 Quoted in Griffin (1962), p. 278.
48 Hooker (1894), p. 635.
49 The price index was formed by taking the Board of Trade wholesale price index for 1871–90, and splicing onto it the Sauerbeck index for 1851–70, the Gayer-Rostow-Schwartz index for the period before that, and an unweighted average of the food and clothing components of the Labour Department's retail price index for the period after 1890. All indices are in Mitchell & Deane (1962), ch. 16.

50 The reasons for the gently declining trend of wages in South Staffordshire up to the 1880s (and again in the 1900s) are probably the high level reached prior to the 1830s, when labour was short (Raybould (1973), p. 187), and the working out or flooding of the Thick Coal after the early 1870s. In Somerset the reason is probably the isolation of a small coalfield and its markets, coupled with lack of entrepreneurial drive by its rather paternalist coalowners, thus perpetuating the eighteenth-century attitudes referred to above.

51 See pp. 174 *et seq.* For the problems and failure of the one union which was something like genuinely national, see Challinor & Ripley (1968).

52 We do not mean to imply that there was not considerable psychological resistance – and not only on the part of workers but of owners too – to the reduction of wages below their previous minimum. But it is the failure to go as far as this – the ratchet effect – which we are here concerned to explain.

53 The nation-wide union of the 1920s did make the attempt, inspired partly by political hopes in 1921, if not in 1926. It gained little, and that at high cost, and one cannot help question the realism and wisdom of its union leaders, if not their courage or inspiration.

54 See, e.g. Cartter (1964), chs. 2 and 3, or, for a more sophisticated and up-to-date approach (far beyond the simple framework used here), Fisher (1971).

55 One must make a specific exception of the earlier stages of the 1873–9 downswing, because of the very large number of new entrants to the labour force in the preceding boom, and because of the exodus of labour up to 1878. That depression, however, was not one in which wage rates failed to fall to their previous trough level.

56 Jevons (1915), pp. 351–2, makes this point, adding that 'on the other hand, where price lists are low there congregate the unskilful, lazy, and irregular men'. He also notes, incidentally, that the places with high basis wages had the best local amenities.

57 That there was such serious resistance in the downswing of the 1870s must be attributed to the unprecedented strength which unions had acquired in both membership and funds during the great boom, to the prolonged nature of the downswing (in some cases – notably the North-East – where the strikes came late), and to the temper of the employers, amongst whom there was still widespread resentment at trade unions having *any* influence.

58 The justification for doing this is that the Scottish price index in the trough of 1858 may be unduly high because of the abnormally high values for exports (on which the index is based at that period) at some East Coast ports.

59 For the Sauerbeck index see Mitchell & Deane (1962), pp. 474–5. If the index is put on the basis 1886 = 100, the rise in the overall index between the two trough was 32 points, 25 of them by 1858.

60 See pp. 312 *et seq.* and 289–90.

61 Saul (1969), pp. 30–4, and the literature cited there.

62 See p. 297.

63 The Sauerbeck index, on an 1886 = 100 basis, fell by 11 points between 1886/8 and 1896/7, and then rose to the succeeding troughs by 12 points and 7 points respectively.

64 See pp. 149 *et seq.*

65 Even then the resulting estimates will not be quite accurate, because some maintenance men worked even when the pits were not open. This is not likely to be a source of much error, however.

66 DC 1907, Q. 6877.

67 DC 1907, Q. 10517–24; CIC, appendix 23.
68 In Anthony Slaven's estimates of weekly average earnings in Scotland (in Payne (1967), p. 222) he has 1878 with as low a number of shifts per week as 1879 (viz. three and a half) and 1877 and 1868 are both lower (viz. three), whilst 1886 is the same as many other years (viz. four). It is hard to accept these very low figures, which he himself describes as 'dubious values'. The Govan Colliery books are only complete for one of these years, 1878, but in that year the pits were apparently open for 300 days; and in 1868 there is only a suggestion of short-time working in the late summer in the fortnightly output figures. Govan may have been untypical, but it is still hard to believe that it was open twice as many days as the Scottish average.
69 In 1886, for example, a Northumberland hewer doing four shifts a week would average 19s 8d plus house and coal, whilst the local farm labourer at that time got 18s a week plus house, garden, 60–80 stones of potatoes, and two bushels of wheat a year (Jeans (1892), pp. 620–51).
70 Charges for house rent, concessionary coal, schools, insurance, medical attention, etc., were also made (as appropriate) in the form of offtakes from pay, but they are not regarded as deductions here.
71 Machin (1958), p. 307.
72 *SP* 1892 XXXVI, Part I, Q. 13627.
73 The word used by the Secretary of the Durham Miners' Association in acknowledging the need for fines to the Royal Commission on Labour (*SP* 1892 XXXIV, Q. 472–7).
74 Machin (1958), pp. 316–17.
75 See para. 2 of the Bentinck Colliery Price List reproduced on p. 166 above for the continued use of forks to keep down output of slack in the twentieth century.
76 The 'big' ton was sometimes adopted, or increased, for another reason, namely to match the London chaldron measure (which was about 26 cwt) in markets where inland coal competed with coal supplied via London. Raybould (1973), pp. 182–9, explains the rise from 21 cwt to a 26 cwt ton in the Earl of Dudley's mines in the 1790s in these terms. Incidentally, matching sales by volume measure, at that time and until the 1830s, piece rates in the North-East were in terms of volume not weight.
77 Williams (1962), p. 131; Machin (19587), p. 352.
78 Morris & Williams (1958), p. 269.
79 SC 1873, Q. 7578. Hostility in the 1870s and 80s was associated with its introduction at the same time as more stringent exclusion of slack from the tubs was being sought. The latter, rather than the Billy Fairplay, was the true object of dislike (Williams (1962), p. 254).
80 E.g. John Dixon in 1873 (SC 1873, Q. 4985) and W.H. Patterson to the Royal Commission on Labour (*SP* 1892 XXIV, Q. 472–7).
81 Quoted in Machin (1958), p. 352.
82 Youngson Brown (1953), p. 49.
83 The manager of Dowlais Iron Company, for instance, wrote to Thomas Guest in 1799 insisting on changing from weekly to monthly pays to 'save us three broken days in the Month besides those drunken combinations' (quoted in Elsas (1960), p. 34). See also *Report* of the Truck Commission (*SP* 1871 XXXVI, Q. 3266–76, 4099–102, 4182, 27538, and 35478).
84 Ashton & Sykes (1929), pp. 143–4.
85 Machin (1958), p. 347.
86 Unwin (1924), ch. 10.

87 See Hilton (1960), p. 10, for discussion of this point.
88 *Report of the Commissioner* . . . (*SP* 1848 XXVI, p. 4); RC on Truck (*SP* 1871 XXXVI), p. vi.
89 Sidney & Beatrice Webb (1902), pp. 317–18.
90 Hilton (1960), pp. 32–4; *Report of the Commissioner* . . . (*SP* 1850 XXIII), p. 22.
91 Evans (1961), p. 11.
92 Hilton (1960), pp. 47–8, draws attention to this, but instead of explaining it, allows himself to be diverted into a discussion of truck as a device for limiting workers' mobility
93 See, for example, the letter to the Dowlais Company shop manager in Elsas (1960), p. 77.
94 See the *First Report* of the Midland Mining Commission (*SP* 1843 XIII), pp. xliii–xliv, xlviii, and 11). Hilton (1960), pp. 34–6, has a good discussion of the subject.
95 Hilton (1960), p. 40. The quotation is from Marshall (1890), p. 580.
96 Hilton (1960), p. 40.
97 *Ibid.*, pp. 3–4.
98 Govan Colliery Letter Book 23/8/1842 to 4/7/1845, letter no. 45 (University of Glasgow Library: UGD/1/56/1).
99 Printed in Elsas (1960), p. 78. Note that 'we, who do not truck' both operated company shops, but priced their goods competitively and exercised no compulsion on their workers to use them.
100 Hilton (1960), pp. 33 and 39, and the sources cited there; Challinor & Ripley (1968), p. 57.
101 SC 1873, Q. 440–7, 550, and 1659; RC on Truck (*SP* 1871 XXXVI), pp. xv–xxv and Q. 8575. Griffin (1962), p. 40, says it was 'dead' in the East Midlands by then, killed by the change from the full to the little butty system.
102 Clapham (1932), p. 456.
103 Hilton (1960), p. 142 n4; Clapham (1932), p. 457.
104 See p. 177.
105 One could cite very many examples, and doubtless there were more which went unrecorded. The following is a selection, mainly from union histories, with the date indicated: Challinor (1972), pp. 37–8 (1843–4) and 164–82 (1881); Evans (1961), pp. 19 (1824), 88–91 (1863 and 1864); Griffin (1962), pp. 77 (1844), and 89 (1872); Machin (1958), pp. 36 (1825), 59–64 (1844), 68–70 (1853–4), 160 and 166–8 (1871–2); Williams (1962), pp. 127 and 131–2 (1871–2), and 269 (1881).
106 Examples can be found in Arnot (1955), p. 59 (1877); Machin (1958), pp. 198–206 (later 1870s); Williams (1962), p. 277 (1883). However, there are many more examples of strikes which were totally unsuccessful in downswings.
107 Using the 1886 Wage Census, changes since that date, and assumptions about absenteeism and the regularity of opening of the pits.
108 Williams (1962), pp. 348–50.
109 See p. 186.
110 Especially chapter 4 and appendix 4.
111 *SP* 1890 LXVIII.
112 The reasons for preferring these figures, given in *SP* 1919 XIII, appendix 49, to others provided by the Coal Controller, are given in Rowe (1923), pp. 54–5.
113 The increases in putters' basis rates were generally a little below those of

hewers, the main exceptions being South Yorkshire, where they were roughly the same, Somerset, where the putters were already in an unusually good relative position in 1886, and Warwickshire, for which the figures are not all that reliable.

114 The grounds for saying this are firstly that there is more detail in the 1886 return, and secondly that the figures for this group in all four of the coalfields concerned are surprisingly low in comparison with wages for the same group in neighbouring or comparable coalfields.

8. Royalties

1 Ashton & Sykes (1929), p. 176, refer to an example of such a lease near Bradford for twenty-one years from 1802.
2 Second *Report* of RC on Mining Royalties (*SP* 1890–1 XLI), Q. 6413–14. There was a similar large difference between cannel and other coal in Lancashire, which was the other main cannel-producing district (see first *Report* (*SP* 1890 XXXVI), appendix B, p. 184).
3 See especially the evidence of W.B. Harrison of Cannock Chase in the first *Report* of the RC on Mining Royalties.
4 See p. 254 and n17 to this chapter.
5 SC 1873, Q. 6601–7 and 6807–11, and second *Report* of the RC on Mining Royalties (*SP* 1890–1 XLI, Q. 6422–33).
6 SC 1873, Q. 2252 and 3196; Ashton & Sykes (1929), pp. 177–82.
7 CIC, p. 653.
8 Watson Collection, North of England Institute of Mining Engineers Library, 3059.
9 *Report* of the (Commons) Select Committee on the Coal Trade, p. 264, and *Report* of the (Lords) Select Committee, p. 37 (both in *SP* 1830 VIII).
10 SC 1873, Q. 3644.
11 Morris & Williams (1958), pp. 116–17.
12 Checkland (1957), p. 131.
13 CIC, p. 658.
14 CIC, p. 312.
15 Morris & Williams (1958), p. 118.
16 J. Barnes of Coalbrook Vale Colliery, Nantyglo, who had made some rather bad bargains in the boom of the early 1870s.
17 Second *Report* (*SP* 1890–1 XLI), Q. 13980 and 13984. This very satisfying result from the economist's point of view was echoed in Merthyr Valley (Q. 13982).
18 Sorley (1889), p. 65.
19 SC 1873, Q. 2923 and 4260; second *Report* of RC on Mining Royalties (Q. 8045).
20 Morris & Williams (1958), p. 118; SC of the Commons on the Coal Trade (*SP* 1830 VIII), pp. 263–4.
21 See the evidence of Alfred De Bock Porter, Secretary to the Ecclesiastical Commissioners, to the RC on Mining Royalties (*SP* 1890 XXXVI).
22 Wayleaves could be used in oligopolistic manoeuvring to limit competition, as is described by Ashton & Sykes (1929), p. 188. Perhaps this is why they retained a significant size in the North-East, where such oligopoly existed until well into the nineteenth century, especially since the bigger coalowners included major landowners, such as the Earl of Durham and the Marquis of

Londonderry. For a particularly clear statement of the effect of wayleaves on the royalty bargain see SC 1873, Q. 2816.

23 See, e.g., Sorley (1889), pp. 90–1, and many statements by J.E.C. Munro, the economist and one of the members of the Royal Commission on Mining Royalties, in his questioning of witnesses before that Commission. That absence of royalty payment *might* have some effect in practice was evidenced by the rather better wages and/or conditions in many pits worked by the landlord himself. (See, e.g., Mee (1975), p. 111, and Ward & Wilson (1971), pp. 104–5.) That general abolition of royalties would have made such conditions universal is almost inconceivable, for the landlord–entrepreneur was using his relatively advantageous position to buy peace, or prestige, or a good conscience; but if no one had a relatively advantageous position of this kind, no one could offer better conditions.

24 The Duke of Hamilton, for instance, averaged £113,793 per year in 1908–17, and the Marquess of Bute £115,772 in 1912–18. (See CIC, pp. 584, 596, 603, 607, 618, 626, 631, 648, and 653 for these and other cases.)

25 Ashton & Sykes (1929), pp. 189–90.

26 For example, at a time when the average Lancashire royalty was probably equal to somewhere between 6d and 9d per ton sold, some payments on a proportional basis were reported from the Oldham district as high as 5s per ton (SC 1873, Q. 2250–1).

27 SC of the Commons on the Coal Trade (*SP* 1830 VIII), pp. 263–4 and 271. R.W. Brandling, incidentally, gave a range of 6d to 1s per *Newcastle* chaldron, which would be only 2¼d to 4½d per ton. He must have meant per *London* chaldron, since the maximum royalty certainly far exceeded 4½d per ton.

28 These calculations are based on the Sunderland price of coal given in Gayer, Rostow, & Schwartz (1953), I, p. 232; transport costs given by Buddle to the SC of the Lords on the Coal Trade (*SP* 1830 VIII), p. 94; and Buddle's estimates of the proportion of large coal (*ibid.*, p. 288).

29 SC 1873, Q. 3195 and 3543. Average pitmouth price in 1870–1 is assumed to bear the same relation to the declared value f.o.b. of coal at the Tyne ports, Sunderland, and Hartlepool as in 1882–3, when both the average pitmouth and f.o.b. prices are available (*Report on Wholesale Prices in the United Kingdom*, *SP* 1903 LXVIII).

30 Ashton & Sykes (1929), p. 182; Ward & Wilson (1971), pp. 68–9 and 93; SC 1873, Q. 6604–7 and 6807–11. It may perhaps be implied from the royalty of 4d per cart (of 13 cwt) at Carnbroe in 1833, cited by Ward, that the royalty level had already come down by the 1830s, though that would really be to put too much weight on a single case.

31 Ashton & Sykes (1929), p. 189.

32 First *Report* (*SP* 1890 XXXVI), appendix B, p. 164.

33 Morris & Williams (1958), p. 117.

34 First *Report* (*SP* 1890 XXXVI), appendix B, pp. 188–9 for the coalowners' and pp. 203 *et seq.* for the landlords' figures. The latter's range of 4d to 9d a ton, whilst probably correct at the bottom end, certainly puts the maximum too low.

35 The owners said it was 85% in 1889. It was certainly a lower proportion earlier in the century, and the change in this respect may have resulted in some slight decrease in the average level of royalty in the coalfield as a whole.

36 Ashton & Sykes (1929), p. 178. To be precise, on the common reckoning of

the day of 100 tons per inch, a 6d a ton royalty implies a rent of £60 per acre for a two-foot seam. From this there had to be a deduction of 10% or more for colliery consumption and the concessionary coal for miners. There were some seams worked in the far west of the Yorkshire coalfield near the outcrop which were a bit less than two feet thick, but it seems a fair general deduction from the figures that the royalties did not exceed 6d per ton.

37 Ashton & Sykes (1929), p. 189.

38 The 1860 figures are from the presidential address of I. Hodges to the Midland Institute of Mining Engineers in 1910 (*TFIME*, XL, pp. 168–75); those for 1867 and 1868 were taken, in the early 1950s, from the company's books, then held at the company's office in Wakefield.

39 Farey (1811), quoted in Ashton & Sykes (1929), p. 178.

40 First *Report* of the RC on Mining Royalties (*SP* 1890 XXXVI), appendix B, p. 205.

41 *Colliery Guardian*, *passim*.

42 First *Report* of the RC on Mining Royalties (*SP* 1890 XXXVI), appendix B, pp. 203 *et seq.*, and Q. 2729–60.

43 CIC, Q. 14809.

44 First *Report* of the RC on Mining Royalties (*SP* 1890 XXXVI), appendix B, p. 203 *et seq.*

45 *The Economist*, 8 March 1879.

46 CIC, Q. 16364–9.

47 Wood (1952), pp. 232–3.

48 See R.F. Pawsey's proof to the CIC, p. 312.

9. Prices, costs, and profits

1 It is not intended to go into the history of this here, for it has been fully covered in Sweezy (1938), supplemented by Taylor (1953). The cartel worked quite effectively, despite occasional periods of 'fighting trade', up to the 1830s, or even the 40s. But it was disrupted by the opening of new collieries on several occasions, and this, together with improved land transport, finally killed it. Even when operating effectively it had always, in the nineteenth century, to bear in mind the danger of encouraging newcomers to the industry, for it never had anything approaching control of entry.

2 *CG*, 3 January 1913.

3 This sort of advertising was confined to local papers, and at no time does there seem to have been any advertising on a national scale.

4 The minutes of the Northumberland Steam Collieries Association, in the Northumberland Record Office, include some good examples of this, with Seaton Delaval the most frequent chiseller, as well as being the most openly defiant. A memorandum of 31 August 1861 by the Secretary, Thomas Doubleday, gives a very good account of how the Association had almost entirely failed to affect prices. For other accounts of the failures of owners' associations, see *CG*, especially 'Leeds Notes' in the mid-1890s and in 1909.

5 *CG*, June 1913, 'District Notes'.

6 And also by individual factors at various times, and quoted in part in some official papers, such as the appendices to the *Reports* of the Select Committees on the Coal Trade of the 1820s and 30s.

7 *CG*, 13 May 1893, p. 865, for example.

8 Discussion in this and the subsequent two paragraphs is based mainly on the

records of a number of individual collieries in all the main and some of the smaller coalfields, on reports in George Johnson's Minute Book and in the records of William Armstrong (both in the Northumberland Record Office), and on the files of the *Colliery Guardian*, which began publication in 1867.

9 Compare the discussion in Morris & Williams (1958), pp. 87–8.
10 See George Johnson's report on Chorley Colliery (*loc. cit.* in n8), and *Statement of the Selling Prices of Coal at the Pit . . . in the Oldham District* (at the John Rylands Library, Manchester).
11 From George Johnson's and William Armstrong's reports (*loc. cit.* n8).
12 Average annual freight rates to 1834 can be found in the appendices to the *Reports* of the SC on the Coal Trade (*SP* 1830 VIII, and *SP* 1836 XI). Later figures are given in George Johnson's Minute Book (*loc. cit.* in n8).
13 RC 1871, Report of Committee E, pp. 1295–6. Sale notes of Gwrhay Colliery, shown to the author in the 1950s by Mr Roger Lewis of Birmingham, whose family once owned the colliery, showed small coal prices from house-coal seams at 24s and large coal at 35s in March 1873. The proportion of large to small coal prices in the house-coal seams would therefore seem to have fallen only a little – from 1.6 to between 1.4 and 1.5 – from 1868 to 1873.
14 RC 1871, p. 1295.
15 Morris & Williams (1958), pp. 104 *et seq.*
16 SC 1873, p. 360.
17 RC 1871, p. 1294.
18 *SP* 1890 XXXVI, appendix B, p. 191.
19 For details of when individual lines were opened see *Railway Returns*, published regularly in the *Sessional Papers* of Parliament from 1847 onwards, and various reports of the Railway Department previously.
20 See the Dowlais Company case referred to on p. 254.
21 *Report* of the House of Commons Select Committee on the Coal Trade (*SP* 1830 VIII), p. 252.
22 The former is from George Johnson's Minute Book (*loc. cit.* in n8), the latter from the records of Dalkeith Colliery in the Scottish Record Office (by kind permission of the Duke of Buccleuch).
23 William Armstrong's Report Book (*loc. cit.* in n8).
24 Colliery records in the Northumberland Record Office.
25 Company records in Birmingham University Library.
26 Griffin (1969), appendix 2.
27 Company records, made available to me in 1953 through the kindness of Mr H. Horspool.
28 The average receipts per ton of the whole of West Yorkshire (given in the *Report* of the RC on Mining Royalties) are very similar to the Briggs firm on its own.
29 *TFIME*, XL, pp. 168 *et seq.* The rise between 1860 and 1885 was 33% in the Stanley Main and 27.5% in the Haigh Moor seam. The rise for all seams was about 17%.
30 William Armstrong's Report Book (*loc. cit.* in n8).
31 Dalkeith records (*loc. cit.* in n22); SC 1873, appendix.
32 Calculated from Slaven (1966), appendix B, table D.
33 See, for example, Buddle's evidence to the SC of the House of Lords on the Coal Trade (*SP* 1830 VIII), p. 38.
34 See, for example, references in *CG*, 23 February 1883 p. 294, 1 August 1884 p. 339, 10 August 1888 p. 201, and 26 February 1892 p. 382.

35 Rimmer (1955), p. 46n.

36 *Report* of the SC of the House of Lords on the Coal Trade (*SP* 1830 VIII), p. 94.

37 CIC, appendix 5.

38 The difference arose entirely in years when wages costs fell, when rises in productivity accounted for almost two-thirds of the change in wages costs. In years of rising wages costs, responsibility for the changes was equally divided between movements in wage rates and in productivity.

39 See ch. 10, especially pp. 313 and 317.

40 If one looks only at the company which seems the most likely of the three to be typical, Briggs, this would certainly be true, wage rates being roughly twice as important as productivity.

41 Again, this is more like the proportion for Briggs.

42 Discussion in this and subsequent paragraphs is based on the same colliery records as in the discussion on wages costs.

43 *The Economist*, 15 March 1873; RC on the Depression of Trade and Industry (*SP* 1886 XXI), p. 643.

44 *TFIME*, XLVIII, p. 149; final *Report* of the Mining Sub-Committee, Coal Conservation Committee (*SP* 1918 VII), pp. 6 *et seq*. For productivity see the next chapter. For the long-run nature of these cost changes see *TFIME*, XLVI, pp. 272–3, and *SP* 1918 VII, pp. 61 *et seq*.

45 See pp. 257 *et seq*.

46 In the case of Briggs, which was probably typical, Workmen's Compensation rose from 0.73d per ton in 1899 to 1.11d per ton in 1910. In 1914, this item, combined with National Health Insurance, cost 1.94d per ton.

47 CIC, Q. 651–4.

48 Stamp (1918), p. 571.

49 This is explained in Stamp (1918), pp. 599–600.

50 Stamp (1916), ch. 8.

51 Stamp claimed, quite correctly, that this did not matter when the object was to examine fluctuations, because non-coal mines were of small and virtually unchanging significance. This is true for the period in which he was interested; but prior to the early 1880s, non-coal mines were of greater importance the further back one goes.

52 An example amongst the firms whose records we have seen is Terry, Greaves & Co. of West Yorkshire. Their profit record in 1883–90 was as follows: 1883 +£167; 1884 −£2,251; 1885 −£1,935; 1886 −£1,147; 1887 +£203; 1888 +£664; 1889 +£1,125; 1890 +£1,403. Earlier profit figures are not known, but it seems likely that the firm's tax assessment was nil for the whole of the 80s, and up to and including 1890/1 (i.e. average profits of 1885–9), whilst its contribution to the actual aggregate profit of the coal industry was, in fact, negative.

53 Certainly they cannot be taken on their own as any sort of reliable guide to the annual profits of the industry. A very good warning against this is the year 1867. When it is first represented in the assessment statistics, in 1868/9, the five-year average profit falls from £5,745 thousand to £5,480 thousand (£2,312 thousand to £2,043 thousand for colliery profits alone), indicating that the profits of 1867 were lower than those of 1862. This is quite incredible. Output in 1867 was 25% higher than in 1862, average price was 10d per ton higher, and such colliery records as we have seen suggest that 1867 was about on a par with the cyclical peak of 1866, whilst 1862 was a rather indifferently poor year in the industry.

54 Coal, copper, iron, lead, and tin. The reason for this limitation, apart from convenience, is that statistics for these minerals are available in a continuous, consistent form, whereas this is not the case for total mineral receipts.
55 Mitchell & Deane (1962), pp. 129 and 159–62.
56 Stamp (1918), pp. 573 and 599–600.
57 If royalties in Scotland were 2d or so higher than the average in the 1880s and earlier, as was concluded above (p. 261), then profits per ton were certainly less than half the UK average throughout the later 1870s and the 80s. Indeed, in 1878–80, there would appear to have been no profit at all, i.e. no single Scottish colliery concern had a taxable average profit to show over the five-year periods ending in those three years. This is unlikely, and almost certainly too much has been deducted for the equally depressed mines producing other minerals. But these are not very large in any case, so that unless our reckoning of royalty incomes is much exaggerated, it really does seem to be the case that coalowners' profits were very small indeed in Scotland at that time.
58 See chapter 8 n47.
59 Rimmer (1955).
60 Challinor (1972), p. 28.
61 Wood (1952), pp. 150–1.
62 Challinor & Ripley (1968), p. 126, say that 'in Northumberland and Durham no time could have been less opportune for a strike than 1844. Trade was suffering from a severe depression.' As was argued above (p. 175), this is probably a considerable exaggeration; but it contains a grain of truth.
63 See pp. 216 *et seq.*
64 The dividend yield on the ordinary shares of the eleven largest UK railway companies between 1880 and 1910 varied from 3.00% to 4.52%.
65 Ashton & Sykes (1929), p. 245.
66 *Report* of the SC of the House of Lords on the Coal Trade (*SP* 1830 VIII), p. 38.
67 J.C. Stamp told the Sankey Commission that profits on industry as a whole 'before the war' were between 9% and 10% (CIC, Q. 838–41).
68 See p. 292.
69 CIC, Q. 838–41.
70 Ashton & Sykes (1929), p. 246.
71 *Report* of the SC of the House of Commons on the Coal Trade (*SP* 1830 VIII), p. 316.
72 Perhaps one individual instance may be quoted by way of example. The long-established Wylam Colliery, of which many records survive and are kept at the Northumberland Record Office, met increasing problems with water in the workings in the 1850s, and was eventually closed in the later 1860s. Its profit record is not known for most years, but the figures for 1855–60 are revealing: 1855 +£1,182; 1856 −£1,370; 1857 −£1,081; 1858 −£2,701; 1859 −£2,203; 1860 +£653. In 1866, shortly before its closure, costs per ton had risen to 6s 3d, having been only 3s 5d in 1859 and 3s 9d in 1864.
73 Richardson & Wallbank (1912), p. 32. The quotation is on pp. 9–10, and refers to Ashley's article in the *Economic Journal* (said to be 1911 but in fact 1910).
74 CIC, Q. 9945.
75 Bwllfa & Merthyr Dare Steam Collieries (1891) Ltd seems to be an example of this. Its paid-up capital remained at £121,160 right through to its becoming a private company in 1908, but the numbers it employed rose from less than

1,500 to nearly 2,000, and its output was well over 500,000 tons. Its dividend payment on its ordinary paid-up capital (all of which was owned by the vendors to the company) reached 155% in 1900, and averaged 68% from 1898 to 1906, a figure quoted in Richardson & Wallbank's appendix. No doubt this was an exceedingly profitable company, but that it genuinely employed only about 4s worth of capital per ton of output, and only about £66 per man, is highly unlikely. Without the company's profit and loss accounts to show development work paid for out of current revenue one cannot say more. When such ploughing back of profits not only occurred but was reflected in the accounts by the issue of bonus shares, there was a curious tendency on the part of some observers to regard it as nothing but a means of disguising the payment of excessively high dividends (e.g. the evidence of A.E. Davies to the Sankey Commission, CIC, Q. 9860 *et seq.*).

76 The principal source is the records of the company in the Northumberland Record Office (especially NCB/BR Box 22), but other information was derived from the *Stock Exchange Official Intelligence*.

77 Only a selected (and rather small) part of the records of defunct companies, sent from the Companies Registration Office to the Public Record Office, is preserved; and, of course, most colliery companies became defunct fairly soon after their nationalisation compensation was settled, only a few continuing in other trades.

10. Productivity

1 See, for example, Fogel & Engerman (1971), pp. 98–9.

2 Domar (1962).

3 A simple exposition of the concept can be found in Fogel & Engerman (1971), pp. 98–9.

4 See pp. 149 *et seq*.

5 If, as in figure 10.2, Derbyshire is shown separately from Nottinghamshire, the trend in its case was slightly upward.

6 It was perhaps for this reason that the district's inspector was the most perfunctory of all in making his estimates of the number of coalminers employed.

7 At the OMY of 1874, the 1871 UK output would have required 500,000 men, instead of the 371,000 returned by the inspectors and the 351,000 given by the census, which, be it noted, still included retired workers in their former occupations. It is impossible that anything near to the necessary number of men was omitted in 1871.

8 For evidence of this in an earlier period see G.G. Hopkinson's contribution to Benson & Neville (1976).

9 Based on estimates presented in chapter 3.

10 Cf. Taylor (1961).

11 SC 1873, Q. 7409.

12 SC 1873, Q. 434.

13 SC 1873, Q. 595, 1227, 1353, 1703–4, 1707, 1788, and 1826–8.

14 SC 1873, Q. 903 and 1015.

15 SC 1873, Q. 2718 and 5111. For other owners' views see Q. 1972, 2103, 3029, 4382, and 6835.

16 SC 1873, Q. 3035 and 3417. J. Whitwell Pease, a member of the Select Committee, testified to a somewhat smaller rise in absenteeism in his collieries (Q. 4237 and 4291).

17 SC 1873, Q. 6501.
18 See his *Report* for 1875.
19 See pp. 134 *et seq.*
20 Joseph Dickinson told the 1873 Committee that 'in the great bulk of collieries in the district [i.e. North and East Lancashire] boys between 10 and 12 years of age have had to give up working altogether' (SC 1873, Q. 232).
21 See pp. 125 *et seq.*
22 See pp. 125–6.
23 Amongst these may be mentioned intensified competition from foreign coal mines, and from oil; the political uncertainty fostered by the Sankey Commission and the attitude of the Coalition Government, as well as by the rise of the Labour Party; and the owners' attempt to return to the *status quo ante bellum* in terms of wage costs, and the abysmal industrial relations which accompanied this.
24 See p. 124.
25 For example, John Watson Ltd, a large Scottish firm, was working thick seams exclusively until shortly after 1900, according to the evidence of its chairman to the Eight Hours Day Committee, but had then had to start working many thin seams. Its OMY had fallen between 1899 and 1905 from 584 tons to 424 tons (DC 1907, Q. 2784–98).
26 Watkins (1934), p. 115. He went on to add that 'a considerable part of the contrast in mining practice in the two countries is due to this single factor age'.
27 See pp. 161 *et seq.*
28 See, for example, Domar (1962) and Gavin Wright in Intriligator (1971), pp. 412–59.
29 Chamberlin (1933), pp. 6–7.

11. Conclusion

1 Taylor (1961).

REFERENCES

Allen, G.C. (1935) *British Industries and their Organization*, London (2nd edition)

Anderson, D. (1964) 'Blundell's Collieries, the Progress of the Business', *Transactions of the Historical Society of Lancashire and Cheshire*, CXVI

Arnot, R. Page (1949–53) *The Miners*, London (2 vols., the second being entitled *The Miners: the Years of Struggle*)

(1955) *A History of the Scottish Miners*, London

(1967) *The South Wales Miners*, London

Ashton, T.S., & Sykes, J. (1929) *The Coal Industry of the Eighteenth Century*, Manchester

Benson, J., & Neville, R.V., eds. (1976) *Studies in the Yorkshire Coal Industry*, Manchester

Bienefeld, M.A. (1972) *Working Hours in British Industry: An Economic History*, London

Bowley, A.L. (1895) 'Changes in Average Wages in the U.K. between 1860 and 1891', *Journal of the Royal Statistical Society*, LVIII

Bowley, A.L., & Wood, G.H. (1905–6) 'The Statistics of Wages in the United Kingdom during the Nineteenth Century: Engineering and Shipbuilding', *Journal of the Royal Statistical Society*, LXVIII and LXIX

Brand, J. (1789) *History of Newcastle*, Newcastle-upon-Tyne (2 vols.)

Brassey, Thomas (1879) *Foreign Work and English Wages*, London (2nd edition)

Broehl, Wayne G. (1964) *The Molly Maguires*, Cambridge, Mass.

Bulman, H.F., & Redmayne, R.A.S. (1906) *Colliery Working and Management*, London (2nd edition)

Burgess, Keith (1975) *The Origins of British Industrial Relations: The Nineteenth Century Experience*, London

Cairncross, A.K. (1953) *Home and Foreign Investment, 1870–1913*, Cambridge

Carter, C.F. & Williams, B.R. (1957) *Industry and Technical Progress*, London

(1958) *Investment in Innovation*, London

Cartter, Alan M. (1964) *Theory of Wages and Employment*, Homewood, Ill.

Challinor, Raymond (1972) *The Lancashire and Cheshire Miners*, Newcastle-upon-Tyne

Challinor, Raymond, & Ripley, Brian (1968) *The Miners' Association: A Trade Union in the Age of the Chartists*, London

Chamberlin, E.H. (1933) *The Theory of Monopolistic Competition*, Cambridge, Mass.

Checkland, S.G. (1957) 'Two Scottish West Indian Liquidations after 1793', *Scottish Journal of Political Economy*, IV

Clapham, J.H. (1932) *An Economic History of Modern Britain*, II, Cambridge

Clegg, H.A., Fox, Alan, & Thompson, A.F. (1964) *A History of British Trade Unions since 1889*, I, Oxford

Colls, Robert (1977) *The Collier's Rant*, London
Cunningham, A.S. (1925) *Mining in Mid and East Lothian*, Edinburgh
Daunton, M.J. (1977) *Coal Metropolis: Cardiff 1870–1914*, Leicester
David, Paul A. (1975) *Technical Choice, Innovation and Economic Growth*, Cambridge
Deane, Phyllis (1968) 'New Estimates of Gross National Product for the United Kingdom 1830–1914', *Review of Income & Wealth*, series 14, II
Deane, Phyllis, & Cole, W.A. (1967) *British Economic Growth, 1688–1959*, Cambridge (2nd edition)
Dennis, Norman, Henriques, Fernando, & Slaughter, Clifford (1956) *Coal is Our Life*, London
Domar, Evsey D. (1962) 'On Total Factor Productivity and All That', *Journal of Political Economy*, LXX
Douglass, David (1972) *Pit Life in Co. Durham*, Oxford (Ruskin College History Workshop Pamphlet No. 6)
Duckham, Baron R. (1969) 'The Emergence of the Professional Manager in the Scottish Coal Industry, 1760–1815', *Business History Review*, XLIII
(1970) *A History of the Scottish Coal Industry, vol. I, 1700–1815*, Newton Abbot
(1973), *Great Pit Disasters*, Newton Abbot
Dunn, Matthias (1944) *View of the Coal Trade of the North of England*, Newcastle-upon-Tyne
(1852) *Winning and Working of Collieries*, Newcastle-upon-Tyne and London
Dyos, H.J., & Aldcroft, D.H. (1969) *British Transport*, Leicester
Edwards, Ness (1926) *The History of the South Wales Miners*, London
Elsas, Madeleine, ed. (1960) *Iron in the Making*, Cardiff
Evans, E.W. (1959) *Mabon*, Cardiff
(1961) *The Miners of South Wales*, Cardiff
Farey, John (1811) *A General View of the Agriculture of Derbyshire*, London
Feinstein, C.H. (1972) *National Income, Expenditure and Output of the United Kingdom, 1855–1965*, Cambridge
Fisher, Malcolm R. (1971) *The Economic Analysis of Labour*, London
Fogel, Robert W., & Engerman, Stanley L. (1971) *The Reinterpretation of American Economic History*, New York
Fynes, Richard (1971) *The Miners of Northumberland and Durham*, Wakefield (originally published in 1873)
Galloway, R.L. (1882) *A History of Coal Mining in Great Britain*, London
(1898–1904) *Annals of Coal Mining*, London (2 vols.)
Gayer, A.D., Rostow, W.W., & Schwartz, A.J. (1953) *The Growth and Fluctuations of the British Economy, 1790–1850*, Oxford (2 vols.)
Goodchild, J. (1970) 'Some Early Yorkshire Coal Masters' Associations', *South Yorkshire Journal*, February
Greenwell, A. (1912) *The Economic Position of the Coal Industry of the United Kingdom*, London
Griffin, A.R. (1962) *The Miners of Nottinghamshire*, London
(1971) *Mining in the East Midlands, 1550–1947*, London
(1977) *The British Coalmining Industry: Retrospect and Prospect*, Buxton
Griffin, C.P. (1969) 'The Economic and Social Development of the Leicestershire and South Derbyshire Coalfield' (unpublished PhD thesis at Nottingham University)
Handy, L.J. (1968) 'Absenteeism and Attendance in the British Coal-Mining

Industry: An Examination of Post-War Trends', *British Journal of Industrial Relations*,VI, 1

Hardwick, F.W., & O'Shee, L.T. (1916) 'Notes on the History of the Safety Lamp', *Transactions of the Federated Institute of Mining Engineers*, LI

Hardy, S.H. (1956) 'The Development of Coal Mining in a North Derbyshire Village, 1635–1860', *University of Birmingham Historical Journal*, V, 2

Harrison, Royden, ed. (1978) *Independent Collier*, Hassocks

Hawke, G.R. (1970) *Railways and Economic Growth in England and Wales, 1840–1870*, Oxford

Hilton, George W. (1960) *The Truck System*, Cambridge

Hobsbawm, E.J. (1964) *Labouring Men*, London

Hodges, T.M. (1947) 'The Peopling of the Hinterland and the Port of Cardiff (1801–1914)', *Economic History Review*, 1st series, XVII, 1

Hooker, R.H. (1894) 'On the Relation between Wages and the Numbers Employed in the Coal Mining Industry', *Journal of the Royal Statistical Society*, LVII

Hughes, H.W. (1904) *A Text-Book of Coal Mining*, London (5th edition)

Intriligator, M.D., ed. (1971) *Frontiers of Quantitative Economics*, Amsterdam

Jeans, J.S. (1892) 'On the Recent Movement of Labour in Different Countries in Reference to Wages, Hours of Work, and Efficiency', *Journal of the Royal Statistical Society*, LV

Jevons, H.S. (1915) *The British Coal Trade*, London

Jevons, W.S. (1864) *The Coal Question*, London

Kerr, G.L. (1900) (1914) *Practical Coal Mining*, London (1st edition or 5th edition)

Liddell, F.D.K. (1954) 'Attendance in the Coal-Mining Industry', *British Journal of Sociology*, VI, 1

Lones, T.E. (1898) *A History of Mining in the Black Country*, Dudley

McCormick, B., & Williams, J.E. (1959) 'The Miners and the Eight-Hour Day, 1863–1910', *Economic History Review*, 2nd series, XII, 2

McCulloch, J.R. (1859) *Commercial Dictionary*, London (4th edition)

MacGeorge, A. (1875) *The Bairds of Gartsherrie*, Glasgow

Machin, F. (1958) *The Yorkshire Miners*, I, Barnsley (vol. II never published)

Marshall, Alfred (1890) *Principles of Economics*, London

Mather, F.C. (1970) *After the Canal Duke*, Oxford

Meade, R. (1882) *The Coal and Iron Industries of the United Kingdom*, London

Mee, Graham (1975) *Aristocratic Enterprise*, Glasgow & London

Merivale, J.H. (1903) *Colliery Manager's Pocket Book*, London

Miller, A. (1864) *The Rise and Progress of Coatbridge*, London

Minchinton, W.E., ed. (1969) *Industrial South Wales, 1750–1914*, London

Miners' Federation of Great Britain (1890) *Statistical Report of the Hours Worked in Every Mining District in Great Britain except Durham and Cleveland*, London

Mitchell, B.R. (1980) *European Historical Statistics, 1750–1975*, London (2nd edition)

Mitchell, B.R., & Deane, Phyllis (1962) *Abstract of British Historical Statistics*, Cambridge

Moore, Robert (1974) *Pitmen, Preachers, and Politics*, Cambridge

Morris, J.H., & Williams, L.J. (1958) *The South Wales Coal Industry, 1841–1875*, Cardiff

(1960) 'The South Wales Sliding Scale: An Experiment in Industrial Relations', *Manchester School*, XXVIII, 2

Munro, J.E.C. (1885) *Sliding Scales in the Coal Industry*, Manchester

(1890) 'Sliding Scales in the Coal Industry from 1885 to 1889', *Transactions of the Manchester Statistical Society* (1889–90)

Mushet, D. (1840) *Papers on Iron and Steel*, London

Nef, J.E. (1932) *The Rise of the British Coal Industry*, London (2 vols.)

Neuman, A.M. (1934) *Economic Organization of the British Coal Industry*, London

Parnely, Caleb (1904) *Colliery Manager's Handbook*, London (5th edition)

Payne, P.L., ed. (1967) *Studies in Scottish Business History*, London

Phelps Brown, E.H. (1960) *The Growth of British Industrial Relations*, London

Pollard, Sidney (1980) 'A New Estimate of British Coal Production, 1750–1850', *Economic History Review*, 2nd series, XXXIII, 2

Porter, G.R. (1847) *The Progress of the Nation*, London (2nd edition)

Price-Williams, R. (1889) 'The Coal Question', *Journal of the Royal Statistical Society*

Raistrick, A. (1953) *Dynasty of Ironfounders*, London

Raybould, T.J. (1973) *The Economic Emergence of the Black Country*, Newton Abbot

Richardson, Thomas, & Wallbank, John A. (1912) *Profits and Wages in the British Coal Trade (1898–1910)*, Newcastle-upon-Tyne

Rimmer, W.G. (1955) 'Middleton Colliery, near Leeds', *Yorkshire Bulletin for Economic and Social Research*, VII, 1

(1960) *Marshalls of Leeds, Flax Spinners, 1788–1886*, Cambridge

Rosovsky, Henry, ed. (1966) *Industrialization in Two Systems*, New York

Rostow, W.W. (1948) *British Economy of the Nineteenth Century*, Oxford

Rowe, J.W.F. (1923) *Wages in the Coal Industry*, London

Saul, S.B. (1969) *The Myth of the Great Depression*, London

Scott, H. (1949) 'Colliers' Wages in Shropshire, 1830–1850', *Transactions of the Shropshire Archaeological Society*, LIII

Scott, William, *et al.* (1972) *Labour Disputes in the Mines: Eight Pamphlets, 1831–1844*, New York

Scrivenor, H. (1854) *A History of the Iron Trade*, London

Slaven, Anthony (1966) 'Coal Mining in the West of Scotland in the Nineteenth Century: The Dixon Enterprises' (unpublished B Litt. thesis, University of Glasgow)

Smart, William (1894) *Miners' Wages and the Sliding Scale*, Edinburgh & London

Smillie, R. (1924) *My Life For Labour*, London

Sorley, W.R. (1889) 'Mining Royalties and their Effect on the Iron and Coal Trades', *Journal of the Royal Statistical Society*, LII

Spicer, R.S. (1928) *British Engineering Wages*, London

Spring, D. (1963) *The English Landed Estate in the Nineteenth Century*, Baltimore

Stamp, J.C. (1916) *British Incomes and Property*, London

(1918) 'The Effect of Trade Fluctuations upon Profits', *Journal of the Royal Statistical Society*, LXXXI

Storm-Clark, Christopher (1971) 'The Miners, 1870–1970: A Test Case for Oral History', *Victorian Studies*, XV, 1

Sweezy, Paul M. (1938) *Monopoly and Competition in the English Coal Trade, 1550–1850*, Cambridge, Mass.

Taylor, A.J. (1953) 'Combination in the Mid-Nineteenth Century Coal Industry', *Transactions of the Royal Historical Society*, 5th series, III

(1961) 'Labour Productivity and Technological Innovation in the British Coal Industry, 1830–1914', *Economic History Review*, 2nd series, XIV, 1

Thomas, D.A. (1903) *Coal Exports, 1850–1900*, London

Thompson, F.M.L. (1963) *English Landed Society in the Nineteenth Century*, London

Tooke, Thomas (1838–58) *History of Prices*, London (6 vols.)

Tunzelman, G.N. von (1978) *Steam Power and British Industrialization to 1860*, Oxford

Unwin, G. (1924) *Samuel Oldknow and the Arkwrights*, Manchester

Walters, R.H. (1977) *The Economic and Business History of the South Wales Steam Coal Industry, 1840–1914*, New York

Ward, J.T., & Wilson, R.G., eds. (1971) *Land and Industry*, Newton Abbot

Watkins, Harold M. (1934) *Coal and Men*, London

Webb, Sidney (1921) *The Story of the Durham Miners, 1662–1921*, London

Webb, Sidney & Beatrice (1902) *Industrial Democracy*, London

(1920) *The History of Trade Unionism*, London (2nd edition)

Welbourne, E. (1923) *The Miners' Unions of Northumberland and Durham*, Cambridge

Williams, J.E. (1962) *The Derbyshire Miners*, London

Wood, Oliver (1952) 'The Development of the Coal, Iron and Shipbuilding Industries of West Cumberland, 1750–1914' (unpublished PhD thesis, University of London)

(1954) 'A Cumberland Colliery during the Napoleonic Wars', *Economica*, new series, XXI, 1

Wrigley, E.A. (1962) 'The Supply of Raw Materials in the Industrial Revolution', *Economic History Review*, 2nd series, XV, 1

(1972) 'The Process of Modernization and the Industrial Revolution in England', *Journal of Interdisciplinary History*, III

Youngson Brown, A.J. (1953) 'Trade Union Policy in the Scots Coalfields, 1855–1885', *Economic History Review*, 2nd series, VI, 1

INDEX